THE INVENTION OF HUMBOLDT

The *Invention of Humboldt* is a game-changing volume of essays by leading scholars of the Hispanic world that explodes many myths about Alexander von Humboldt and his world.

Rather than 'follow in Humboldt's footsteps,' this book outlines the new critical horizon of post-Humboldtian Humboldt studies: the archaeology of all that lies buried under the Baron's epistemological footprint. Contrary to the popular image of Humboldt as a solitary 'adventurer' and 'hero of science' surrounded by New World nature, *The Invention of Humboldt* demonstrates that the Baron's opus and practice was largely derivative of the knowledge communities and archives of the Hispanic world. Although Humboldtian writing has invented a powerful cult that has served to erase the sources of his knowledge and practice, in truth Humboldt did not 'invent nature,' nor did he pioneer global science: he was the beneficiary of Iberian natural science and globalization. Nor was Humboldt a pioneering, 'postcolonial' cultural relativist. Instead, his anthropological views of the Americas were Orientalist and historicist and, in most ways, were less enlightened than those of his Creole contemporaries.

This book will reshape the landscape of Humboldt scholarship. It is essential reading for all those interested in Alexander von Humboldt, the Hispanic American enlightenment, and the global history of science and knowledge.

Mark Thurner is Distinguished Professor of Anthropology, History and Humanities at FLACSO-Ecuador, and Emeritus Professor of History at the University of Florida. He was Professor of Latin American Studies at the Institute of Latin American Studies, University of London, until the Institute's forced closure in 2021. He is the author of *History's Peru: The Poetics of Colonial and Postcolonial Historiography* (2011) amongst numerous other publications. He is the editor of *The First Wave of Decolonization* (Routledge, 2019).

Jorge Cañizares-Esguerra is the Alice Drysdale Sheffield Professor of History in the Department of History, University of Texas at Austin. He is the author of *Nature, Empire, and Nation: Explorations of the History of Science in the Iberian World* (2006) amongst numerous other publications. He is editor of *Entangled Empires: The Anglo-Iberian Atlantic, 1500–1830* (2018).

Routledge Studies in Global Latin America

Routledge Studies in Global Latin America publishes critical, post-Area Studies scholarship that connects local histories with the global history of modernity. The editors are keen to publish in those areas where Latin or Iberian America has played a pioneering role in global history.

Series Editors: Peter Burke (University of Cambridge), Jorge Cañizares-Esguerra (The University of Texas at Austin), Irina Podgorny (CONICET), Mark Thurner (FLACSO)

The First Wave of Decolonization
Edited by Mark Thurner

A History of Book Publishing in Contemporary Latin America
Gustavo Sorá

The Invention of Humboldt: On the Geopolitics of Knowledge
Edited by Mark Thurner and Jorge Cañizares-Esguerra

For more information about this series, please visit: https://www.routledge.com/Routledge-Studies-in-Global-Latin-America/book-series/RSGLA

THE INVENTION OF HUMBOLDT

On the Geopolitics of Knowledge

Edited by Mark Thurner and
Jorge Cañizares-Esguerra

Routledge
Taylor & Francis Group

NEW YORK AND LONDON

Designed cover image: © LAGLOBAL

First published 2023
by Routledge
605 Third Avenue, New York, NY 10158

and by Routledge
4 Park Square, Milton Park, Abingdon, Oxon, OX14 4RN

Routledge is an imprint of the Taylor & Francis Group, an informa business

© 2023 Taylor & Francis

Library of Congress Cataloging-in-Publication Data
Names: Thurner, Mark, editor. | Cañizares-Esguerra, Jorge, editor.
Title: The invention of Humboldt : on the geopolitics of knowledge / edited by Mark Thurner and Jorge Cañizares-Esguerra.
Description: New York, NY : Routledge, 2023. | Series: Routledge studies in global Latin America | Includes bibliographical references and index. | Contents: The Apotheosis of Humboldt during the Nineteenth Century / Leoncio López-Ocón -- A Sense of Place: Early Modern Roots of Humboldt's Natural History Practices / Florike Egmond -- Six Days on Tenerife: The Making of Humboldt's Tropical Antique / Peter Mason -- Caldas and Humboldt in the Andes: Who Invented Biogeography? / Alberto Gómez Gutiérrez -- An Archaeology of Mutis' Disappearing Gift to Humboldt / José Antonio Amaya -- Incas, Pyramids and Amazons: Notes on Humboldt's Equatorial Encounters / Neil Safier -- Humboldt's Magic Mountain / Juan Pimentel -- Peruvian Desencuentro: Humboldt's Fog, Unanue's Light / Mark Thurner -- Air in a Flask: The Mexican Making of Humboldt's Objects of Knowledge / Miruna Achim and Gabriela Goldin Marcovich -- Humboldt's Misreading of the Mercantilist Face of New Spain / José Enrique Covarrubias -- Bonpland's Cactus, Or Trafficking in Exotics and Ignorance / Irina Podgorny -- Humboldt's Columbus, Or the Iberian Worlds that Humboldt Ignored / Jorge Cañizares-Esguerra. |
Identifiers: LCCN 2022031308 (print) | LCCN 2022031309 (ebook) | ISBN 9781032139173 (hardback) | ISBN 9781032139166 (paperback) | ISBN 9781003231479 (ebook)
Subjects: LCSH: Humboldt, Alexander von, 1769-1859--Travel--Latin America. | Humboldt, Alexander von, 1769-1859--Influence. | Natural history--Latin America--History--19th century. | Latin America--Historiography. | Latin America--Colonial influence.
Classification: LCC Q143.H9 I58 2023 (print) | LCC Q143.H9 (ebook) | DDC 509.2--dc23/eng/20220705
LC record available at https://lccn.loc.gov/2022031308
LC ebook record available at https://lccn.loc.gov/2022031309

ISBN: 978-1-032-13917-3 (hbk)
ISBN: 978-1-032-13916-6 (pbk)
ISBN: 978-1-003-23147-9 (ebk)

DOI: 10.4324/9781003231479

Typeset in Bembo
by SPi Technologies India Pvt Ltd (Straive)

To the brave students of the 2019 Chimborazo Summer School.

CONTENTS

FIGURES

TABLES

ACKNOWLEDGEMENTS

This book is the fruit of *The Invention of Humboldt* workshop held in Quito, Ecuador, in August 2019. We are most grateful to then director Juan Ponce of FLACSO for his decisive support of the workshop and the Chimborazo Summer School of which it formed part, and to Mireya Salgado, then coordinator of the Doctoral Programme in The History of the Andes of the Department of Anthropology, History and Humanities at FLACSO, Ecuador, and of the FDA IP-1073 research grant. We also wish to take this opportunity to acknowledge the Leverhulme Trust (IN-2015-056) for its initial support of the LAGLOBAL network during 2016–2019. Finally, we collectively give thanks to all those FLACSO students and staff, members of the LAGLOBAL network, and the wider Ecuadorian intellectual community, without whom this book would have been impossible.

SERIES EDITORS' INTRODUCTION

Routledge Studies in Global Latin America

The *Invention of Humboldt* is a game-changing collection of essays by leading scholars of the Hispanic world. It dispels many myths about Alexander von Humboldt. Despite all the solemn portraits of Humboldt set in New World greenery, it is clear the Baron was not a solitary 'genius' in the wilderness of American nature. Instead, at every step the Prussian relied on the knowledge, networks, and archives of Spain and Hispanic America. Humboldt did not 'invent nature,' nor did he pioneer globalization. Instead, he was the direct beneficiary of Iberian globalization and natural science. Humboldt has been lionized as a 'second Columbus' who 'truly discovered America.' In truth, Humboldt self-fashioned himself in the image of a fabricated Columbus who never existed. Nor did Humboldt pioneer interdisciplinary research that bridged the arts and the sciences: such work was commonplace among Hispanic American polymaths of the period. Humboldt also did not invent biogeography. Instead, Francisco José de Caldas and Humboldt co-invented the field in the Andes. Humboldt did not discover the 'Humboldt Current.' Instead, the more aptly named 'Peruvian Current' was described earlier by José Hipólito Unanue. Humboldt was also not a 'postcolonial' cultural relativist. Instead, his anthropological views of the Americas were consistent with those of European Orientalism and historicism.

In short, *The Invention of Humboldt* outlines a new task for post-Humboldtian Humboldt studies: The archaeology of all that lies buried under Humboldt's epistemological footprint.

CONTRIBUTORS

Miruna Achim is Associate Professor in the Humanities at the Universidad Autónoma Metropolitana-Cuajimalpa, in Mexico City. She is the author of *From Idols to Antiquity: Forging the National Museum of Mexico* (Omaha: University of Nebraska Press, 2017). She is the co-editor, with Susan Deans-Smith and Sandra Rozental, of *Museum Matters: Making and Unmaking Mexico's National Collections* (Tucson: University of Arizona Press, 2021); with Laura Cházaro and Nuria Valverde, *Piedra, papel y tijera. Instrumentos en las ciencias en México* (Mexico City: UAM, 2018); and, with Irina Podgorny, *Museos al detalle: colecciones, antigüedades e historia natural* (Rosario: Prohistoria, 2014).

José Antonio Amaya is Associate Professor of History in the Department of History at the Universidad Nacional de Colombia. He is the author of *Enrique Umaña Barragán: Ciencia y política en la Nueva Granada* (Bogotá: Editorial Maremagnum, 2014) and co-editor, with Olga Restrepo, of *Ciencia y representación: Dispositivos en la construcción, la circulación y la validación del conocimiento científico* (Bogotá: Universidad Nacional de Colombia, 1999).

Jorge Cañizares-Esguerra is the Alice Drysdale Sheffield Professor of History in the Department of History, University of Texas at Austin. He is the author of *Puritan Conquistadors: Iberianizing the Atlantic, 1550–1700* (Stanford, CA: Stanford University Press, 2006); *Nature, Empire, and Nation: Explorations of the History of Science in the Iberian World* (Stanford, CA: Stanford University Press, 2006); and *How to Write the History of the New World: Histories, Epistemologies, and Identities in the Eighteenth-Century Atlantic World* (Stanford, CA: Stanford University Press, 2001). He is editor of *Entangled Empires: The Anglo-Iberian Atlantic, 1500–1830* (Philadelphia: University of Pennsylvania Press, 2018) and co-editor, with James Sidbury and Matt Childs, of *The Black Urban Atlantic in the*

Era of the Slave Trade (Philadelphia: University of Pennsylvania Press, 2013), and co-editor, with Erik R. Seeman, of *The Atlantic in Global History, 1500–2000* (New York: Routledge, 2006).

José Enrique Covarrubias is Research Professor at UNAM, Mexico. He is the author of *En busca del hombre útil: Un estudio comparado del utilitarismo neomercantilista en México y Europa, 1748–1833* (Mexico City: UNAM, 2005), *La moneda de cobre en México, 1760-1842: Un problema administrativo* (Mexico City: Instituto Mora, 2000), and *Visión extranjera de México, 1840-1867: El estudio de las costumbres y la situación social* (Mexico City: UNAM, 1998). He is co-editor, with Matilde Souto, of *Economía, ciencia y política. Estudios sobre Alexander von Humboldt a 200 años del Ensayo político sobre el reino de la Nueva España* (Mexico City: UNAM, 2012).

Florike Egmond is Associate Fellow at the LUCAS Research Institute, University of Leiden. She is the author of *Eye for Detail: Images of Plants and Animals in Art and Science* (London: Reaktion, 2016), *The World of Carolus Clusius: Natural History in the Making, 1550-1610* (New York: Routledge, 2010). She is editor of *Conrad Gessners Thierbuch. Die Originalzeichnungen* (Darmstadt: Wgb, 2018), and co-editor of *Correspondence and Cultural Exchange in Early Modern Europe, 1400-1700* (Cambridge: Cambridge University Press, 2007).

Alberto Gómez Gutiérrez is Professor of Human Genetics at the Pontificia Universidad Javeriana and Member of the Colombian Academy of Exact, Physical, and Natural Sciences. He is the author or co-editor of the following works, all published by the Editorial Pontificia Universidad Xaveriana in Bogotá: *Scientia Xaveriana: Los jesuitas y el desarrollo de la ciencia en Colombia, siglos XVI–XX* (2008), *Medicina científica mutisiana* (2008), *Filosofía natural mutisiana* (2009), *A impulsos de una rara resolución: El viaje de José Celestino Mutis al Nuevo Reino de Granada, 1760–1763* (2010), *Academia mutisiana* (2011), and *Humboldtiana neogranadina* (2018).

Leoncio López-Ocón is Research Professor in the Department of the History of Science, Institute of History, CSIC, Madrid. He is the author of *Botánicos y biólogos en el Ecuador* (Quito: Banco Central del Ecuador, 2010), *The Scientific Commission of the Pacific: From Expedition to Cyberspace* (Madrid: CSIC, 2006), and *Los americanistas del siglo XIX* (Madrid: Iberoamericana Vervuert, 2005) and co-editor, with Jean-Pierre Chaumeil and Ana Verde, of *Breve historia de la ciencia española* (Madrid: Alianza, 2003) and, with Carmen Mª Pérez-Montes, of *Marcos Jiménez de la Espada: Tras la senda de un explorador* (Madrid: CSIC, 2000).

Gabriela Goldin Marcovich is a postdoctoral fellow in the Instituto de Investigaciones Históricas at the Universidad Nacional Autónoma de México, working under the direction of Estela Roseló Soberón, PhD. She received her PhD in 2020 in history at the École des Hautes Études en Sciences Sociales, Paris. Her unpublished dissertation, *Creole Voices: New Spain's Intellectuals between Mexico City* *and the Italian Exile (1767–1814)*, reconstructs the transatlantic production of knowledge of a generation separated by the expulsion of the Jesuits in 1767.

Peter Mason lectures widely on the history of art and visual culture in Europe and the Americas. He is the author of The Modernists that *Rome Made: Turner and other Foreign Painters in Rome, XVI–XIX Centuries* (Rome: Gangeme, 2020), *El Drago en el Jardín del Edén: Las Islas Canarias en la circulación transatlántica de imágenes en el mundo ibérico, siglos XVI y XVII* (London: Reaktion, 2018), *The Ways of the World: European Representations of Other Cultures from Homer to Sade* (Canon Pyne: Sean Kingston, 2015), and *The Colossal: From Ancient Greece to Giacometti* (London: Reaktion, 2013).

Juan Pimentel is Research Professor in the Department of History of Science, Institute of History, CSIC, Madrid. He is the author of *Fantasmas de la ciencia española* (Madrid: Marcial Pons, 2020); *The Rhinocerous and the Megatherium* (Cambridge: Harvard University Press, 2017); *Testigos del mundo: Ciencia, literatura y viajes en la Ilustración* (Madrid: Marcial Pons, 2003); and *La física de la Monarquía: Alejandro Malaspina, 1754–1810* (Madrid: Doce Calles, 1998).

Irina Podgorny is Research Scholar at CONICET, Argentina, and Director of the Historical Archive at the Museo de la Plata, University of La Plata. She is the author of *Florentino Ameghino & Hermanos: Empresa Argentina de Paleontología Ilimitada* (Buenos Aires: Edhasa, 2021), *Los argentinos vienen de los peces: Ensayo de filogenia nacional* (Buenos Aires: Viterbo, 2021), *La momia que habla: Microensayos de historia natural* (Rosario: CB, 2020), and, with Margaret Lopes, of *El desierto en una vitrina* (Rosario: Protohistoria, 2008).

Neil Safier is Associate Professor in the Department of History at Brown University. From 2013 to 2021, he was Beatrice and Julio Mario Santo Domingo Director and Librarian of the John Carter Brown Library. He is the author of *Measuring the New World: Enlightenment Science and South America* (Chicago: University of Chicago Press, 2008).

Mark Thurner is Distinguished Professor of Anthropology, History and Humanities at FLACSO-Ecuador and Emeritus Professor of History at the University of Florida. He was Professor of Latin American Studies at the Institute of Latin American Studies, School of Advanced Study, University of London, from 2014 to 2021. He is the author of *History's Peru: The Poetics of Colonial and Postcolonial Historiography* (Gainesville: University Press of Florida, 2011) and *From Two Republics to One Divided: Contradictions of Postcolonial Nationmaking in Andean Peru* (Durham: Duke University Press, 1997). He is editor of *The First Wave of Decolonization* (New York: Routledge, 2019) and co-editor, with Juan Pimentel, of *New World Objects of Knowledge: A Cabinet of Curiosities* (London: University of London Press, 2021) and, with Andrés Guerrero, *After Spanish Rule: Postcolonial Predicaments of the Americas* (Durham: Duke University Press, 2003).

INTRODUCTION

Under Humboldt's Footsteps

Mark Thurner and Jorge Cañizares-Esguerra

> What one senses and hears [here in the Cortes of Cádiz] is an almost servile insistence on following in the footsteps of the Baron von Humboldt. But the blood of good Americans boils when he is cited as an authority on some species relative to America, not only for the abundance of errors but for the gross and detestable calumnies that he vomits on those ill-fated natives upon whom so much impunity is visited …
>
> – Manuel Olaguer Feliú[1]

Alexander von Humboldt's most enduring invention was probably himself. The proof lies in the fact that so many have for so long wished to 'follow in his footsteps.' Widened and deepened by generations of foot-stepping Humboldtians, Alexander's *huellas* have become veritable craters, inescapable footprints in the epistemological landscape.[2] Yet another reminder of this geohistorical fact was 2019. By virtue of yet another product of globalisation, the 250th anniversary of the Baron's birth was plagued by COVID-19, from which at the time of this writing the world still ails. Perhaps one day this book will be read as a one-shot vaccine for the 'Humboldt-19' virus, or as a seismograph for detecting the next Humboldtian eruption. But we doubt it. The Baron's footprints are simply too deep and too wide, and too much is concealed beneath them.

'Humboldt studies' is a crowded field with an irresistible life of its own driven by insatiable popular consumers and captive academic readerships. Behind these consumers and readerships stand formidable national and international cults institutionally embodied in memorials, museums, universities, schools, and clubs; behind these stand dozens of biographies aimed at both adult and child readerships; behind these stand hundreds of monographs, thousands of articles and chapters, a massive paper and digital archive, a host of TV and online documentaries and, not least, proliferating open access journals, many of which emanate from Germany, where Alexander

DOI: 10.4324/9781003231479-1

(together with his brother Wilhelm) is the object of a longstanding national cult with cosmopolitan pretensions. Most Humboldtiana is published in German and English (but curiously not in French, the print language that Alexander notably favoured), languages in which Humboldt has enjoyed yet another popular revival, in part thanks to the reception of Andrea Wulf's prize-winning *The Invention of Nature: The Adventures of Alexander von Humboldt, the Lost Hero of Science.*

Surely it is eternally useful and profitable for an editor or author to claim that the subject of their book is otherwise lost or forgotten. But there is little that is 'lost' about Humboldt. He is one of the most celebrated figures of modern science and was so long before Wulf's biography 'found' him yet again. The sheer mass of Humboldtian commentary is more than sufficient proof. That is not where the lost part is to be found. Instead, what is really lost about Humboldt is to be found under his footsteps and those of his 'Humboldtian' followers.

Wulf tells her grateful readers toward the end of her meandering, epic biography that she was inspired to write because Humboldt was a 'founding father' to ecologists and writers and because he bridged the arts and sciences. What she does not tell her readers is that many of Humboldt's Hispanic and Hispanic American contemporaries had routinely bridged the arts and sciences too and that some of them were 'founding fathers' for Humboldt. But the larger point is not to take away from Wulf's popularising narrative achievement. The success of *The Invention of Nature* is not an isolated event that may be traced to the shortcomings of one author. That success is underwritten by an epistemological mindscape and geopolitics of knowledge in which the otherwise ignorant South (in this case, the transoceanic Hispanic world) offers exotic objects of study to the North's enlightened 'discoverers' and 'heroes.' Although for any serious student of the Hispanic American Enlightenment this tired trope is plain to see in the English or German editions of Wulf's book, the Spanish translation makes it painfully obvious. Here, innocent Spanish readers, fanned by a fawning local press eager to 'follow in Humboldt's footsteps,' are coaxed into an Anglo-Germanic hero myth that almost completely erases the Hispanic American making of Humboldt's 'New World.'[3]

If Wulf's popular book is another best-selling brick in the wall, one more huge crater in the global epistemological landscape, we may be thankful that the considerable scholarship on Humboldt is more critical and sophisticated. On the teeming academic planet of Humboldt studies, Nicolaas Rupke's brilliant *Alexander von Humboldt: A Metabiography* (Peter Lang, 2005; Chicago, 2008) remains the most insightful and useful historical survey. Notably, its purview is predictably limited to the German and, to a lesser extent, English inventions of Humboldt. Surveying this field, Rupke deftly demonstrates that, in accordance with institutional, political, cultural, and calendrical cycles in Germany's modern history, various 'Humboldts' have been periodically refashioned to fit new agendas. He thus finds romantic, abolitionist, Nazi, communist, cosmopolitan, global, postcolonial, green, and gay 'Humboldts,' depending on the source, place, and period. Rupke's history of Humboltian history in Germany should be sufficient to dispel any notions of a heroic Humboldt of the kind Wulf paints. But alas, Wulf's success demonstrates that clearly it is not enough.

For us, the limit of Rupke's metabiographical approach lies not in the ironic, reflexive reading it performs – the author is keenly aware that his reading of German inventions is also an invention – but in its limited vernacular reach. 'Humboldt' is, of course, not only a German or an English phenomenon. Other 'Humboldts' are out there. Although it is not his focus or competence, Rupke does indeed take note of the reported existence of a 'Humboldt cult' in Latin America, mainly because that cult is in part the work of the agents of German cultural diplomacy. This 'cult' bloomed on cue for Alexander's 250th birthday party. German diplomats, accompanied by Humboldt experts and armed with books and euros and backed by a German media blitz of TV documentaries broadcast on the local networks, 'followed in Humboldt's footsteps,' albeit in jets and helicopters, spreading funds and commemorative plaques as they went.[4] The flowers of the cult's latest manifestation were many, and they are still being counted.

Our own party took place in Quito and on the slopes of Chimborazo during the happy, pre-pandemic month of August 2019. It was organised by LAGLOBAL and funded by the Ecuadorian campus of FLACSO (Latin American Social Science Faculty), a network of centres for postgraduate study founded in 1957 with the aim of decolonising the social sciences in the region. 'The Invention of Humboldt' workshop formed part of the inaugural, 2019 edition of the Chimborazo Summer School, an ongoing LAGLOBAL-FLACSO postgraduate training initiative in the global history of knowledge.[5] The Chimborazo Summer School seeks to recast 'Chimborazo' not as a metonym for Humboldt or Humboldtian science (see the Pimentel chapter in this volume) but as a critical vantage point from which to reshape the global epistemological landscape. One of the first fruits of the Chimborazo Summer School is the publication of this book, notably delayed due to the local and global repercussions of the COVID-19 pandemic.

Although the Latin American 'Humboldt cult' has surely resonated with the more well-funded, celebrated, studied and promoted German and Anglo cults, there is more than sufficient reason to suspect that Humboldt's moon over Hispanic America reflects other suns.[6] To be sure, nearly everywhere in the region Humboldt is periodically adulated and not without reason, but in Cuba, Mexico, and Ecuador, he is not quite the deified ancestor that he is in Germany, as several contributions to this volume reveal; in Colombia, the reception of Humboldt is more nuanced, in part perhaps because such rival savants as Caldas, Mutis, or López Ruiz remain thorns in his side (see the Amaya and Gómez chapters in this volume).[7] In Peru, Humboldt's legacy has been even more ambivalent. Nor, in the case of Peru, is this ambivalence new. Suspicion and criticism of Humboldt have made their way into print since at least 1811, when an exasperated Ramón Olaguer Feliú, one of Peru's elected representatives at the Cortes de Cádiz and who was in Lima when Humboldt visited, slammed the Baron and those who gleefully 'followed in his footsteps.' Olaguer's dissenting comment, discussed in Thurner's chapter on Peru, reveals that Humboldt's dim view of the intellectual and artistic capacity of the natives was used at the Cortes of Cádiz (1810–1813) by those who argued against granting them citizenship.

Rupke notes in the closing pages of his book that in more recent years German Humboldt scholars have sought to engage Latin Americanists, the result being an opening in the digital horizon. This is surely true, and it has produced and will continue to produce important insights and exchanges. But there is also considerable literature in the Spanish language that dates to the period of Humboldt's voyage. It is for the most part independent or critical of northern European trends. With notable exceptions, some of which are discussed in the chapters that follow, much of this literature tends to be national in scope and audience, often focusing on Humboldt's travels in, or writings on, 'my country.' As a result, there are at least as many literary shades of 'Humboldt' in Spanish as there are national territories carved from what was then the Hispanic 'Empire of the Indies.' In short, probably most 'Humboldts' are American, not German or English.

The cubist portrait or montage that emerges in the pages that follow reflects something of Humboldt's varied reception across the region but in no way exhausts it. It conjures not a solemn image of a heroic, solitary genius in the jungle of the kind that routinely adorns the covers of most books on Humboldt, Wulf's and Rupke's included, but instead a montage of ambiguities. The 'Humboldts' that emerge here include an anti-slavery liberal and champion of yeomanry at home among prominent slaveholders in Cuba, Venezuela and the United States; a liberated man who contracted a mestizo slave ('José') to lug his barometer up and down the Andes; an incessant polyglot scribbler who published his illustrated travelogues in cosmopolitan French but not in the vernaculars of the host countries where he learned the most; the Prussian who denounced Inca despotism and Creole gambling in Lima but later published coffee-table albums of antique exotica for Parisian salons; the privileged guest who shared his host's state secrets with rival powers, all supposedly in the name of Enlightenment; the consummate networker, gatekeeper, and promoter who often failed to give due credit to his sources, collaborators, and rivals.

Despite all the enticing cover portraits of a serene Humboldt set in New World nature, it is abundantly evident that the Prussian traveller was not a solitary 'genius' who invented 'nature' on the slopes of Chimborazo or in the jungles of Venezuela and New Granada. 'Nature' was on old concept whose modern global study was not only, as is often claimed, a northern European invention but was instead gainfully pursued across the early-modern transoceanic Hispanic world that, incidentally, included much of Italy, as Egmond's contribution to this volume attests. Humboldt did not 'introduce' the 'New Continent' to enlightened natural science. Enlightened science was already everywhere he went because it was based in part on the findings and interpretations of early modern Hispanic science, as the contributions of López-Ocón and Cañizares-Esguerra demonstrate. At each step, Humboldt followed in the networked footsteps of Hispanic American intellectuals, informants, and archives. But Humboldt's footprints have largely erased theirs.

Humboldt was granted a royal passport not to venture into an unexplored 'New Continent' but instead to move about a modern empire driven by science and whose workings he only gradually and imperfectly came to understand. It was by no means an unprecedented event: the Hispanic Crown sponsored more scientific expeditions in the eighteenth century than any other world empire. The Crown also supported numerous American-based institutions and academies for the specialised study of mining, botany, zoology, statistics, and the history and customs of local societies. Many of these initiatives were part of an eighteenth-century project of imperial reform in political economy that in many cases responded not only to designs conjured in Madrid but mostly to local petitions, interests, and insurrections, which in turn had built upon a long tradition of scientific inquiry and governance by petition that had begun in the sixteenth century. In this enlightened milieu of reform and experimentation, everywhere Humboldt went he found willing and capable interlocutors. But the Prussian also encountered learned individuals and communities duly sceptical of his motives, methods, and findings. They were often well positioned to critique his work as derivative if not potentially manipulative, self-aggrandising, and biased against Spain and its Empire of the Indies.[8]

Decades before Humboldt's arrival on the scene, La Condamine and his French academicians had visited the equator in what was then the southern province of the New Kingdom of Granada to settle scores between the Cartesian and Newtonian positions on the question of gravitation. It was a European debate that took place on Ecuadorian terrain under the auspices of the Hispanic Crown. The terrain and its inhabitants were also protagonists. To begin, the Hispano-Franco expedition was deliberately concerted by the Crown to prevent the epistemological erasure of Hispanic science, albeit to little avail. Enlightened Hispanic naval officers assigned to the expedition, trained in Madrid, Seville, and Cadiz, fully expected the French academicians to produce reports prejudicial to the reputation of the Empire, silencing or underplaying Hispanic contributions to knowledge. As expected, as soon as the French arrived in Quito, they made evident their unwillingness to respect Hispanic institutions and learned traditions. Among other acts, La Condamine and his peers had commemorative pyramids erected in the places where they had developed baselines of calculation, adorned with Latin texts that omitted Hispanic contributions. After a fray, local authorities had the pyramids destroyed.[9] The Crown invested in new printing presses and types to transform the reports of Antonio de Ulloa and Jorge Juan into digestible, alternative accounts that could circulate in learned circles. Indeed, upon their return to Europe, Ulloa and Juan were eventually inducted into the British Royal Society and the French Academy of Sciences, largely due to the success of these reports. These treatises were mostly ignored by Humboldt and, subsequently, by the historiography of science, as Neil Safier notes in his contribution. Sadly, traces of this erasure are still evident today at the 'Mitad del Mundo' theme park in Ecuador.[10]

The notable fact that Humboldt's two-man expedition was *not* state-sponsored, organised, or deliberate, may partially explain why the Crown did not assign a competent scholar of the likes of an Ulloa or Juan to protect its epistemological authority and secure its reputation. As Humboldt moved through Hispanic America, local authorities and Creole polymaths seriously doubted whether he would fully acknowledge their perspectives and contributions.[11] Francisco José de Caldas was proposed for the job in New Granada but to no avail. Humboldt had other companions in mind. Several chapters in this volume serve to demonstrate that intellectuals across the region found occasion to question, critique, and offer alternatives to many of Humboldt's observations and claims. Thus, Humboldt's *Géographie des plantes* (1806) was promptly translated and printed in the *Semanario de Nueva Granada* in 1809 with abundant critical notes by Caldas, who identified numerous errors. As Gómez Rodríguez demonstrates in his contribution to this volume, Caldas had developed a more accurate biogeography on the slopes of Imbabura. In Peru, as Thurner shows in his chapter, Humboldt's Orientalist, antiquarian and ethnographic speculations about the Inca and Peruvian civilisations were rebuked by erudite Creoles whose research eventual eclipsed the Prussian's dangerous missteps in this key area, although he did not cease to attract followers for that reason. Another notable example of contesting Humboldt was Vicente Talledo y Rivera, a Spanish engineer and cartographer detailed to New Granada who supported the royalist cause in the wars of independence.

A highly accurate map of the course of the Magdalena River from its mouth in the Atlantic to the Province of Velez was published in London in 1820.[12] It was not Humboldt's. Talledo's map superseded Humboldt's chorographic map of the river, drawn in 1801. It was part of an effort to re-engineer the flow of the river at dangerous rapids and straights.[13] Along with his commission to map the river, Talledo received a copy of Humboldt's map from the viceroy, with an order to confirm the Prussian's observations. He remained in New Granada for sixteen years, roaming up and down the rugged course of the Magdalena River countless times while mapping the viceroyalty. He witnessed the unfolding of the revolutionary wars, taking an active military role in the defence of the Province of Santa Marta, until he finally surrendered its fortification to patriot armies in 1812, whence he was removed to Panama. Talledo returned to Spain in 1818, but not before selling a copy of his detailed map of the Magdalena to the British agent of the London map publishers Laurie and Whittle. Humboldt first accessed this map in 1821.[14] In a note added to his *Personal Narrative*, Humboldt used Talledo as a foil.[15] Humboldt concluded that Talledo had merely copied his own calculations of the longitude and latitude at every spot on the map, from Turbaco to Nare. In a brief but devastating addendum, he ridiculed Talledo for having used his 1801 astronomical observations. Why had Talledo ignored his more recent and more accurate *Collection of Astronomical Observations*? Humboldt's published innuendo on Talledo became a diatribe in his diary, where he made more explicit the cause of his outrage: an ignorant 'Catalan' had dared to question his work.[16]

Talledo was not 'Catalan' but instead born in Valencia to a Cantabrian family. He did not simply copy Humboldt's observations. His map was also an effort to correct the many errors that plagued Humboldt's map. For example, he located the port of Guarumo in Boyaca 11′ 20″ of a degree further to the west than did Humboldt.

Unsure whether his observations were the result of his own lack of instrumentation (not having a chronometer), Talledo took a long detour to Muzo, where he could observe Guarumo from the Andes and thus triangulate the port's position using known mountain elevations. Once he had confirmed his original measurement of Guarumo, he then returned to Bogotá to measure the capital city's latitude and longitude, suspecting that Humboldt had also got them wrong. After measuring the immersion of Jupiter's satellites using Bogota's observatory and a chronometer, Talledo corrected Humboldt once again. Bogotá was located 10′ of a degree west of Humboldt's reading.[17]

Humboldt's reaction to Talledo is not untypical of his attitude toward the Hispanic science produced in the places he was allowed to visit and study, particularly when it took issue with his observations or claims. More, the story of Humboldt and Talledo is emblematic of the modern invention of Humboldt. It was not only that Talledo assembled more accurate empirical observations or that he did so at the behest of the Crown's imperial navy. Talledo had also spent thirty more weeks exploring the Magdalena than did the Prussian. Humboldt's six weeks there, however, left enduring 'footprints' that others, blissfully or wilfully ignorant of the works of the likes of Talledo, would follow. These followers 'in the footsteps of Humboldt' often ignore that part of the reason for their ignorance of Talledo's map is that Humboldt cast doubt on its maker.

As López-Ocón notes in his contribution to this volume, the transatlantic invention and apotheosis of Humboldt began in the context of the Age of Revolution and the Hispanic American wars of independence. The wars in Europe steered the course of Humboldt's haphazard voyage, and in America, they prompted many partisans of the new republics to recast 'Spain' as an empire not of science and exploration but of inquisition and ignorance. But Spanish scientists who read Humboldt's work later came to question the Humboldtian narrative of heroic invention, noting that the Prussian had built up his opus on the back of an earlier Hispanic science. Nation-builders on both sides of the Atlantic had uplifted a few heroes of science and philosophy as founding fathers. The new Latin American 'Galileos' included, in the case of the new state of Colombia, men such as Columbus, Caldas, and Humboldt.[18] Notably, Bolivar christened Humboldt a 'second Columbus' and the 'true discoverer of America.' But as Thurner and Cañizares-Esguerra both note in their respective chapters, Humboldt was neither except in an ironic sense that most commentators have failed to grasp.

Many creole patriots invented their own Humboldt cult, in the process belittling or eliding the Hispanic Enlightenment. Claiming Humboldt as heir to the knowledge assembled by eighteenth-century institutions sponsored by the Bourbons would have implied acknowledging the massive Spanish investments in enlightened cosmography, medicine, and natural history. Such recognition would undercut patriot claims to Spanish ignorance, misrule, and inquisitorial persecution, primary justifications for independence. Humboldt was also a foil for those local 'Galileos' who now claimed to have been marginalized and ignored by an obscurantist, 'medieval' empire. Beyond the region, creole patriots seeking diplomatic recognition in European courts found Humboldt's work timely and useful as well. (Anglo-American empire-builders also found Humboldt useful in their efforts to expand at the expense of Spain and Mexico.) The Prussian 'Columbus' had assembled massive amounts of data culled from local scholars and imperial bureaucracies while visiting Spain, Cuba, Venezuela, New Granada (Colombia and Ecuador), Peru, and Mexico. Humboldt's profusion, mostly published in the newly prestigious French language, was now cast as the product of a singular, cosmopolitan genius. His work offered a visually impressive, seemingly encyclopaedic compilation that, unknown to many of its readers, was based upon the eclipsed scientific tradition of 'ignorant' Spain. Humboldt's work could now serve to demonstrate to wary British capitalists that financial support and political recognition would be handsomely rewarded with merchant, mining, and agricultural contracts. Creoles could have their cake and eat it too: Humboldt's opus offered both the erasure of Spain and recognition of the new republics.[19] The acrobatics transformed Humboldt into a nineteenth-century demigod, the greatest polymath scientist of the century, the fitting intellectual parent of Darwin, and today's hero of the Anthropocene.

The image of Humboldt as lone genius was not only invented by anti-Hispanists but also by chance. The 'Second Columbus' who inspired Darwin arrived in America by accident, and once there his travels were governed mostly by happenstance and improvisation. In most cases, the plan or narrative of the voyage was construed retrospectively, after the fact. Planning would have forced Humboldt to visit the Indies with a coterie of rival Hispanic scientists who could have secured recognition of Hispanic contributions to knowledge. The young Humboldt sought to circumnavigate the globe and survey the Pacific, as his friend and mentor Johann Georg Forster had on Cook's second voyage, but international events intervened. When the option to travel to Spain and Hispanic America presented itself, Humboldt's plan (or at least that is what he tells us, retrospectively, in his rewritten travelogue) was still to rendezvous in Lima's port of Callao with the voyage of French captain Nicolas Baudin. If not for this desire to meet up with Baudin, it is unlikely that Humboldt would have made his way to the Andes to meet Mutis, let alone attempt to climb Chimborazo. In the end Baudin took another route, leaving Humboldt high and dry. Months later Alexander departed from Lima's port of Callao on a royal frigate for Guayaquil, whence he hoped to board a commercial vessel to Acapulco. While in Guayaquil the Cotopaxi volcano erupted high in the sierras. Humboldt rushed back up the Andean foothills to witness the event but, he tells us, was forced to turn back at Babahoyo when word arrived of the imminent departure of the awaited frigate. From Acapulco Humboldt could have sailed east to the Philippines and beyond but instead decided to stay in New Spain, where he spent a very productive year, thanks to his Mexican hosts who, as in Colombia, were in several notable cases not justly rewarded for their generosity. When Humboldt finally left Mexico behind, departing from the Gulf port of Veracruz, he sailed not to Spain as expected but instead north to Philadelphia, where he shared with Jefferson and others state knowledge he gathered from Mexicans and Mexican archives. From the US, Humboldt embarked not for Spain but Bordeaux and from there made his way to Paris. Despite his heralded return to Paris, Humboldt's luck as a traveller did not improve much. He made short but important trips to Italy and England, but his dogged efforts to reach India were blocked, this time by London. His legendary carriage ride across the Russian steppe was an alternative, last-ditch effort to reach the mountain ranges of the East so that he could 'compare' them with the Andes, now as the czar's mine inspector. *Kosmos* was the final product of his itinerant voyage, and the Americas is written all over its face. Despite its American writing, Humboldt's face would be portrayed and remembered as that of a solitary genius pioneering the science of nature in the American wilderness. The massive scientific output of the Hispanic Enlightenment would be assigned to the dustbin of the Age of Revolution. The geopolitics of knowledge in that Age and in our own has served to erase or marginalise that output. The chapters in this volume are only a modest beginning in the scholarly and geopolitical work of restoring, recognising, and rewriting the critical place of the Hispanic world in the global history of knowledge.

Chapter Synopses

In 'The Apotheosis of Humboldt,' Leoncio López-Ocón explains how Humboldt rapidly became a demigod on both sides of the Atlantic. The iconic invention of Humboldt as a saint of science and progress was the product of transatlantic scientific and political networks, which served to cultivate local legacies. In Spain, notable scientists negotiated Humboldt's legacy, arguing that the Baron's findings and indeed what we now call 'Humboldtian science' at large, were deeply indebted to an earlier, Hispanic science. Humboldt himself occasionally recognised this debt, although never fully or consistently.

In 'A Sense of Place,' Florike Egmond argues that Humboldt's fieldwork practices in geology and botany were shaped by an early modern natural history tradition that, in its continental European iteration, often privileged Italian sites of knowledge. We know that mining in the German lands was Humboldt's first 'working world' and one source of his keen interest in botanical and geological 'layers' or 'verticality,'[20] but Egmond's chapter pushes our understanding of Humboldt's practice back to the early modern period and beyond the German lands. Peter Mason's 'Six Days on Tenerife' takes a more punctual look at Humboldt's travelling scientific practice, tracing the emergence of Humboldt's 'tropical antique' to his stopover in the Canary Islands and a subsequent trip to Rome, where the Prussian traveller stitched together his botanical, geological, and antiquarian interests.

In 'Caldas and Humboldt in the Andes,' Alberto Gómez Gutiérrez presents a comparative chronology of the emerging biogeographical practices of Francisco José de Caldas and Alexander von Humboldt. His detailed analysis of notes, drawings, citations, and publications restores Caldas to his rightful place in the history of science as co-founder of what is now called 'biogeography.' It can no longer be said that Humboldt is the sole inventor of this field, as was proclaimed in 2019 at the International Biogeography Society meeting convened in Quito to celebrate

Humboldt's 250th anniversary.[21] In 'An Archaeology of Mutis's Disappearing Gift to Humboldt,' José Antonio Amaya unearths the hidden traces of José Celestino Mutis's generous but now mostly lost gift of an unprecedented collection of botanical drawings and specimens. This gift was privately entrusted to Humboldt and Bonpland and intended for the National Museum of Natural History in Paris. Amaya provides a detailed reconstruction of the ways in which the work of the Crown-sponsored but locally run 'Botanical Expedition' in Bogotá was absorbed by Humboldt and Bonpland without providing due citation of the source.

Revisiting Humboldt's 'Incas, Pyramids and Amazons,' Neil Safier cracks open the 'black box of Humboldtian science' and 'Humboldtian writing,' where he finds the looming narrative shadow of La Condamine, which, in turn, tended to blot out other narrative possibilities, including those brilliant ones of the Hispanic savant Antonio de Ulloa. Next, Juan Pimentel reclimbs 'Humboldt's Magic Mountain,' reminding us that Humboldt's failed attempt to reach Chimborazo's summit in June 1802 occupied, and continues to occupy, a privileged place in his own self-invention. Such heroic legends of fame and destiny are built on necessary ellipses and oblivions, and these, in turn, Pimentel reminds us, intervene in the history of knowledge. In 'Peruvian *desencuentro*,' Mark Thurner retraces Humboldt's ambivalent 'footsteps' in Peru, focusing on enlightened debates on the origins of Inca civilisation. The debate was won by Peruvian polymath Joseph Hipólito Unanue and his followers, who collectively outshined the Baron's dark, Orientalist footprint in Peru.

In 'Air in a Flask,' Miruna Achim and Gabriela Goldin Marcovich explore the historical making of three of Humboldt's key objects of knowledge: Erythronium, a metal; Cochineal, the dye made from the body of an insect; and Xochicalco, an archeological site to the south of Mexico City. This potent trilogy of 'Humboldtian' objects of knowledge was, the authors reveal, made in erasure, fragmentation, loss, and the mistranslation of Mexican sources. On the political economy side of things, in 'Humboldt's Misreading of the Mercantilist Face of New Spain,' José Enrique Covarrubias argues that Humboldt's Physiocratic-Smithian approach both took advantage of, and turned a blind eye to, certain key features of the *novohispano* or Mexican economic regime. Unlike many accounts that celebrate Humboldt's *Political Essay on the Kingdom of New Spain*, Covarrubias explores its conceptual assumptions and practical limitations, in particular Humboldt's influential approach to demography, transportation, and agricultural labour.

Moving south to Argentina and to Humboldt's companion Bonpland, who took up residence in the River Plate, Irina Podgorny argues that the Frenchman's botanising responded to French aristocratic tastes, widespread botanical ignorance, and the creation of markets in exotic plants on both sides of the Atlantic. In short, 'Bonpland's Cactus' tells a prickly story of wide dimensions. Finally, in 'Humboldt's Columbus' Jorge Cañizares-Esguerra argues that Humboldt's romantic and enlightened invention of Columbus was an exercise in self-fashioning built on and at the same time ignorant of the deep archival scholarship of Spanish historians. Humboldtian ignorance of the worlds that made the Hispanic archive continues to underwrite those who would 'follow in his footsteps.'

In summary, the chapters of *The Invention of Humboldt* outline a new task for post-Humboldtian 'Humboldt studies': The archaeology of all that lies under Humboldt's footsteps.

Notes

1 'Discurso del Señor Feliú en que hace la apología de los indios contra las imputaciones del Barón de Humboldt,' *Noticias del Perú*, n. 8, 1–12. Excerpt translated by Mark Thurner.

2 Here we play on the double meaning of the Spanish word 'huella,' used by Olaguer Feliú and others to characterise Humboldt's legacy. 'Huella' may be translated both as 'footstep' and 'footprint.' The point is that by following in his footsteps, Humboldtians have left deep footprints in the epistemological landscape. In addition, Humboldt himself left many footprints of his footsteps in the form of a visual and textual archive of vast proportions.

3 See Mark Thurner and Jorge Cañizares-Esguerra, 'La invención de Humboldt y la destrucción de las pirámides de La Condamine,' *Procesos* 51 (June 2020), 201–204.

4 The manner of Wulf's triumphal arrival in Ecuador in 2019 was another example: the author accompanied the German president Frank-Walter Steinmeier in a helicopter tour and media blitz of the country.

5 For further information on LAGLOBAL, FLACSO, the Invention of Humboldt workshop, and the Chimborazo Summer School, please visit https://flacso.edu.ec/laglobal/.

6 Nicolaas A. Rupke, *Alexander von Humboldt: A Metabiography* (Chicago, IL: University of Chicago Press, 2008).

7 Although more historical research is needed, German cultural diplomacy combined with the groundwork of the region's German schools, communities, and 'Humboldt clubs,' accounts for a good measure of the celebratory manifestations of the 'Humboldt cult' in Latin America, as was plainly evident in 2019.

8 Jorge Cañizares-Esguerra, *How to Write the History of the New World* (Stanford, CA: Stanford University Press, 2001).

9 Neil Safier, *Measuring the New World: Enlightenment Science and South America* (Chicago: University of Chicago Press, 2008); Antonio de la Fuente and Antonio Mazuecos, *Los caballeros del punto fijo: ciencia, política y aventura en la expedición geodésica hispanofrancesa al virreinato del Perú en el siglo XVIII* (Madrid: CSIC, 1988).

10 Thurner and Cañizares-Esguerra, 'La invención de Humboldt.'

11 José Antonio Amaya, 'Cuestionamientos internos e impugnaciones desde el flanco militar a la Expedición Botánica,' *Anuario Colombiano de Historia Social y de la Cultura* 31 (2004), 75–118.

12 This account of Talledo is based on Sergio Mejía, *Cartografía e Ingeniería en la Era de las Revoluciones. Mapas y obras de Vicente Talledo y Rivera en España y el Nuevo Reino de Granada, 1758-1820* (Madrid: Ministerio de Defensa, 2021).

13 See Manuel Lucena de Giraldo, 'Ciencia y crisis política. La creación de la Escuela Náutica de Cartagena de Indias (1810-1822),' *Revista de Historia Naval* 30 (1990), 31–38.

14 Mejía, *Cartografía e Ingeniería*, passim.

15 Alexander von Humboldt and Aimé Bonpland, *Personal Narrative of Travels to the Equinoctial Regions of the New Continent during the years 1799–1804* (London: Longman, 1829), v. 5, part 2, annex, 1–3.

16 Alexander von Humboldt, *Aus seinen Reisetagebüchern zusammengestellt und erläutert durch Margot Faak; mit einer einleitenden Studie von Kurt-R. Biermann* (Berlin: Akademie-Verlag, 1986), 115–116.

17 Mejía, *Cartografía e Ingeniería*, 146.

18 Lina del Castillo, 'Caldas as Galileo: Republican Print Culture Invents an Obscurantist Monarchy to Legitimate Rule,' *Anuario Colombiano de Historia Social y de la Cultura* 45:2 (2018), 89–111. For the connection between science and Independence, see Thomas F. Glick, 'Science and Independence in Latin America (with Special Reference to New Granada),' *Hispanic American Historical Review* 71:2 (May 1991), 307–334; José Luis Peset, *Ciencia y libertad: El papel del científico ante la Independencia americana* (Madrid: CSIC, 1987). See also Miguel de Asúa, *La ciencia de Mayo: La cultura científica en el Río de la Plata, 1800–1820* (Buenos Aires: Fondo de Cultura Económica, 2010).

19 Lina Del Castillo, 'Entangled Fates: French-Trained Naturalists, the First Colombian Republic, and the Materiality of Geopolitical Practice, 1819–1830,' *Hispanic American Historical Review* 98:3 (2018), 407–443.

20 See Patrick Anthony, 'Mining as the Working World of Alexander von Humboldt's Plant Geography and Vertical Cartography,' *ISIS* 109:1 (2018), 28–55.

21 International Biogeography Meeting, Universidad San Francisco de Quito, Quito, Ecuador, August 5–9, 2019.

1

THE APOTHEOSIS OF HUMBOLDT DURING THE NINETEENTH CENTURY

Leoncio López-Ocón

TRANSLATED BY MARK THURNER

Few scientists have accumulated as many honours and recognitions as Alexander von Humboldt, both on Earth and beyond. An asteroid is named after him (54 Alexandra) and on the Moon his footprint is present in the *Humboldtianum Sea*. On our planet, Humboldt's name has served to designate animal species such as *Conepatus humboldtii* (Humboldt skunk) or *Broteas humboldtii* (Venezuelan scorpion) and plant species such as *Quercus humboldtii* (South American oak) and *Lilium humboldtii* (Humboldt lily). His name also adorns national parks and monuments (such as the caverns in the Venezuelan state of Monagas), nature reserves conveniently situated in various American countries, and far-flung climatic phenomena such as the Humboldt Current, or geological features as far away as China, Australia, New Zealand, New Caledonia, or the Canary Islands.

But the trace of the Prussian savant is evident not only in the eponymy of the components of nature but also in various elements of American culture, be it in the name of educational and scientific institutions, or as the subject of inspiration for artistic works such as the drama of the Venezuelan author Ibsen Martínez, *Humboldt y Bonpland taxidermistas*. His likeness also occupies city parks and squares, and his image also has a volatile presence, as in the stamps distributed among the philatelic.

Many elements may explain what can be called the 'Humboldt phenomenon.' Many observers have alluded to his grand communicative capacity. When in company, he was a great conversationalist; when alone, a great writer. He published more than 600 books and articles over nearly six decades since his first monograph on the mineralogy of basalt appeared in 1790. And he wrote thousands of letters. If that were not enough, since his death in 1859, a huge literature on his life and work has been generated. The Humboldt Research Centre in Berlin tallied 5,000 publications at the turn of century, and 453 publications per year in 1959, when the first centenary of his death was celebrated. This vast bibliographic production,

DOI: 10.4324/9781003231479-2

accentuated in the commemorative years of his vital trajectory, is what has led some authors to underline the existence of a 'Humboldt industry.' This industry has amplified Humboldt's own self-promotion, which included an active policy of publishing multiple editions and translations of his travel diaries and scientific works in the three most prestigious European languages of his time: French, German, and English, as Nicolaas A. Rupke has noted.[1]

That the Baron should become a hero of science during his own lifetime was more than a question of self-promotion, however. After his American journey and a half-century-long career as a researcher, Humboldt made fundamental contributions to several branches of knowledge. As Goethe noted, Humboldt was a fountain with many spouts.[2] Among these, his contributions to biogeography or the geography of plants, which he considered an essential part of 'General Physics' stand out as particularly noteworthy.[3] Here Humboldt was interested in drawing systematic relations between the organic and the inorganic, in particular, the influence of clime and elevation on the regional 'type' or landscape of vegetation.[4]

Thanks to his skills as a communicator and his studies on the 'physics of the globe' presented in his *magna obra*, *Kosmos*, Humboldt is also credited with creating a school or style of collective thinking that historians of science have coined 'Humboldtian science' to denominate a method or programme of research and dissemination.[5] In this effort, Humboldt sought to build a transatlantic scientific network, connecting the scientific and technical capacities on both sides of the Atlantic. This transatlantic network was later used by different societies and scientists of Europe and the Americas as a 'site of scientific memory,' sometimes shared among different countries but also appropriated by each of the American states forged during the nineteenth century, albeit to different political ends.

Little wonder, then, that after giving a series of lectures in Berlin in 1827–1828, Humboldt was awarded a medal with an image of the sun and the caption: 'Illustrans totum radiis splendentibus orbem.'[6] It is well known that in the ancient world knowledge of the past, present, and future was attributed to the Sun. The fact that his compatriots compared him to the sun king is incontestable proof of his *apotheosis*, that is, of his having received honours proper to the gods. This apotheosis was part of a long German process of mythologising von Humboldt that continues under various guises to this day.

In the pages that follow, I underline two key aspects of the process of deification of Humboldt. First, I suggest how Humboldt traced his path towards the Pantheon by way of his capacity to deploy and 'globalise' the knowledge he amassed on his American journey, thanks in large part to the networks spanning the Atlantic space. Second, I highlight how these networks shifted over time, such that in the decades after his death, Humboldt's texts and science continued to interpolate multiple readers in several European and American societies, thereby contributing to various national scientific traditions on both sides of the Atlantic. On this score, I devote particular attention to the nineteenth-century reception of Humboldtian science in Spain.

Building a Transatlantic Scientific Network

The global circulation of scientific and technical knowledge occurs via networks.[7] Created by scientific practice, networks act in a tentacular way, and their function is twofold: to ensure the circulation of scientific and technical objects and/or the flows of authority that legitimise the continuity of scientific practices. Since the texts produced by scientists differ depending on the target audience, several kinds of distribution networks for the circulation of scientific-technical objects are necessarily formed. These include *networks of co-responsibility* that group scientists with peers or colleagues, generating flows in which the authority of a work is configured; *patronage networks* that bind scientists to political or economic powers that may provide financial resources for their activities; and *popularisation networks* that link scientists with citizens, facilitating the organisation of broader audiences or readerships for their work. In this case, Humboldt's ability to exploit potent and heterogenous networks of communication served to render consistent a scientific practice that gradually acquired increasing authority in both scientific and non-scientific circles.

Several studies have emphasised the rhetorical skill, literary drive, and communicative power of Humboldt's writings, present especially in the narrative of his transatlantic voyage, in the descriptions of nature, and in his portraits of the vistas of the mountains and ancient monuments, which moved readers to contemplate in the mind's eye a harmonious whole of natural and cultural objects of intertropical America. In this regard, he is rightly considered one of the great nature writers, a point underlined by Juan Pimentel.[8]

The historiography has stressed Humboldt's care to measure and represent reality with various instruments and visual devices. With these devices, he composed an accurate atlas of the territory he visited and explored, generating admirable fragments of natural reality worthy of study and aesthetic delight, and in which it is possible to visualise Humboldt's scientific achievements. In this fashion, Humboldt managed to create one of the most powerful icons of geoscience: the *Tableau physique des Andes et pays voisins*, which was inserted in the *Essai sur la géographie des plantes* published in Paris in 1805. The *Tableau* was published shortly after his return from America and his installation in the French capital, at the time perhaps the primary node of the scientific world.[9] Sylvia Romanowski described it as 'a structure, in which each element depends on the other elements, and is inseparable from the whole' while Malcolm Nicolson sees it as a holistic vision of a unified landscape.[10] This elaborate engraving of a cross section of the Andes from the Atlantic to the Pacific Oceans at the latitude of Chimborazo presented a maximum amount of information with a determined objective and in a concise form, providing at a glance a complete impression of the 'equinoctial region' of South America. The tableaux maps the elevations at which certain species of plants grow, animals live, and kinds of agricultural crops are found while also noting the underlying geological structures and all manner of measurable parameters.

Attention has also been drawn to Humboldt's prodigious, life-long ability to exchange information with hundreds of interlocutors through thousands of letters, indeed more than 35,000 according to Charles Minguet.[11] What remains to be investigated is how this vast correspondence was put to the service of Humboldt's transatlantic cultural strategy. This strategy was, as noted, aimed at intensifying cultural and scientific relations between Europe and the Americas, capturing powerful allies, and seducing the public to support his work and research program. This program was transatlantic first, but ultimately, it should not be forgotten, global, since Humboldt's tree of knowledge culminated in *Kosmos* conceived as a 'physics of the terrestrial globe.'

The correspondence was one of the tools that allowed for the circulation of the scientific objects that Humboldt created over the course of his fruitful career, enabling his research program to progressively acquire volume and consistency. His ambition to become a *passeur* or intermediary between European and American science, and vice versa, can be seen clearly in one of the organs of expression of transatlantic science then being built in the transit between the eighteenth and nineteenth centuries.[12] I refer to the notable Madrid scientific journal *Anales de Historia Natural*, published from 1799 to 1804, that is, precisely during the years of Humboldt's American journey.[13] The journal was driven by a group of scientists committed to the project of making Madrid a hub for transatlantic scientific communication. Given the precarious situation of the Hispanic Monarchy at a historical moment when the crisis of the *ancien régime* and the dismantling of the empire loomed on the horizon, the project suffered ups and downs. At this moment, several scientific institutions in Madrid dedicated themselves to the processing, systematising and analysis of vast quantities of information gathered by the members of a vast scientific expeditionary movement that, in the second half of the eighteenth century, had participated in what has been called 'the reconquest of America.' These institutions included The Hydrography Depository, the Royal Cabinet of Natural History, the Royal Study of Mineralogy, and the Royal Botanical Garden.[14]

In the wider context of Spain's enlightened expeditionary movement, it is unsurprising that the journal followed with great interest the scientific pilgrimage of Humboldt in America. In numbers 2 and 6, which correspond to December 1799 and October 1800, the journal published extracts of Humboldt's letters to Baron Forell, Plenipotentiary of Saxony at the court of Madrid, dated in Cumana on July 16, 1799, and in Caracas on February 3, 1800, respectively.[15] The translator and editor, the mineralogist Cristiano Herrgen, was one of the promoters of the *Anales*. A professor at the Royal Study of Mineralogy trained by the prominent mineralogist and geologist Abraham G. Werner (1749–1817) at the Freiberg School of Mines where Humboldt also studied in 1791–1792, Herrgen was one of the many voices or spokesmen that the Prussian naturalist engaged in Europe to help spread the news of his American scientific discoveries. Such spokespersons played a key role in the deployment of Humboldt's transatlantic network.

How did Herrgen carry out this work as spokesperson for Humboldt? Essentially by exercising the function of messenger. He transported via translation Humboldt's science in action, and he began to practice a species of exegesis on the early results of the expedition to America. Not only did Herrgen pour into Castilian Humboldt's report on the release of caloric considered a geognostic phenomenon, published in German in 1799 in the *Annals of the Baron de Moll*, or translate some of Humboldt's American letters, as noted earlier; he used the pages of the *Annals of Natural History* to discuss and record work that Humboldt had remitted to scientific institutions in Madrid.[16] One such case involved the American rock collection Humboldt sent to José Clavijo, director of the Royal Cabinet of Natural History. Clavijo had been decisive in the Madrid court's decision to facilitate the passage to America of Humboldt and his partner Bonpland.[17] The rock shipment had as its primary objective 'the news communicated to the Baron von Forell about the disposition and direction of geological strata in South America, and its comparison with those of the Old World; an interesting problem, which I think I shall someday treat with greater clarity when I have examined more lands,' as noted in note 6 of the *Annals of Natural History* published in October 1800.[18]

Humboldt's transatlantic network of communication posed risks. At times, the readers and receivers of Humboldt's scientific adventures were so hungry for news that results were published prematurely. As a result, the author did not always have control of product quality. Indeed, Humboldt would present himself to us as a prisoner of the expectations raised. Humboldt thus appealed to the visible head of the Madrid circle of scientific communications, the powerful director of the Royal Botanical Garden, Antonio José Cavanilles.[19] In a letter sent from Mexico on April 22, 1803, and printed in n. 18 of the *Anales*, Humboldt begged that greater care be taken in the editing and circulation of the scientific information he transmitted. After acknowledging the praise showered upon him in note 15 of the *Annals*, demonstrating once again that scientific news did indeed move back and forth across the Atlantic, Humboldt called for the rectification of inaccurate data presented in a table of elevations, expressed in fathoms, of twenty-one sites in the Viceroyalty of New Granada between Cartagena and Santa Fe.[20] Here we sense the tension in Humboldt's communication strategy between the desire to accelerate access to his findings and the scientific exigency of precision in the titanic task of measuring American reality:

> If the frankness with which I have without reservation communicated my findings of plants, animals, geographic charts, and observations, permitting everyone who so desired to make copies, gave rise to the mentioned equivocation, it is also true that the occasion has allowed me to rectify the locations of several important sites, provided to me by knowledgeable persons. I would prefer that you print only what I, myself write in my letters or memoirs, because everyone knows that initial ideas are only a sketch to be concluded, and that the calculations and measurements always require further and more detained examination. The savants Condamine and Bouguer are proof of this truth. They thought their measurements were precise and conclusive, and so upon leaving Quito had the longitude inscribed in stone at the Jesuit College, even though that measurement differs by one degree from the one adopted in Europe.[21]

Despite the risks involved, the communication strategy of availing various publics of his findings accompanied Humboldt throughout his life. And this risky strategy partly explains why his scientific conduct became a role model for the legion of romantic European naturalists who, following in his footsteps during the first half of the nineteenth century, constructed and disseminated a vast inventory of nature and culture in the Americas.

Thanks to these nineteenth-century naturalists, the networks that Humboldt put into circulation were displaced and renewed in time and space. These follow-ers applied the quantitative and visual methods that, following a long tradition of Euro-American knowledge production, Humboldt had raised to a high degree of perfection, as any reader could see in such masterworks as the *Tableaux de la Nature*, the thirty volumes of the *Voyage aux régions du Nouveau Equinoxiales au Monde*, or the political and economic essays on Mexico and Cuba.

Notable Central European savants who followed in Humboldt's steps included Karl Wilhelm Koppe (1777–1837),[22] Carl Christian Sartorius (1796–1872),[23] and Karl Sapper (1866–1945),[24] among others, who followed the trail of Humboldt in Mexico and Central America, while Eduard Poeppig (1798–1868)[25] and Johan Jakob von Tschudi (1818–1889)[26] were inspired by the Baron in Peru, Bolivia, and Chile, and Prince Maximilian zu Wied (1782–1867),[27] the zoologist Johann Baptist von Spix (1781–1826), and the botanist Carl von Martius (1794–1868), in explorations of southeast Brazil and the Brazilian Amazon, respectively, all imitated Humboldt's methodological approach.[28]

Humboldt also encouraged many European naturalists to do what he had not done (although his companion Bonpland, did take the plunge), namely, resettle to the Americas and, working from her natural laboratories, build institutions of natural science and geography in the new republics. The work of many European naturalists who settled in the US or various Latin American countries during the middle decades of the nineteenth century – including Burmeister in Argentina, Philippi in Chile, Ernst in Venezuela, Raimondi in Peru, Codazzi in Colombia and Venezuela, Claudio Gay in Chile, Agassiz in the United States, and Ramón de la Sagra in Cuba – would be unintelligible without the powerful transatlantic influ-ence that Humboldt exerted.

But Humboldt's long tentacles did not reach America only via these central European intermediaries. He had a direct line to the political leaders of the emerg-ing republics, from Bolivar,[29] Jefferson, and Vicente Rocafuerte[30] to Santa Anna[31] and Lucas Alamán.[32] His influence was also enormous in the incipient scientific and technical communities of the recently emancipated republics of America, as well as in the Empire of Brazil. The agendas of scientific societies such as the Geographic and Statistical Society of Mexico, or the Historical-Geographical Institute of Brazil were influenced by Humboldt's work. Humboldt advised the team appointed by Bolivar to create an Institute of Sciences in Bogotá; distinguished members of Bolivar's Bogotá team included the French agronomist and chemist Boussingault and the Peruvian mineralogist Rivero.

By means of these allies, supporters, and interlocuters, Humboldt shaped a school of scientific practice in the Americas, on which pivoted the cult that was being built around his figure and that reached its paroxysm in the later years of his life, when Latin American intellectuals visited his palatial home at Tegel as if it were a sacred sanctuary. There are vivid descriptions of the rapture such visitors felt upon contemplating the savant in his *sancta sanctorum* or cabinet in Berlin, including a notable one by the Chilean historian and statesman Benjamín Vicuña Mackenna.[33]

Some governments in Latin America, most notably the Mexican, showed great interest in canonising Humboldt during his lifetime. The Mexicans sought to immortalise his name, graciously paying homage for services provided in the construction of 'national scientific culture.' In September 1827, for example, the Congress of the State of Mexico declared Humboldt and Bonpland honorary citizens of the state.[34] In 1854, General Urraga, emissary of Santa Anna, awarded Humboldt in Berlin with the Mexican Order of Guadalupe.[35] On September 14, 1857, Mexican president Ignacio Comonfort signed a decree to found three new cities in the Isthmus of Tehuantepec, to be named, respectively, after Columbus, Iturbide, and Humboldt.[36]

Shortly before his death Humboldt showed signs of fatigue at so much fame. He lamented the burden of his success, manifested in the crushing mass of correspondence that was by now impeding his activities. Charles Minguet, editor of Humboldt's correspondence, reproduces the dramatic appeal issued by Humboldt on March 15, 1859. Given the eloquence of this autobiographical testimony, which confirms once again that networks of scientific communication may cease to be operational by virtue of saturation, I quote at length:

> Crushed under the weight of an ever-increasing correspondence of on average 1,600–2,000 items per year (including letters, publications on topics totally unrelated to my interests, manuscripts whose authors seek my opinion, travel plans and colonial expeditions, models, machines and natural history objects, questions about air travel, enrichment of autograph collections, offers to take care of me, to entertain me, etc.), I beg publicly, once again, of all those who honour me with their favours, to please persuade those of both continents to shower less attention upon me and not use my house as a mailbox. In this way, I will be able to devote myself with gusto and tranquillity to my research, despite the decline of my physical and intellectual forces. May this call for help, to which I have resolved with remorse and much too late, not be interpreted as a sign of hostility.[37]

It is worth noting, however, that at the time he wrote this urgent and desperate public appeal to his correspondents, he was about to deliver the fifth and final volume of *Kosmos*. Shortly before this, Humboldt gave scientific instructions for the circumnavigation of the globe by the frigate *Novara* of the Austro-Hungarian empire.[38] In turn, that voyage would inspire, to some extent, the organisation of the Spanish expedition known as Scientific Commission of the Pacific.

A Sage for All Tastes

The cult of Humboldt was neither paralyzed nor diminished by the occasion of his death on May 6, 1859. On the contrary, as of that date both in Europe and America various initiatives would sustain, via various ritual actions, strong interest in his figure and work for the remainder of the nineteenth century. The scientific networks created by Humboldt continued to move *postmortem* through time and space, for two reasons. First, the very nature of the networks. As is known, networks built from laboratories extend across space and time. The production of facts is a continuous process and the mobilisation of the networks, based on various negotiations, is permanent.

The extent and strength of Humboldt's literary and iconographic inscriptions, taken from and initially processed in the natural laboratories through which he travelled and which sustained his communication strategy, were decisive in gaining the recognition of peers and, perhaps just as importantly, in his ability to conquer and move masses of readers in different societies. His knowledge thus circulated and expanded across national borders, becoming 'globalised.' But it also spread among diverse audiences, becoming popular. His writing, backed by powerful images, appealed to diverse tastes, and it continued to interpolate diverse audiences and far-flung geographical settings for decades. Many of these audiences kept Humboldt in the pantheon of science, through a series of recurring signs.

Everywhere politicians appealed to artists to sculpt his memory in stone. No sooner having learned of the savant's death, President Juárez, amidst of tribulations to consolidate his power, signed a decree on June 29, 1859, declaring Humboldt 'Benemérito de la Patria.' He further ordered that a life-sized marble statue be sculpted in Italy, to be installed in the School of Mines in Mexico City, as a 'public testimony to the esteem in which Mexico, and everyone else, holds him.' Humboldt's studies 'on nature and the products of our soil, on our political economic elements and on so many useful subjects' had brought 'honour and advantage to the Republic.'[39]

Many publicists worked in the 1860s and 1870s to sustain Humboldt's cult among the citizenry, evoking periodically his merits and exploits in scattered publications at various locales. For example, the Spanish promoter of science, José Genaro Monti, considered Humboldt a model to be imitated. In his article 'Astronomy' published in the Madrid magazine *El Museo Universal* on December 28, 1867, he pointed out to his readers that Humboldt had opened new avenues of scientific communication. 'The sciences' he noted

> did not become popular until Alexander von Humboldt, in our century, gave a new and concrete direction to all forms of knowledge by routing the sciences down never trodden paths. He stripped scientific observation of its aridity, presenting the results of the physical contemplation of the world to the intelligence of all in a clear and synthetic exposition.[40]

When a few years later in 1874 the Colombian consul in Le Havre, Adriano Páez, published the *Revista Latinoamericana* in Paris, Humboldt's imprint was there. In its pages, the Venezuelan physician, author, and promoter of science Arístides Rojas (1826–1894) published an article under the title 'Memories of Humboldt in America,' dedicated to Adolfo Ernst – the German naturalist considered to be the founder of the Venezuelan positivist school,[41] and as a gesture of his admiration for the travelling Prussian naturalist.[42]

Many scientists emulated and used Humboldt as a fundamental reference for their research programmes. Scientific associations surged and expanded in Latin America during the romantic and positivist eras. In many cases, they did so under the influence of Humboldt and in a few cases with explicit reference to his name.[43] The New Granada Society of Naturalists was founded in Bogotá in the year of Humboldt's passing (1859). The society intended to systematise indigenous knowledge about the flora of New Granada, resuming the research on useful plants carried out by the botanical expedition of Mutis, as well as by Humboldt and Bonpland.[44] The group of Venezuelan scientists who edited the journal *Vargasia* decided in 1869 to dedicate a special issue to him on the centenary of his birth. When a group of students from the Seminar on Mining in Mexico chose to become a scientific society in the 1880s, they agreed to adopt the name of 'Humboldt Society.'

European naturalists and expedition members who travelled to the Americas in the 1860s and the decades following continued to follow in his footsteps, adopting Humboldt as a guide; subsequently, and out of a desire to emulate, they strove to clarify and improve on his observations. This was the case, for example, of the *Commission scientifique du Mexique* organised by Napoleon III. Humboldtian influence was evident both in the works that were supplied to the commission and in the working materials as well as the instructions elaborated by the four committees in charge of the organisation of this complex mobilisation of French imperial science, which included the natural and medical sciences; physical and chemical sciences; history, linguistics, and archaeology; political economy, statistics, public works and administration.[45] The botanical expedition's instructions specified that the main objective was to make botanical maps in which one could appreciate the relations between geological and botanical phenomena, 'since the world's vegetation constitutes a whole whose parts are linked to one other by general causes that are none other than those large geological phenomena printed on the surface of our planet.'[46]

The Spanish naturalist Marcos Jiménez de la Espada was another example. A principal member of the Spanish *Scientific Commission of the Pacific* that traversed much of the Americas between 1862 and 1865, Jiménez sought to emulate Humboldt both in his investigations as a naturalist and in his historical works. Jiménez sought to ascend the 'Höhepunkte' or sacred places of naturalists, some of which Humboldt had sanctified and mythologised, such as the Chimborazo volcano in Ecuador.[47] He observed American nature and cultures with a totalising and harmonising Humboldtian gaze, while his historical works on Peruvian antiquities continued along the path blazed by Humboldt. One of the ways he paid homage to the man he called 'the Prussian Aristotle' was by publishing in 1889 in the *Bulletin of the Geographical Society of Madrid* details of Humboldt's journey from Quito to Lima, in the company of Bonpland and Carlos Montufar.[48]

Jiménez de la Espada was also keen to correct some of Humboldt's erroneous observations. For example, he studied the formation and characteristics of the eruptive materials at Lisco and Ansango, in the proximity of the Antisana volcano, revising Humboldt's findings.[49] He also inspected the crater of the Pichincha volcano. He used the occasion to point out that, contrary to the Prussian savant's claims, La Condamine was not the first to ascend Pichincha. Instead, it was the Licentiate Francisco de Uncibay, Oidor of the Audiencia of Quito, who had first climbed the volcano overlooking Quito in the late sixteenth century. Jimenez also seized upon his visit to correct some misleading, European perceptions of the volcano:

> Science is also martial, although without the trappings, gadgets and noise of the warrior; and although I do not recall receiving a promotion for my actions at Pichincha, I cannot complain about the booty I won: some boulders torn from the eruptive cone, showing the transformation of trachyte into pumice; two or three specimens, like honeycombs, of a very beautiful micaceous or scaly sulfur that I took with my hands from the fiery and spongy sulfuric beds that girdle the promontory; and, finally, a nest with its eggs, taken in one of the bushes on the central mound of the crater, proof of the security with which animal and plant life may subsist at the bottom of the abyss where La Condamine and Humboldt only saw the image of chaos.[50]

Although an admirer and to a certain extent disciple of Humboldt, whose footsteps he followed in his American exploration, and particularly so when in the Andes, Jiménez chose to grant more credibility to the power of his senses. The 'testimony of the eyes that see and the hands that feel' trumped 'the authority of the words universally believed and accepted as oracles' of truth.[51]

In some European countries such as France and Germany, Humboldt was worshipped because he was seen by some as part of their own scientific legacy and by others as a shining star that had passed through their respective scientific constellations. The cults were manifested in a permanent dialogue with Humboldt's legacy among German scientists like Ritter, or among Frenchmen such as Boussingault. This engagement, which included perfecting or increasing Humboldt's stock of knowledge, can be seen in the editions of correspondence promoted in Paris by La Roquette in the early 1860s or by Hamy in the first decade of the twentieth century, or in the biographies written by German scholars, such as the three volumes edited in Leipzig by Karl Bruhns in 1872.

In the Spanish case, it can be asserted that the cult of Humboldt did not produce great achievements in the form of a better study of his work. But it is true that the translation of some of his works, such as Bernardo Giner de los Rio's *Cuadros de la Naturaleza* (1876), in this case thanks to the initiative of the Gaspar press, demonstrated notable interest in the reception of Humboldt's scientific legacy in Spanish society toward the last third of the nineteenth century.[52] A group of scientists then involved in building a 'third way' in the so-called polemic of Spanish science, which I have analysed in another work, did pay attention to that legacy.[53] They endeavoured to enrich and value that legacy as one more proof that on the question of the development of Spanish science it was not necessary to show either 'a foolish pride or a sad discouragement,' according to the formulation of one of the theorists of that third way, the zoologist Laureano Pérez Arcas. This attitude is what made it possible for some of the defenders of this third way, grouped in the Spanish Society of Natural History founded by Jiménez de la Espada, to edit various materials related to Humboldt's American voyage. Among these is the notable work published between 1871 and 1872 by José María Solano Eulate in the Proceedings of the Society, titled 'Cartas ineditas del baron Alexandro de Humboldt con un facsimile' and chemist José Rodríguez Carracido's reflections on 'Alejandro Humboldt y la ciencia hispano-americana' in his *Estudios historico-criticos de la ciencia espanola*.[54]

Rodríguez Carracido's tribute to 'a glory of humanity' served to recall 'the glories of Hispanic American science.' The future rector of the Central University built his argument around three axes. He evoked the scientific culture that Humboldt found in the Hispanic world on both sides of the Atlantic to highlight how and in what ways it had influenced the research programme of the Prussian savant. In this regard, he highlighted how Humboldt had valued the work done in the botanical garden of Mexico that he had visited. He also noted the praise he had showered on the professors of the Royal Seminary of Mining of Mexico, Fausto Elhuyar and Andrés Manuel del Río, where Humboldt had found 'all the necessary elements to design the Mexican Geographical Chart.' Humboldt had found such 'a great wealth of data on the geology of the region and such a great abundance of scientific material for the study of Mineralogy and Chemistry' that he considered the seminar to be 'a second School of Freyberg.' Rodríguez presented Humboldt as a 'paladin of our intellectual interests.' He 'demonstrated in all of his publications a firmness of mind and a righteous sense of justice sufficient to return to us that part in the opus of civilization that belongs to us.' Rodríguez gave several examples, taken from his *Ensayo histórico sobre el desarrollo progresivo de la idea del Universo*, particularly in chapter VI, where he explains the 'development of the idea of the Cosmos in the XV and XVI centuries,' of how Humboldt helped to defend the contributions of Hispanic scholars to the knowledge of American nature and culture.

In this regard Rodríguez highlighted how

> Humboldt elevates Hispanic American science to the point of presenting it as a precursor to his own work when he declares that 'the fundamentals of what today we call the physics of the globe, dispensing for the moment with mathematical considerations, is contained in the book by the Jesuit José de Acosta, entitled *Historia natural y moral de las Indias*, and in Gonzalo Fernandez de Oviedo, published twenty years after the death of Columbus.'

And this 'at a time when we only heard loud accusations of ignorance, greed, and fanaticism raining down on our history with unanimous reprobation.' Humboldt had defended some of the first Spaniards who stormed the American continent,

> enraptured by the vehemence of curiosity, they threw themselves into the realization of their desires, excavating mountains, collecting plants and insects, learning languages of the indigenous people, studying their institutions and customs, and even observing the strange differences in that celestial vault; and all this … translated in works composed of elements of reality, conceived in the light of its many aspects, and written with the love of one who feels new revelations that dilate the horizons of human life.[55]

Rodríguez Carracido anticipated in nineteenth-century idioms what, *grosso modo*, later investigations, particularly those carried out in the last three decades, have highlighted. To wit: that science and technology were key elements in the construction of the Hispanic Monarchy throughout the early modern and modern eras; in short, that science and the Hispanic empire went hand in hand.[56] For Rodríguez Carracido, the knowledge gained and deployed throughout the colonial era was reappropriated by Humboldt, as he himself recognised to some extent.

Apropos of Humboldt's debts with the scientific activities carried out in colonial America, it should be added, finally, that his research programme is unintelligible without his interactions with the principle scientific projects of the American viceroyalties toward the end of the colonial period. Unfortunately, this fact has been passed over in Eurocentric narrative and historiography.[57] I refer to the scientific projects organised in the metropole under the reign of Charles III when expeditions were organised to 'reconquer' the colonies. The idea was to bolster the proper scientific traditions of the viceroyalties, promoted by the viceroys and aided by military engineers. This was done for the purposes of organising the defence of the territory, fomenting public works, and caring for public health. This scientific work included undertaking expeditions to border areas with rival empires while attending to the education and cultural demands of the American population.[58] On a few occasions, Humboldt recognised his debts with these late colonial Hispanic and American scientific projects, as in the pages of his *Political Essay on the Kingdom of New Spain*.[59] On other occasions, he occluded or undervalued them, as in the case of the geobotanical knowledge he gained from the Creole botanist Francisco José de Caldas.[60]

Notes

1 Nicolaas A. Rupke, *Alexander von Humboldt: A Metabiography* (Frankfurt am Main: Peter Lang, 2005), 20–27.
2 Rupke, *Alexander*, 46–47.
3 For a characterisation of the 'general physics' see Susan Faye Cannon, *Science in Culture: The Early Victorian Period* (New York: Science History Publications; Folkestone, England: Dawson, 1978), 73–110.

4 Malcolm Nicolson, 'Alexander von Humboldt, Humboldtian Science and the Origins of the Study of Vegetation,' *History of Science* 25:2 (1987), 167–194.

5 Cannon, *Science*; Michael Dettelbach, 'Humboldtian Science,' in *Cultures of Natural History*, eds. Nicholas Jardine, James A. Secord and Emma C. Spary (Cambridge: Cambridge University Press, 1996), 287–304; Gregory T. Cushman, 'Humboldtian Science, Creole Meteorology, and the Discovery of Human-Caused Climate Change in South America,' *Osiris* 26 (2011), 19–44.

6 Julius Löwenberg, Robert Avé-Lallemant and Alfred Dove, *Life of Alexander von Humboldt*, vol. 2, Karl Bruhns, ed. (trans. Jane and Caroline Lassell) (Cambridge: Cambridge University Press, 2012 [1873]), 120.

7 See in this regard Michel Callon, ed., *La science et ses réseaux. Genèse et circulation des faits scientifiques* (Paris: Editions La Découverte / Conseil de l'Europe / UNESCO, 1989).

8 Juan Pimentel, *Testigos del mundo. Ciencia, literatura y viajes en la Ilustración* (Madrid: Marcial Pons Historia, 2013), 179–210. Also see Ottmar Ette, 'Un espíritu de inquietud moral,' in *Humboldt Writing: Alexander von Humboldt y la escritura en la modernidad*, Special Issue of *Cuadernos Americanos*, XIII, 4/76 (July–August 1999), 16–43.

9 Michel Serres, '1800,' in Michel Serres, ed., *Historia de las ciencias* (Madrid: Cátedra, 1991), 381–410.

10 See Alexander von Humboldt and Aimé Bonpland, *Essay on the Geography of Plants*, edited with an introduction by Stephen T. Jackson, trans. Sylvie Romanowski (Chicago, IL: University of Chicago Press, 2009), 162, cited in David Oldroyd, 'Humboldtian Science,' *Metascience* 20 (2011), 581–584. And Nicolson, 'Alexander von Humboldt,' especially 178.

11 Alejandro de Humboldt, *Cartas americanas*, Charles Minguet, ed. (Caracas: Biblioteca Ayacucho, 1980), ix.

12 See Diana Cooper-Richet, 'La figure du passeur dans l'histoire de la diffusion transnationale de la théorie des *Cultural Studies* (années 1990),' *História* 32:1 (June 2013), 190–197. Also see Serge Gruzinski, *Les Quatre parties du monde. Histoire d'une mondialisation* (Paris: Editions de La Martinière, 2004).

13 See Joaquín Fernandez, 'Estudio preliminar,' *Anales de Historia Natural 1799-1804*, vol. 1 (Aranjuez: Doce Calles, 1993), 13–130. In January 1801 the journal's name was changed to *Anales de Ciencias Naturales*. See also Jan-Henrik Withaus, 'América como espacio exploratorio en los *Anales de Historia Natural*,' in *Cuadernos de Ilustración y Romanticismo: Revista Digital del Grupo de Estudios del Siglo XVIII* (Cadiz: Universidad de Cádiz, 2010), n. 16.

14 For a summary of this movement, see Leoncio López-Ocón, *Breve historia de la ciencia española* (Madrid: Alianza editorial, 2003), 156–204. Also see Angel Guirao de Vierna, 'Análisis cuantitativo de las expediciones españolas con destino al Nuevo Mundo,' in *Ciencia, vida y espacio en Iberoamérica*, ed. José Luis Peset, vol. 3 (Madrid: CSIC, 1989), 65–94.

15 See *Anales de Historia Natural* 1:2 (December 1799), 125–127 and 2:6 (October 1800), 251–261.

16 *Anales de Ciencias Naturales* 6:17 (June 1803), 246–258.

17 See M. Villena, J. S. Almazán, J. Muñoz and F. Yagüe, *El gabinete perdido. Pedro Franco Dávila y la Historia Natural del Siglo de las Luces* (Madrid: CSIC, 2009), 925–944. On the relation between Humboldt and Clavijo, see pages 936–938.

18 'Extracto de otra carta del Barón de Humboldt escrita al Sr. D. Joseph Clavijo, Director del Real Gabinete de Historia Natural,' *Anales de Historia Natural* 2:6 (October 1800), 262–271.

19 See Antonio González Bueno, *Antonio José Cavanilles (1745–1804). La pasión por la ciencia* (Aranjuez: Doce Calles, 2002).

20 See *Anales de Ciencias Naturales* 5:15 (November 1802), 231–233.

21 Extract of letter from Humboldt to D. Antonio Josef Cavanilles, *Anales de Ciencias Naturales* 6:18 (October 1803), 284–285.

22 Karl Wilhelm Koppe, *Cartas a la patria. Dos cartas alemanas sobre el México de 1830*, trans. Juan A. Ortega y Medina (México: UNAM, 1955). And *Drei Berichte des General-Kapitains von Neu-Spanien Don Fernando Cortes an Kaiser Karl V. Aus dem Spanischen übersetzt, mit einem Vorworte und erläuternden Anmerkungen von Dr. Carl Wilhelm Koppe … Mit einer Karte und einem Fragment des in Hieroglyphen abgefassten Alt-Mexikanischen Tribut-Registers* (Berlin: Theodor Chr. Fr. Enslin, 1834).

23 See Wilhelm Pferdekamp, *Auf Humboldts Spuren. Deutsche im jungen México* (Munich: Max Huber Verlag, 1958) and José Enrique Covarrubias, 'Carl Christian Sartorius y su comprensión del indio dentro del cuadro social mexicano,' in *La imagen del México decimonónico de los visitantes extranjeros: ¿un Estado-nación o un mosaico plurinacional?* (México: Universidad Nacional Autónoma de México, Instituto de Investigaciones Jurídicas, 2002), 217–236. Sartorius was the author of *Mexico. Landschaftsbilder und Skizzen aus dem Volksleben* (1852).

24 See Guillermo E. Alvarado and Percy Denyer, eds., *Karl T. Sapper (1866–1945): geólogo pionero en América Central* (San José: Universidad de Costa Rica, 2012).

25 Author of *Reise in Chile, Peru und auf dem Amazonenstrome 1827–1832* (Leipzig: Fleischer, 1835–1836), and *Un testigo en la alborada de Chile, 1826–1829* (Santiago: Zig-Zag, 1960), with a prologue by the translator, Carlos Keller. Also see Verónica Ramírez Errázuriz, 'Ciencia y literatura: Eduard Poeppig y su representación de la Araucania (siglo XIX),' *Cuadernos de Historia Cultural* 6 (2017), 41–65.

26 Von Tschui studied at the University of Zurich where Humboldt's work exercised notable influence. He became interested in zoology, languages, and medicine. In 1843, he met with Humboldt in Berlín. He prepared his grand opus on Peruvian fauna and in 1851 co-authored with the Peruvian Mariano Eduardo Rivero, *Antigüedades peruanas*. See Peter Kaulicke, ed., *Aportes y vigencia de Johann Jakob von Tschudi, 1818–1889* (Lima: Pontificia Universidad Católica del Perú, 2001).

27 Humboldt was his mentor. He explored the Brazilian southeast between 1815 and 1817. Author of *Reise nach Brasilien in den Jahren 1815 bis 1817*. 2 Bände (Frankfurt: Verlag Heinrich Ludwig Brönner, 1820–1821) and *Beiträge zur Naturgeschichte Brasiliens*, 4 vols. (1824–1833).

28 See Karen Macknow Lisboa, *A nova Atlântida de Spix e Martius: naturaleza e civilizaçao na Viagem pelo Brasil, 1817–1820* (Sao Paulo: Editora Hucitec, 1997).

29 For correspondence with Bolivar, see de Humboldt, *Cartas*, 196–197; 220–221, 266–267.

30 For correspondence with Rocafuerte, see Humboldt, *Cartas*, 270–271.

31 On December 22, 1854, Humboldt wrote to Santa Anna to thank him for the decoration of the Mexican Order of Guadalupe. See Humboldt, *Cartas*, 240–241.

32 For correspondence with Alaman, see Humboldt, *Cartas*, 205–206, 269–271.

33 See Benjamín Vicuña Mackenna, 'Una visita a Humboldt en 1855,' *Anales de la Universidad de Chile*, 1960. Centenarios 1950-1960, 93–95.

34 Humboldt, *Cartas*, 272–273.

35 Humboldt, *Cartas*, 240.

36 Humboldt, *Cartas*, 276–277.

37 Humboldt, *Cartas*, ix.

38 See Karl Scherzer, *Reise der Österreichischen Fregatte Novara um die Erde in den Jahren 1857–1859 unter den Befehlen des Commodore B. von Wüllerstorf-Urbair. (Physicalische und geognotische Erinnerungen von A. von Humboldt), Beschreibender Theil*, Hof – und Staatsdruckerei (Viena: Bände, 1861–1862). An English translation appeared shortly: *Narrative of the Circumnavigation of the Globe by the Austrian Frigate Novara … in the Years 1857, 1858 & 1859*, 3 vols. (London: Saunders, Otley and Co., 1861–1863). See also G. Treffer, ed. *Die Weltumseglung der Novara, 1857–1859* (Vienna: Molden, 1973).

39 Humboldt, *Cartas*, 277–278.

40 José Genaro Monti, 'Astronomía,' *El Museo Universal* XI:52 (Madrid 28 diciembre 1867), 410.

41 Arístides Rojas, 'Recuerdos de Humboldt en América,' *Revista Latino-Americana*, París, I (1874), 192–210.

42 See Gregory Zambrano, 'Arístides Rojas y la memoria colectiva venezolana,' *Revista de Teoría y Didáctica de las Ciencias Sociales* 12 (2007), 215–234. See also Eduardo Röhl, *Humboldtianas* (Caracas: Tip. Vargas, 1924).

43 See Horacio Capel Sáez, 'El asociacionismo científico en Iberoamérica. La necesidad de un enfoque globalizado,' in *Mundialización de la ciencia y cultura nacional*, eds. Antonio Lafuente, Alberto Elena and María Luisa Ortega, Actas del Congreso Internacional Ciencia, Descubrimiento y Mundo Colonial (Aranjuez: Doce Calles, 1993), 409–428.

44 Diana Obregón, 'La sociedad de naturalistas neogranadinos y la tradición científica,' *Anuario Colombiano de Historia Social y de la Cultura* 18–19 (1991), 101–123.

45 See Leoncio López-Ocón, 'La Comisión Científica del Pacífico (1862-1866) y la Commission Scientifique du Mexique (1864-1867). Semejanzas y diferencias de dos viajes colectivos de naturalistas europeos a tierras americanas,' in *Viajeros y migrantes franceses en la América española y portuguesa durante el siglo XIX*, eds. Chantal Cramaussel y Delia González (Zamora: El Colegio de Michoacán, 2007), 201–212.

46 Joseph Decaisne, 'Instructions sommaires sur la botanique,' in *Archives de la Commission scientifique du Mexique*, t. I (Paris: Imprimerie Nationale, 1865), 31–37.

47 Pimentel, *Testigos del mundo*, 179–210.

48 Marcos Jiménez de la Espada 'Viaje de Quito a Lima de Carlos Montúfar con el Barón de Humboldt y don Alexandre Bompland,' in *Boletín de la Sociedad Geográfica de Madrid*, vol. XXV, eds. Imprenta de T. Fortanet (Madrid: Sociedad Geográfica de Madrid, 1889), 371–389. Also see Teodoro Hampe, 'Carlos Montúfar y Larrea (1780–1816), el quiteño compañero de Humboldt,' *Revista de Indias* LXII:226 (2002), 711–720. It was newly transcribed, based on the original manuscript held at the Lilly Library of University of Indiana by Reinhard Andres and Silvia Navia, 'Das Tagebuch von Carlos Montúfar: Faksimile und neue Transkription,' *Humboldt im Netz (HIN)* XIII:23 (2012), 24. https://verlagsarchivweb.ub.uni-potsdam.de/5924/html/andress-navia.htm. See also Reinhard Andress, 'Alexander von Humboldt und Carlos Montúfar als Reisegefährte: ein Vergleich ihrer Tagebücher zum Chimborazo-Aufstieg,' *Humboldt im Netz (HIN)* XII (2011), 22.

49 Marcos Jiménez de la Espada, 'El volcán de Ansango,' *Anales de la Sociedad Española de Historia Natural* I (1872), 49–76. Reprinted in Leoncio López-Ocón y Carmen María Pérez Montes, eds., *Marcos Jiménez de la Espada (1831–1898). Tras la senda de un explorador* (Madrid: CSIC, 2000), 225–242. Here he challenged data published by Boussingault in *Annales de Physique et Chimie* LII (1833), 18 and by Humboldt in his chapter dedicated to the 'Lavas of American volcanos' in *Cosmos*, IV.

50 Marcos Jiménez de la Espada, 'Una ascensión a El Pichincha en 1582,' *Boletín de la Institución Libre de Enseñanza* XI (1887), 345–351, reprinted in López-Ocón, *Marcos Jiménez*, 321–329. The quotation appears on page 329.

51 Jiménez, 'El volcán,' in López-Ocón, *Marcos Jiménez*, 227.

52 See Leoncio López-Ocón, 'Notas sobre la recepción de Humboldt en España. Maneras de leer a un sabio a lo largo de dos décadas (1851-871),' in *Estudios de Historia das Ciencias e das Técnicas*, eds. María Alvarez Lires et al., vol. I (Pontevedra: Diputación Provincial de Pontevedra, 2001), 335–347.

53 Leoncio López-Ocón, 'Ciencia e historia de la ciencia en el Sexenio democrático: la formación de una tercera vía en la polémica de la ciencia española,' *Dynamis* 12 (1992), 87–103.

54 See José Rodríguez Carracido, *Estudios histórico-críticos de la ciencia española*, eds. Antonio Moreno González y Jaume Josa Llorca (Barcelona: Editorial Alta Fulla, 1988).

55 Rodríguez Carracido, *Estudios*, 108–120.

56 An extensive literature exists on the topic. See Jorge Cañizares-Esguerra, *Nature, Empire and Nation: Explorations of the History of Science in the Iberian World* (Stanford, CA: Stanford University Press, 2006); María Portuondo, *Secret Science. Spanish Cosmography and the New World* (Chicago, IL: University of Chicago Press, 2009); Antonio Barrera-Osorio, *Experiencing Nature. The Spanish American Empire and the Early Scientific Revolution* (Austin: University of Texas Press, 2006); Arndt Brendecke, *Imperio e información. Funciones del*

saber en el dominio colonial español (Madrid-Frankfurt: Iberoamericana, 2012); Jesús Mª Carrillo Castillo, *Naturaleza e Imperio. La representación del mundo natural en la Historia general y natural de las Indias de Gonzalo Fernández de Oviedo* (Madrid: Ediciones Doce Calles, 2004); Mauricio Nieto Olarte, *Las máquinas del imperio y el reino de Dios: Reflexiones sobre ciencia, tecnología y religión en el mundo atlántico del siglo XVI* (Bogotá: Universidad de los Andes, 2013); José Sala Catalá, *Ciencia y técnica en la metropolización de América* (Madrid: Consejo Superior de Investigaciones Científicas, 1994); Juan Pimentel, *Fantasmas de la ciencia española* (Madrid: Marcial Pons, 2020); and Leoncio López-Ocón, 'Quarta Pars. El impacto de un nuevo mundo en la ciencia europea de principios del siglo XVI según Alejandro de Humboldt y Marcos Jiménez de la Espada,' in *De la unión de Coronas al imperio de Carlos V*, ed. Ernest Belenguer, vol. II (Madrid: Sociedad Estatal para la Conmemoración de los centenarios de Felipe II y Carlos V, 2001), 371–388 and *Breve historia*, 27–105.

57 Clear examples include Andrea Wulf, *La invención de la naturaleza. El Nuevo Mundo de Alexander von Humboldt* (Madrid: Taurus, 2016), as noted by Mark Thurner and Jorge Cañizares-Esguerra, 'La invención de Humboldt y la destrucción de las pirámides de La Condamine,' and by Michael Zeuske, 'La invención de Humboldt,' *Procesos. Revista Ecuatoriana de Historia* 51 (January–June 2020), 201–204 and 205–211, respectively.

58 For more detailed analysis see Antonio Lafuente and Leoncio López-Ocón, 'Tradiciones científicas y expediciones ilustradas en la América hispana del siglo XVIII,' in *Historia social de las ciencias en América latina*, ed. Juan José Saldaña (México: UNAM-Porrúa, 1996), 247–281. For an English translation, see Juan José Saldaña, ed., *Science in Latin America: A History* (Austin: University of Texas Press, 2007).

59 I outlined this argument in 'Dos maneras de ver un pasado científico: de los historiadores humboldtianos a los americanistas positivistas,' paper presented to the LAGLOBAL workshop, 'Sites of Invention: Latin America and the Global History of Historical and Anthropological Knowledge,' London, 9 June 2016.

60 See Leoncio López-Ocón, *Botánicos y biólogos en el Ecuador*, vol. I. (Quito: Corporación Editora Nacional-Banco Central del Ecuador, 2010).

2

A SENSE OF PLACE

Early Modern Roots of Humboldt's Natural History Practices[1]

Florike Egmond

> Dalle viscere lor Pichinca ed Etna/Non più ceneri ardenti, ed infuocati Fiumi di zolfo, e liquefatte pietre/Pei culti campi spargeranno; immense Catene di Montagne, Isole, e Regni/Dal loro seno sorgeranno; e ad esse Base saranno ampie Città sepolte./Chi allor di nostrafama, e delle nostre Cosi basso giacenti ossa petrose/Ardira sospicar? chi fia, che voglia Riconoscerle mai sotto gl'immensi/Letti alternati di marine spoglie, E di lave pesanti, o la nel Fondo/Del avare miniere, ove pur troppo Spesso penetra insaziabil', empia/ Sete d'oro fatal? ...[2]
>
> —Alberto Fortis

These lines are excerpted from *De' Cataclismi sofferti dal nostro pianeta* (On the Natural Disasters Suffered by our Planet) a long, scientific *poem-essay* by the Paduan naturalist and geologist Alberto Fortis (1741–1803), published one year before the birth of Alexander von Humboldt. Fortis was an Augustinian cleric, writer, folklorist, fieldworker, expert on fossil fuels, and elected fellow of the British Royal Society.[3] It is noteworthy that Fortis's poem places side by side the New World volcano Pichincha (Quito, Ecuador) and the Old World Etna (in Sicily). Our author also indirectly refers to the catastrophic burial of Pompei and Herculaneum, where excavations had recently begun (in 1738 in Herculaneum and 1748 in Pompei) under the direction of Bourbon military engineers, who, at the time, were also engaged in hydrological explorations.[4] Interesting parallels and connections with the Hispanic world (via the Bourbons) and with Humboldt do not end here. Fortis operated in court society both as a scientist and technical mining advisor, and he was actively involved in mining and mineralogical exploration. In the years between 1796 and 1801, he formed part of the cosmopolitan, Parisian scientific elite.[5] Like Humboldt, Fortis emphasised autopsy and observation-based experience and experiment over theoretical or philosophical schemes.

DOI: 10.4324/9781003231479-3

Ferri has pointed out that 'the convergence of archaeological and geological investigations in Fortis's poem-essay, and the analysis of strata that they both required, introduced an equivalence between the ideas of spatial and temporal depth,' and that 'to the eighteenth-century antiquarian-naturalist, the ability to trace and read these changes, and thus to decipher the fossils and antiquities embedded in superimposed strata, is the key to historical knowledge.'[6] At a less abstract level, Fortis's line about alternating beds (strata) of marine remnants and heavy lava is a crucial clue to his ideas about earth history. Fortis introduces two significant themes here, namely, stratification and the relevance of place. These same themes are also central to two works that Humboldt regarded as among his most important publications.[7] In Humboldt's iconic *Geography of Plants* (publ. [1805] 1807), a single site of knowledge (the Ecuadorian volcano Chimborazo) visually and textually 'condenses' and rhetorically anchors observations and measurements from multiple sites. Layers, strata, rendered in colours characterise Humboldt's famous representation of Chimborazo; they represent the vegetation and habitation zones (or snow) that are visible on the mountain's surfaces. The same stratified patterns return in Humboldt's *tableaux* of other mountains. Almost in counterpoint, his *Geognostical Essay* (1823) names and compares hundreds if not thousands of locations of geological and mineralogical relevance from several continents. The essay gathers together Humboldt's accumulated experience and observations from the preceding thirty years. Its central theme is the relative position of the earth's strata. As the dissimilar fates of these two major works show, the relative unity of place in the former, supported in a major way by the tableau of Chimborazo, was key to its rhetorical and dramatical effect. The *Geography of Plants* can still be read, enjoyed, and understood, whereas the largely non-visual *Geognostical Essay* is hard to get through, at least for non-specialists. There, Humboldt's ideas are buried under a prosaic avalanche of names and places.[8] It is thus not so difficult to understand why the *Geography of Plants* was published at least six times in the twentieth century, whereas the *Geognostical Essay* was published only in the early nineteenth century.[9]

The notable parallels between the projects and styles of the earth science of Fortis and Humboldt point to a shared tradition. In what follows, I trace some of the early-modern European roots or sources of that shared tradition. My focus is not the history of ideas, scientific theories, or natural philosophies, which have been amply and ably studied by others. Instead, I am interested in a particular style of doing science, one linked to methodologies in which autopsy, fieldwork, and site-specific expertise played key roles. These practices, I suggest, were linked to 'a sense of place' and all were as much part of Humboldt's style and practice as was his emphasis on measuring and the use of instruments.[10] My focus on practices and places inevitably involves paying attention to the connection between Humboldt's writings and practice-based forms of knowledge about plants and the earth, including his own knowledge, as well as that of others.

While innovative research by experts on South America is unravelling the American contributions to and inspirations of Humboldt's scientific practice and knowledge (as the contributions to this volume attest), it seems worthwhile to also take a closer look here at the scientific practices in Europe that inspired and contributed to Humboldt's work in botany and geology. My attention goes especially to the northernmost part of Italy, that is, the pre-Alps and Alps of the Veneto and the Trentino. This site of expert natural knowledge possessed a centuries-long tradition with indirect and direct links to Humboldt's practice and style that takes us back into the early modern period. This chapter therefore has its own temporal stratification which, like geological strata, does not always present an orderly aspect. We move in an exploratory fashion between the sixteenth and late eighteen and early nineteenth centuries and within Humboldt's lifetime between the years well before and long after his American journey.

Plants and Rocks

Humboldt often reflected in print on his own intellectual development. In *Geography of Plants*, he explicitly places the origins of his study of the connections between altitude, climate, temperature, and botanical strata to 1790, although it is clear that the substance of his work in this area may be traced to his famous American journey (1799–1804).[11] Much later, in his *Geognostical Essay* (1823), Humboldt wrote in retrospect: 'I have been guided by the idea of the relative age of rocks, in this yet very imperfect labour which was begun long before my voyage to the Cordilleras of the New Continent, in the year 1792, when, upon leaving the school of Freyberg [*sic*], I was appointed to the direction of the mines in the Fichtelgebirge.' In the same publication he speaks of his early fascination with patterns in the orientation of mountain ranges, which 'became one of the principal reasons for my voyage to the equator'; four years of travel in the Andes had corrected his early views which, he writes, had been rather 'vague and less accurate.'[12] Both between 1805 and 1807, aged thirty-six to thirty-eight, and in 1822–1823, aged about 54, Humboldt placed the roots of his geological and biogeographical ideas in his very early twenties, just before and during the time (1791) he started studying at the famous mining academy in Freiberg under the direction of the most influential geologist-mineralogist in Europe, Abraham Gottlob Werner (1749–1817). Clearly, Humboldt rewrote his past with a dose of personal teleology. As several contributions to the present volume make clear, Humboldt arrived in the New World largely by chance. Major travel became possible for Humboldt only after his mother's death in November 1796 when he came into a notable inheritance. Politics and warfare then blocked him from travelling to North Africa. A journey to America only became feasible when the Spanish King gave his special permission in May 1799. The main research themes developed by Humboldt during the years 1790 to 1796 cannot possibly have been focused on America, therefore, but, if we are to take him at his retrospective word, were general enough to be adapted and moulded to *any* part of the world. That flexibility marked Humboldt's scientific practice from early on.[13]

Humboldt's writings and activities of 1789 to 1798 suggest that his interest in both geological investigation and habitat-oriented plant research date back before his student days at Freiberg. Gómez Gutiérrez and Leitner have carefully retraced the appearance in these early years of ideas about plant geography in Humboldt's letters and published works. Their finds include not only a botanical article published in early 1789, a text about basalt formations (1790), and an explicitly habitat-oriented publication (1793) on the subterranean vegetation of the Freiberg mines but also a notable letter dated August 1794 and addressed to Friedrich Schiller.[14] The study of the cryptogams of the Freiberg mines was, in retrospect, already a typically 'Humboldtian work' in terms of its composition, with a descriptive first part based on his personal research in the mines, and a second part that addresses more theoretical and general questions.[15] In 1793–1794, Humboldt was also thinking of a publication on underground and light-shunning plants which, if finished, would have been dedicated to Goethe.

The latter had first met the Humboldt brothers in March 1794, and they certainly discussed Goethe's ideas about morphology and *Urpflanzen* in that year.[16] As is well known, Goethe was one of the important influences on the young Humboldt, particularly in the areas of botany, mineralogy, and geology, lifelong interests of both men.[17] Goethe's studies of plant shapes and development (*Urpflanzen, Metamorphosen*) and his vast mineralogical collection embodied the close link at the time between what we now call plant and earth sciences. As Hamm has shown, Goethe started collecting minerals and rocks in the late 1770s and never stopped. At Goethe's death in 1832, his mineralogical collection comprised almost 18,000 items and was housed in its own garden pavilion. There were only two comparable private mineralogical collections in the whole of Germany; one of those belonged to Werner, the Freiberg mining academy director. Interestingly, Goethe's collection did not focus on unusual or particularly expensive items but was a true research collection. In 1783, Goethe ordered his three early collections (mineral, rock and suite collections; fossils followed) according to the principles of Wernerian classification. By then Goethe had been appointed by Karl August, Duke of Saxe-Weimar, to oversee the re-opening of the copper-silver mine in Ilmenau, a project with which he was passionately engaged, and it was the Ilmenau project that brought him in direct contact with Werner and the Freiberg academy. What is more, Goethe's collecting inspired his Weimar circle: by 1780, several men and women of his acquaintance were enthusiastically collecting rocks and minerals.[18] Goethe was following, rather than setting, a European trend in this respect. Various other and earlier examples are known of men and women from the European elites who collected minerals, stones, rocks, and fossils and became known for their considerable expertise and interesting attempts at ordering their collections.

Among the cultural roots of Goethe's, and by implication Humboldt's, interest in minerals, stones, and geological formations we should not only look at formal educational institutions but also at the history of collecting and the particular types of expertise and interest it generated. Collecting was practised mostly but not only by elite men and women. Among the better-known collectors of the period was Margaret Cavendish, Second Duchess of Portland in England, who assembled a stunning naturalia and porcelain collection (ca. 1760–1785) and achieved great botanical expertise, exchanging ideas with Rousseau and one of Linnaeus's students.[19] A lesser-known female collector in Italy, the Marchesa Margherita Sparapani Boccapaduli (1735–1820), is an illustrative case and deserves to be mentioned here because her interests, travels, and contacts reveal many similarities with those of Goethe and Humboldt. The multilingual Marchesa spent most of her life in Rome. Among her many social and scientific activities, she created a museum in her palace (from about 1771) with particularly important sections on minerals, stones, and fossils. Among her stronger interests were electricity, mineralogy, and volcanism. Of course, she ascended Vesuvius herself, partly on foot, and had her walking stick burnt by viscous lava.[20] Given her many international scientific and artistic contacts, it is hard to imagine that she would not have encountered at least Goethe (in the course of 1786–1788) or Wilhelm von Humboldt in Rome (ca. 1802–1808), and perhaps even both Wilhelm and Alexander during the latter's visit to Rome in the summer of 1805.

Training at elite institutions was also a mark of distinction. Humboldt's training first at Goettingen University (1789–1790), and afterwards (for eight months in 1791–1792) at the Freiberg Mining Academy (founded 1765), places him in two of the top educational institutions in Europe where theoretical and applied science went hand in hand.[21] Werner was at the time the most influential earth scientist in Europe, and his staunch promotion of Neptunism made it for many decades the dominant earth theory. Werner held that the various layers of the earth were deposits on top of a solid (non-hot) earth core, sediments of a primal ocean that had gradually retracted, revealing the continents. All stones, including basalt and granite, were crystallisations of minerals and sedimentary in origin, and all volcanic activities were superficial phenomena. Earth strata were the documents of successive phases in the deep time of the watery earth that could be 'read.' Most of Werner's students followed his theories. Many, including Humboldt, modified or adapted his views later in their lives, allowing volcanoes and volcanism a far more important role in the creation of rocks, minerals, mountains, and landscapes. Werner's ideas, however influential, were not the only ones available in Europe, and the heated debates of the late eighteenth and early nineteenth centuries reveal that a very wide range of positions existed, regarding the age of the earth, the role of the oceans, sediments, and volcanoes, the interpretations of fossils in earth layers, or the nature of the core of the earth.[22] Basalt and the question of whether it was volcanic or sedimentary in origin became a test case, and this was true of fossils as well. Some locations, and in particular Italian ones, presented fine samples of both phenomena and thus were apt to become special sites of natural knowledge.[23] Werner taught more than neptunism, and one of his most important lessons actually came from his expertise in the identification of minerals on the basis of morphology.[24] His empirical style of mineralogy deeply influenced the Freiberg academy, Goethe's approach to minerals and collecting, and Humboldt's *geognosy*. A considerable part of the expertise that Humboldt and his fellow students acquired at the Freiberg Academy[25] was actually practice-based and served both scientific and practical purposes. The students spent much of their time on location in the mines learning about

> shaft timbering, tunnelling, the methods of transporting ores, the machines for lifting water underground and so on. He also observed veins, minerals, mine airs, plants growing underground, and beyond this he learned the methods of mining with his own hands.[26]

Hamm has pointed out that Werner actually presented a very fine-grained method for describing minerals, which relied upon sight, taste, touch, smell and even hearing (the tone rocks made when they were struck). It was a system, in effect a technology, that relied upon carefully disciplined senses and made the body an instrument for identifying minerals in the field.[27] Notably, Humboldt's observations, whether concerning animal electricity, the colour of the sky on Chimborazo and Teide, the smell and touch of minerals in mines, the smell of gases, the taste of tropical plants, frequently depended on his five senses, however strong his emphasis on measuring and instruments.

As Klein notes, Humboldt, in his five years (1792–1797) as the leading Prussian mining official, systematically 'used inspection tours for mineralogical and geological observations. He transformed mines into chemical laboratories, and he transmitted both knowledge and material samples from his natural inquiries in mines to the institutions of learned men.'[28] Humboldt himself made no secret of the eminently practical purposes of his science and travels in this and later periods of his career.[29] In a letter to Simon Bolivar, Humboldt would write:

> Usted hallará también en él un gran conocimiento práctico en el trabajo subterráneo del minero y en el arte de la fundición de metales. Él adquirió esta práctica en las montañas, y como yo mismo adquirí una vez la mía en la dirección de las minas de una región célebre por sus explotaciones en Alemania …[30]

However, the importance of practical mining knowledge to geological science was not only a German but a European, if not worldwide, phenomenon. As Vaccari has noted, the 'interaction between geology and mining was a decisive element for the development of stratigraphy during the eighteenth century in Germany, Sweden, England, and also Italy' and that it was crucially important to the theoretical work on the classification of mountains by various Italian geologists: 'from the early eighteenth century, mines and quarries became some of the best places for observing rock strata and sometimes for elaborating general theories about them.'[31]

Fieldwork and a Sense of Place

Humboldt's famous tableau of Chimborazo (first published 1805–1807; manuscript versions 1803[32]), the rather less famous one of Teide on Tenerife (1817) and the comparative one of Mont Blanc, Chimborazo and the Lapponian Sulitelma (1817) are hybrid visual representations,[33] part table, part schemata, and part naturalistic.

In essence, they depict an approximate distribution in more or less irregular strata or layers (*Schichten* in German) of major vegetation types in relation to altitude zones on mountains, while textual information links this information with temperature, colour of the sky, and numerous other factors. The layers depicted in these hybrid images represent an extension of the notion of plant habitat beyond an individual plant species to a vegetation type.

Various scholars have pointed out that Humboldt was certainly not the first to study supra-local patterns in the connection between vegetation, altitude, and climate. However, he certainly went furthest in the range of his comparison, the extent of measuring, the concept of vegetation types, and the number of interconnections that he integrated.[34] The tendency to combine the study of plants, climate, stratification, and altitude had been climbing fast on the international scientific research agenda since the late seventeenth century. Indeed, the integrated study of botany and mineralogy-geology, the importance of both aesthetics and measuring in the service of scientific research and representing, the extremely close interaction between utilitarian (and practice-based) expertise and science, all became common features in the European natural sciences of the late seventeenth and eighteenth centuries.[35] Indeed, Humboldt sometimes explicitly recognized the work of his European counterparts and predecessors.[36] A brief examination of two such predecessors from the early eighteenth century will lead us back into the early modern period: Carl Linnaeus and Count Luigi Ferdinando Marsili.

Linnaeus (1707–1778) is remembered especially for his botanical classification system. Humboldt used it together (or in alternation) with his own morphology-based division of plants into fifteen vegetation types. Linnaeus's research concerned from the very beginning not only plants but also animals, minerals, and stones. As Beretta has shown, Linnaeus stood in a long tradition of Swedish excellence in mineralogy.[37] During his first field expeditions (1724–1727), the young Linnaeus gathered minerals and fossils and recorded in detail the geological context. He visited the Swedish mining districts, studied mining techniques, extraction methods, the assaying of minerals, and so on and actively involved himself in the complicated issue of classifying minerals.[38] Like Humboldt some seventy years later, he combined training in practical mining techniques with a scientific interest in geology and mineralogy and a vast knowledge of both plants and rocks that were strongly informed by fieldwork and personal observation, which always comprised the contextual setting of naturalia. Stratification entered into this pattern as well: in Linnaeus's view, still following Beretta, 'minerals were subject to modification over time. Hence rocks, especially those to be found at the foot of mountains and other geological elevations, were the result of processes – aggregating and giving rise to diversified strata.'[39] His knowledge of specifics led him to construct systems that aimed to go beyond classification, or as Beretta puts it:

> He combined his interest in systematics with the constant observation of specimens *in situ*, convinced from the very outset that these signs should be regarded as filaments in a thread of Ariadne whose meaning would unwind in a geographic map as extensive as the universe itself.[40]

Fieldwork for Linnaeus was much more than a simple means to obtain specimens. As a research method, it combined personal observation of the context and a valuation of that context as an essential part of the interpretation of the naturalia observed, whether for plants, animals, or minerals. As Hodacs has noted, fieldwork was also one of the more important teaching methods of Linnaeus: to turn students into naturalists they needed to develop observational skills that enabled them to quickly identify relevant features and naturalia in a landscape while passing (slowly) through it.[41]

In the remarkable case of the Italian natural scientist Count Luigi Ferdinando Marsili (also Marsigli, 1658–1730) from Bologna, we find different ideas, but some notable similarities with Humboldt in terms of social status, a plurality of scientific activities, and the capacity to reason from large quantities of highly specific and localised personal observations to far more general patterns in nature. The count was a high-ranking military officer in Habsburg service with a technical background in mining and engineering. He spent two decades in Hungary (1682–1702) during the war against the Turks, and was a meticulous and widely travelled field observer, who lived and worked or travelled in Italy, France, Switzerland, central and eastern Europe, and Constantinople. During his Austro-Hungarian years, Marsili directed mining exploitations, the excavation of channels, and the construction of fortifications, always linking these with geographic, cartographic, and geological observations, underpinned by precise measurements. He collected many specimens of naturalia, as well as antiques, prints by Dürer, and above all drawings of perishable naturalia. Marsili's interest in the interconnected domains of nature was already pronounced in his teenage years. He continued to investigate all these aspects, as is evident for instance from his monograph about Lake Garda: a perfect example of an integrated approach to the study of one site of knowledge, combining descriptions of its geography, the organic structure of the lake, the lake-bed springs, the rivers ending in it, the botany of its surroundings, its winds and currents, fish, and plants. Especially during his years in Marseille, Marsili conducted an enormous series of marine investigations that ranged from salinity to marine creatures (esp. coral) and from currents and tidal movements to wave heights, water temperature, and the shape and profundity of the seabed. On the basis of his 1725 treatise on oceans, Marsili is regarded as the founding father of oceanography.[42]

While Marsili's pattern of combining interests and methods resonated with Renaissance traditions, Olmi has emphasised that Marsili's astonishing capacity of uniting very different activities and duties should not be interpreted as a sign of the eclecticism of a 'virtuoso,' in the sense of superficial brilliance combined with dilettantism. Marsili was a professional soldier as well as a professional scientist. He systematically collected, observed, measured, identified, described, and illustrated, besides organising a botanical garden, acquiring astronomical instruments, and founding the (still extant) Bolognese Institute of Sciences and Arts (1715).[43] Olmi has further stressed, and here we find another parallel with Humboldt, that for Marsili there was little distinction between science and practical use. This stance not merely implied that a journey undertaken for military purposes also served to investigate the terrain, its botany, geology, and zoology. Marsili's wide-ranging notion of geography, like Humboldt's, included economic, statistical, geological, cartographical, mining, botanical, and many other aspects. It was a kind of 'continua e divisa anatomia,' a metaphor in use among Italian geologists to describe their analysis of the layered structure of the earth's body with its veins of ores and liquids, bowels, and stony skeleton, as the Fortis poem-essay that serves as an epigraph to this chapter illustrates.[44] Marsili's scientific expertise was recognised internationally; he published and was in touch with the Scheuchzer brothers in Switzerland, met Newton in England and Boerhaave and Leeuwenhoek in Holland, was elected a member of the British Royal Society (1691), and became a foreign associate of the Parisian Academy of Sciences (1715). His scientific status was connected with his wide-ranging experience, his use of scientific instruments and measuring, and the way in which he, like Humboldt, made his detailed field observations and practical experience speak to wider issues. Noting the correspondence between the rocky strata on the opposite slopes of various Swiss and North Italian lakes stimulated him to further analyse their morphology and formulate a theory about the structure of mountains. The structures of lakes, sea beds, and mountains were all relevant to the development of his theory of the 'organic structure of the Earth.'[45] As Vaccari phrased it, 'Marsili believed that the structure of the Earth could be interpreted through the 'lines' and the 'correspondences' that may be found in its geological features,' and was well aware that his special attention to precisely these features was grounded in his mining experience.[46]

A final parallel with Humboldt should be mentioned here: Marsili was a gifted draughtsman, who personally made plans of fortifications and cities; beautiful drawings of minerals and stones, fish and other aquatic creatures; and detailed graphic representations of mountain profiles, quarries, mines, and rocky stratifications with details about their lithological, mineralogical, and fossiliferous contents.[47] Like many other naturalists of sufficient means, he also employed professional painters, such as Raimondo Manzini, a Bolognese architectural draughtsman and painter who made many of the drawings for Marsili's work on the Danube, and the Swiss painter Felix Meyer from Winterthur, who created (ca. 1703–1705) drawings as well as unusual, science-based oil paintings of the Swiss mountains and glaciers for Marsili that emphasised geological characteristics.[48]

Early Modern Fieldwork

Marsili takes us to some key sites of natural knowledge in Europe that lead both back into deeper layers of historical time, and forward toward Humboldt. One of Marsili's shorter publications concerns the marble caves and their stratification in Monte Baldo, a 2,218-m-high mountain about 50 km to the northwest of Verona that forms part of the eastern shore of Lake Garda.[49] Nearby, in 1725, Marsili explored the Monti Lessini, a section of the Venetian pre-Alps that forms an arch roughly between Verona and Vicenza. These mountains were famous for their curious rock formations, complex stratifications, and, in particular, the fish fossils of Bolca and the formations of basalt columns some 10 km to the south of Bolca.[50]

Marsili made a detailed map of the so-called *Lastrara* of Bolca, where thin slabs of limestone could (and still can) be opened like books to reveal spectacular fish and plant fossils now known to date back approximately 50 million years.[51]

Monte Baldo and the Bolca zone have been two of the most iconic sites of nature knowledge in Europe since at least the early sixteenth century. Monte Baldo was already famous in the fifteenth century for its rich variety of plants. By the late sixteenth century, groups of Italian and foreign naturalists visited it almost every year to herborise.[52] The 1530s and 1540s saw the very first university botanical gardens in Europe in Padua and Pisa, while botanical fieldwork came to be quite generally recognised in Europe as a scientific research practice and teaching method.[53] Much sixteenth-century fieldwork focused on the mapping and identification of European (and some non-European) flora. Naturalists like Fuchs, Brunfels, Bock, Gessner, Mattioli, Anguillara, Clusius, Dodoens, Lobel, and Bauhin, the biggest names of the century, were engaged in a kind of collective enterprise to survey or inventory the plants of Europe, identify them (often with the help of vast 'databases' of coloured drawings), and reach at least some provisional agreement about plant names using this growing visual archive.[54]

While Humboldt's interest in plant stratification linked to altitude concerned whole vegetation groups, most of the sixteenth-century naturalists focused on individual plant species or on very small groups. Still, the increasing emphasis on direct observation in the field (which all sixteenth-century naturalists confronted with book learning and the classical authorities), stimulated their interest in habitat and ecological context. In some cases, this led to an explicit awareness of and interest in a correlation between plant geography or vegetation patterns, altitude, and climate.

The Swiss naturalist and humanist Conrad Gessner (1516–1565), 'father' of zoology and one of the great botanists and mineralogists of his time, was as iconic for sixteenth-century natural science as Humboldt became for *his* period. From 1540 or even earlier, Gessner climbed one or more mountains in the Swiss Alps every year, partly for pleasure, partly for a systematic exploration of the plants. One of his best-known trips was his ascent of the 2,128-m-high Mount Pilatus (near Lucerne) in 1555. Gessner's published description includes a list of the plants that he identified on the mountain, using their Swiss German names, with short descriptions per plant species of its habitat that focus on humidity, light or shaded position, and altitude. Gessner paid special attention to plants growing in grottoes. While Gessner did not link these observations to the distribution of *groups* of plants, he compared his findings with those of several other naturalists in the Alpine zone. He also summarised his findings on the combination of vegetation, altitude, and climate zones by superimposing a temporal mind map of seasonal change on the geographical map of altitude:

> So, I would call the highest part of mountains of this kind, which is around the peak, wintery. There permanent winter and snowdrifts prevail … and cold and winds. The second, which lies below the peak, could be called spring-like, although the winter is very long there and the spring short. The

third is autumnal because it also has an autumnal part as well as a spring-like and a wintery one. The bottom one is summery.[55]

The most famous early-modern excursion of naturalists to Monte Baldo took place in 1554, just one year before Gessner's ascent of Pilatus. A group of eminent naturalists in Italy travelled from Verona to Monte Baldo.[56] The published account of this journey, Francesco Calzolari's *Il viaggio di Monte Baldo* (1556), describes the mountain's flora in a narrative that follows the itinerary up the mountain and records the plants as they grew in their specific location on the slope. Speaking of the rich variety of its plants, Calzolari draws a direct link with its many microclimates:

> but what shall we say about its diversity of air – it is truly marvellous that whoever walks from one location to another here, even at a small distance, feels an enormous difference which gives one the impression of changing not merely region but also clime … and this diversity of sites must be the cause that such a great variety of plants can be found here, of hot and cold nature, woodland plants, house and garden plants, water and mountain plants.[57]

A second and much bigger publication in 1617 by Giovanni Pona (1565–1630), an apothecary-naturalist from Verona just like Calzolari, consecrated Monte Baldo as *the* European site of botanical knowledge par excellence. Pona describes the ascent of the mountain, compressing numerous field trips into one account. He mentions the individual plant species with their principal characteristics (flowering period, evergreen or not, annual or perennial, taste) and gives long plant lists per location. To avoid repetition, he presents plants in the places where they were first found on the climb up the mountain or were most noticeable. For the highest summits, therefore, only those plants that had not been encountered lower down are mentioned. Pona thus devised a textual system to convey plant stratification and habitat on the mountain. He did not, however, visually represent that pattern on paper.[58]

An awareness of habitat among sixteenth-century naturalists passed directly from personal observation in the field into published works on botany. It had practical consequences as well. The university garden of Montpellier, created in the years from about 1593 to 1605 by the botanist-physician Pierre Richer de Belleval (ca. 1564–1632), was probably the very first botanical garden to be organised on a habitat basis. It had a wet and shaded zone with water plants and at least three further sections with different aridity or humidity and soil types.[59] Richer had systematically explored the regional flora of the Languedoc during field trips, sometimes with groups of students, and he observed that the variety in plant species was directly related with ecological diversity, or as he called it 'différence de territoir.'[60] In his hortus, he tried to re-create the habitat he observed in the field.

Fieldtrips during the period from about 1520 to 1550 covered large tracts of the mountainous terrain between Trent, Lake Garda, Verona, and Vicenza, including both Monte Baldo and Bolca. The publications of the German mineralogist-humanist Georg Agricola (1494–1555), for example, were based in part on his studies, work, and travel in the Veneto during the 1520s. In a successive period (late

1520s–early 1530s), he travelled and acquired practical knowledge in the mountains and mines of the Saxon-Bohemian border area, some 250 years before Humboldt worked there. Agricola learnt much about ores and refining metals from the mining technicians. His *De re metallica libri XII* (1556) covers a wide range of practical aspects of mining sciences and metallurgy, and it remained a standard work for 200 years.[61]

Another, even more famous naturalist who investigated Trento and the pre-Alps of the Veneto was Pietro Andrea Mattioli (1501–1577). He spent the years from 1527 to 1541 as personal physician of Bishop Cles, convener of the Council of Trent. During these years Mattioli surveyed and climbed many mountains in the region. This fieldwork was of essential importance, as Ciancio has argued, to the making of Mattioli's commentaries on Dioscorides's *De Materia medica*, the most widely distributed, translated, and reprinted publication on natural history in the sixteenth century. Mattioli's work dealt especially with plants and medicine but included sections on animals and minerals too. It appeared in many successive and ever-expanding editions in various languages and sold well over 30,000 copies during the author's lifetime.[62] Mattioli is also the very first author to mention Bolca and its fish fossils in print, this in 1550. By then the fossils were already collector's items. Mattioli himself saw Bolca fish fossils in Venice in the collection of the Spanish diplomat Diego Hurtado de Mendoza, famous for his great library, collection of antiquities, and ancient Greek manuscripts.[63]

Early in the sixteenth century, the fish and plant fossils of the pre-Alps between Verona and Vicenza, especially those from Bolca, had already taken up a place as key evidence in the debate on the origin of fossils and its implications for the history of the earth. This debate would continue well beyond Humboldt's lifetime.[64] Indeed, some of the positions in the sixteenth-century debate look remarkably similar to those held in Humboldt's age. It is relevant to note that personal observation on location was an important method of the naturalists on both sides of these sixteenth-century discussions. Jan van Gorp (better known as Goropius Becanus, 1519–1572), the Flemish court physician of Emperor Charles V, climbed several Trentino mountains in search for fossils, partly to *disprove* the ideas of Girolamo Fracastoro (1476/1478–1553) of Verona, who acted as physician to the Council of Trent in the mid-1540s. Fracastoro held that the many fossils that emerged during excavations for city repairs of Verona in 1517 were remains of living creatures and, therefore, neither created as fossils by God, nor a *lusus naturae*, nor the result of Noah's biblical flood. Notably, his reasoning was based on the geographical and stratigraphical locations of the fossils, and on the presumed time span involved in their sedimentation and fossilisation.[65]

Sites of Knowledge

Fieldwork practices, the interest of naturalists in geological stratification and plant habitat linked to altitude and climate, as well as fossil-hunting, all reveal continuities between the early 1500s and Humboldt's time. The continuities are no less impressive when it came to the preferred sites of knowledge and fieldwork across Europe and the

Mediterranean. Particular mountains, such as Mount Ida on Crete, the Schneeberg not far from Vienna, and several mountains in Switzerland, remained favourites for many centuries, in part because they combined relatively easy access with a wide variety of microclimates and plant species on a relatively small land surface. Such lines of continuity also apply to Monte Baldo and Bolca. Famous since the early sixteenth century, their reputation increased especially during the eighteenth century.[66] The fish fossils of Bolca and the basalts of the nearby valley of Roncà link the methodology of fieldwork and the debates on geology and earth history from the sixteenth to the nineteenth centuries while throwing light on the ways in which local knowledge entered mainstream scientific discussions before and during Humboldt's own times.

As we have seen, in the early 1700s Count Marsili explored Monte Baldo for both botanical and geological reasons. And in 1707, Bartolomeo Martini, a surgeon and botanist from Verona, created an herbarium in four volumes with about 200 dried plants collected on Monte Baldo, bound together with a copy of his printed catalogue of plants found on that mountain. Notably, he grouped the species topographically in this herbarium, after the locations on the mountain where they were collected.[67] Linnaeus, too, was connected with Monte Baldo although he never visited it in person. Linnaeus received detailed information about the Baldo plants (1755–1757) from the French naturalist and collector Jean-François Séguier (1703–1784), who operated from Verona, regularly explored Baldo's flora, and collected fossils in the region.[68] Significantly, via a remarkable correspondence (1762–1772) between the physician-botanist Antonio Turra (1736–1797) of Vicenza and Linnaeus himself, Monte Baldo's plants served to correct the Swede's plant classifications. Turra's epistolary description of a four-day botanising expedition to Monte Baldo in the summer of 1764 (*Dei Vegetabili di Monte Baldo*), which the author explicitly placed in the tradition consecrated in the mid-sixteenth century by Italian naturalists, is accompanied by a catalogue of 240 plants. There, as Ciancio writes, Turra, 'painstakingly indicated what he believed to be errors in the classification of certain plants by Linnaeus in *Species plantarum*.'[69] Turra was much more than a local enthusiast and 'mere' contributor to Linnaeus's investigations. His publications contributed to the acceptance of Linnaeus's binomial system in Italy, and he was closely connected with the distinguished geologist and mining specialist Giovanni Arduino (1714–1795) from the region around Verona. Arduino was in contact with Linnaeus as well, and he developed an early and very influential classification of geological time based on the geology of northern Italy.[70] All this makes it easy to understand why Turra was the very first naturalist whom Goethe met during his Italian journey, on September 21, 1786.[71]

Interest in Bolca, its fish fossils and the nearby basalt formations grew exponentially in the eighteenth century.[72] Here, it is the history of their collecting that provides crucial information about local expertise and the way it, and the fossils themselves, entered wider debates. The presence of Bolca fossils in naturalia collections can be traced with a high degree of probability from the mid-sixteenth-century collection of Hurtado de Mendoza (which Mattioli saw in Venice), and with certainty from the collection of the Veronese pharmacist Calzolari, who published on the Monte Baldo excursion of 1554. As noted, Count Marsili mapped Bolca, and in 1709 the fossils figured in his friend Johann Jakob Scheuchzer's work. A considerable number of Bolca fossils entered the collection of the Italian naturalist Antonio Vallisneri (1661–1730) in 1707. The source was the local Veronese collector and one-time owner of the Bolca site, Marquess Scipione Maffei (1675–1755). The latter also published his theories on how the fossils had originated and how they ended up on the tops of mountains. Linnaeus studied Bolca fossils and identified some further fish genera (adding to Scheuchzer's identifications), while the same Jean-François Séguier who provided Linnaeus with information about the plants on Monte Baldo also wrote a beautifully illustrated but unfinished work (1740s) on the petrifications of the Verona area. Like Maffei, Séguier had many Bolca fossils in his private museum, and his manuscript is accompanied by some seventy plates of Bolca fossil fish and plants.[73] A peak of interest in Bolca fossils (and in fossils more generally) occurred between circa 1780 and the 1820s. Geological travel in the Alps and Apennines, often sponsored by governments or private entrepreneurs with mining interests, increased during the 1780s.[74] During the last decades of the eighteenth century an interesting cross-fertilisation emerged at various levels, from local collectors and naturalists to natural history museums of international fame, between collecting and the development of ideas about earth stratification, mineralogy, earth history, and the relevance of fossils. That cross-fertilisation is particularly clear in the Bolca case. The local fossil and geological expertise of the Cerato family, who have managed the fossil exploitation of Bolca since 1777, entered the top levels of natural science, including many European museum collections from the early 1780s on and increasingly so over the course of the nineteenth century.[75] In Humboldt's time, it was the local Veronese society of naturalists that systematically studied the Bolca fish fossils in various private collections. That society commissioned a major study (begun in 1789 and finished in 1809) on the subject from the geologist Abbot Giovanni Serafino Volta (1754/?1764–1842) of Mantua, based on these collections. His *Ittiolitologia Veronese* (1796), described 123 species and contained 76 beautiful copper engravings. It was regarded as the very first systematic work on palaeoichthyology.[76]

A rather different way in which Bolca fossils entered international collections and scientific debate was via the confiscations of the French revolutionary and Napoleonic armies. In 1797, when French troops occupied Verona, the entire collection of Count Giovanni Battista Gazzola (1757–1834), which included more than 1,200 items and a large component of Bolca fossils organised according to Linnaean principles, was confiscated. The collection entered the Musée d'Histoire Naturelle in Paris in 1798.[77] Many years later, in 1831–1832, the Gazzola fossils were studied intensively in that same museum by the famous Swiss geologist and biologist Luis Agassiz (1807–1873). The next collection in importance in the Parisian museum was, according to Agassiz, the Humboldt collection of fossils from Mansfeld, the Saxon copper-mining town where Martin Luther grew up in the last years of the fifteenth century as the son of a miner and ore smelter.[78] It is hard to believe that Humboldt would not have looked at the Gazzola group of Bolca fossils when he left his own collection to the Parisian Museum in 1804.

Meanwhile, in the 1780s and 1790s, scientific debate around the Bolca fish fossils and the question of the volcanic or aquatic origins of basalt formations was heating up.[79] On the Italian side, a long *querelle* in print between three abbots-geologists centring on Bolca fossils, vulcanism, stratification and sedimentation, and the time depth of earth's history pervaded the years between 1789 and 1795. They included Giovanni Volta, then working on the *Ittiolitologia*, Domenico Testa, and Alberto Fortis, whose poem has served as the epigraph to this chapter. In spite of their shared background as church officials, the three abbots held strongly divergent views. Their *querelle* highlighted explicit contrasts in terms of methodology as well. Fortis strongly criticised armchair naturalists (in casu the Roman Abbot Testa) who never examined naturalia on location. He also showed little respect for the superficial and often inaccurate observations of naturalists who travelled rapidly but failed to study their material in depth.[80]

Fortis applied his own rules of in-depth investigation, and not only in the field. His study *Della Valle vulcanico-marina di Roncà* (1778) focuses on a small territory less than 20 km south of Bolca and likewise part of the Monti Lessini. This area was and still is famous for its complex stratification with alternating basalt (volcanic) and sedimentary formations, the latter of which contain masses of fossils.[81] As Ciancio has shown in a brilliant analysis, Fortis's theoretical ambitions in what we might well call a geological micro-history were proportionally inverse to the size of the territory studied. On the basis of detailed topographical descriptions, with particular attention to the position of basalt in relation to limestone strata, and to the stratification on both sides of the valleys in which the erosive force of the stream had cut open, anatomised as he put it, the earth and revealed its structure, 'he systematically refuted all of the principal geological hypotheses that circulated at the time.'[82] Fortis emphasised that both sedimentation in water *and* volcanic activity were relevant while pointing to the slowness and extreme *longue durée* of earth's history. He thus refused to be pulled into either of the opposing camps of neptunists and plutonists (also vulcanists). Fortis's nuanced theoretical position was explicitly grounded in detailed observation during intensive fieldwork at a micro-site. This unity of place was rhetorically effective. Fortis's publication rapidly resonated also outside Italy. It was well known in Paris (where Fortis, as we have seen earlier, formed part of the Parisian scientific elite in the years 1796–1801), and a German language translation was reprinted several times between 1779 and 1792. It reached an international audience of geologists and palaeontologists, including Werner and other members of the Freiberg school during the decade before Humboldt became a student there in 1791.[83]

Humboldt's Personal Links with the Early Modern Scientific Tradition

But is it possible to go beyond morphological resemblances and to connect Humboldt in a more direct, specific, personal way with the crucial practices and sites of knowledge of the North Italian pre-Alps? Humboldt travelled through this area in the summer of 1795. He was also connected with the region via two men who strongly influenced his early formation as a mining expert and mineralogist: Goethe and Werner. A third figure, Leopold von Buch (1774–1853), a fellow student at the Freiberg Mining Academy and probably the most important influence on Humboldt's later geological views, was also connected with this region. Goethe and Werner were well aware of Fortis's publications and the importance of the geological formations in the Verona–Vicenza region. Goethe possessed and annotated an exemplar of the German translation of Fortis's work on the Roncà valley.[84] As Marchi has shown, Goethe also personally collected Bolca fish fossils since at least 1782 and continued to do so, mainly through acquisitions, until at least the early 1820s. In 1782–1783 Goethe acted as an intermediary and contact person for a gift of boxes or crates of Bolca fossils from a great fossil collector in Verona to Goethe's patron, the Duke of Saxe-Weimar. A part of this consignment was meant for Goethe personally. Four years later (1786), during the first part of his Italian journey, Goethe visited the Canossa collection in Verona, where he remarked on the beauty of the Bolca fossils.[85] A considerable number of petrified fish from Monte Bolca figure in the inventory of Goethe's collections, which also contained so-called green earth (a pigment also known as Verona green) and fossilised animals from Monte Baldo that he had acquired during his Italian journey.[86] By the time Goethe and Humboldt met for the first time, in 1794, Goethe was therefore quite familiar not only with the fossils and stones of the Bolca and Monte Baldo area but also possessed items, had seen them in a major collection in Verona, had travelled through the area, and knew Fortis's thoughts on the geological formations of the region and their implications for neptunism and plutonism.[87]

It is unsurprising, therefore, that on the day before his departure for the Alps and Northern Italy, the young Humboldt wrote to Goethe about his plan to travel to Venice, through the *Säulengebirge* of Vicenza to Milan, and then on to Switzerland: 'I wish to see the Alps of Tirol, Lombardy, and Switzerland in context [*in Zusammenhang*].'[88] Johann Carl Freiesleben (1774–1846), another student at the Freiberg Academy and a lifelong friend of Humboldt's, who accompanied Humboldt on the last part of this Alpine journey, related that Humboldt was primarily interested in the *Lagerungsverhältnisse der Gebirge und die Pflanzenwelt* (the depositional relationships of mountain strata and the world of plants).[89] It was only in the autumn of 1822, however, that Humboldt personally explored the pre-Alps between Verona and Vicenza, together for part of the way with his old friend the geologist and palaeontologist Leopold von Buch. In the course of their sixty or more years of friendship, Von Buch deeply influenced Humboldt's geological perceptions and his shift away from Werner's neptunism, while Humboldt in his turn stimulated Buch's comparisons with other regions and the latter's research on the volcanism of the Canary Islands.[90] Shared fieldwork experiences were at least as important to their mutual influence as the reading of each other's publications and letters. As young men, in November 1797, Humboldt and Von Buch had explored together the geological formations and salt mines of Styria. While Humboldt left for America, Von Buch extended his field work to Northern Italy in the spring of 1798. Detailed autopsy in the same area that had already triggered Fortis's nuanced position made Von Buch revise his ideas on the origins of basalt, from aqueous to volcanic. A later (1802) examination of extinct volcanoes and strata of basaltic lava of the Puy de Dôme in the south of France distanced Von Buch even further from Werner's theories.[91]

Meanwhile, in America, Humboldt went through a similar but very gradual process of learning, recognising the importance of volcanism in which basalt formations played a key role, and gaining new insights into the geological stratifications that he encountered there. Shortly after his return to Europe, Humboldt and Von Buch spent time together in Rome and climbed Vesuvius just five days after its eruption on August 12, 1805, which they first witnessed from the house in Naples of Baroness Elisa von der Recke.[92] By 1809–1810, Humboldt's increasing awareness of the importance of volcanism was becoming more explicit.[93] It certainly cannot be a coincidence that Humboldt's life-size portrait in oil painted (1813) in Paris by Karl von Steuben under supervision of Gérard, depicted him leaning against a low group of basalt columns.[94] Ten years later, in his *Geognosy*, Humboldt unequivocally recognised the forces and importance of volcanism, which rather displeased his friend and former mentor Goethe.[95]

The joint field trip of Humboldt and Von Buch in the mountains and valleys of Monte Baldo, Bolca and the basalt area of Roncà in October 1822, when both men were on the verge of producing major works in which they proclaimed their belief in the importance of volcanic forces in the formation of the earth (Humboldt's *Geognosy* of 1823; Von Buch's standard work on the Canary Islands of 1825), undoubtedly cemented this approach. Personal observation, together, in the mountains must have helped. Apart from the local experts of the Verona-Vicenza area, few geologists would have known this area better than Von Buch. Humboldt, in fact, made fun (in a letter to his brother) of Von Buch's eccentricity and described him as a loner who spent five months hiking without company in the North Italian mountains, armed with an umbrella, a coat, and with his pockets full of books and geological instruments, walking up to fourteen hours in a day. But there is no doubt that Humboldt also had the very highest respect for the man whom he called (in the same letter) 'le premier géognoste de notre siècle.'[96]

Humboldt travelled from Germany to Verona through the Alps in late September 1822 and met Von Buch in a village between Trento and Verona, just a few kilometres to the east of Monte Baldo. A few days later they met again in Verona, where they had several long conversations about parts of the text for Humboldt's *Geognosy*, then in the final phase of writing-editing.[97] In Verona, Humboldt also saw what was left of the Gazzola museum and some of the Bolca fish fossils. He was bad-tempered, as little impressed by Verona itself as by its fossils.[98] However, the many pages that he devoted to fossils in his *Geognosy* (1823) and his remarks there on 'the celebrated impressions of fish of Monte Bolca,' show that he was more than aware of their importance.[99] Humboldt and Von Buch then undertook an excursion of several days on foot 'dans les vallons basaltiques de San Giovanni et de Ronca,' the valley on which Fortis's microstudy with its much wider implications had focused.[100] For Humboldt, not only the days with Von Buch but his whole trip southwards through the Alps and into the Monti Lessini appear to have been challenging in a geological sense. While crossing the Alps on his way to Verona

(September 30, 1822), Humboldt had also stopped in Predazzo (Val di Fiemme-Val di Fassa), southeast of Bolzano, to see yet another iconic site of knowledge with his own eyes.[101] Near Predazzo, the botanist, geologist and mining inspector Count Giuseppe Marzari Pencati (1779–1836) from Vicenza, whom both Von Buch and Humboldt had met in Paris, had discovered granite on top of stratified limestone rocks during field trips in 1818–1819.[102] That discovery plainly contradicted neptunist ideas which held that granite was much older than sedimentary layers such as limestone and should therefore be found underneath them; it eventually helped end the long phase in which neptunism had been a dominant theory. Humboldt felt the geological ground literally shift under his feet and realised the possible implications. As he so often did, he immediately linked the Old and the New World, writing to his brother Wilhelm (Milan, October 11, 1822):

> Cette vallée est une continuation de celle de Fassa, elle est récemment devenue célèbre par les observations du Comte Marzari Pencati, qui a vu du granite au dessus d'un calcaire coquilleux devenu grenu par l'éruption granitico-volcanique. C'est comme tu vois, une fière atteinte contre la légitimité du granite. Nous vivons dans un siècle ou rien ne reste plus à sa place. Le Mont Blanc n'est pas seulement déchu, avec le Chimborazo même, de son antique grandeur, on en a encore miné les fondamens.[103]

Humboldt's Style of Doing Science

Much work on Humboldt focuses on his innovative approach to the natural sciences, his modernity, and his role as a precursor of trends in present-day science. These works often highlight in particular his emphasis on interconnections; his global comparative approach; the originality of his visual representations; and the combination of a Romantic-aesthetic vision of nature with measuring and quantification, resulting in what has sometimes been called a fusion of Goethian and Humboldtian science. In this chapter I have preferred to look not toward the present but in another direction, back in time to the early modern past, at how Humboldt's style of doing science fitted into long-standing European traditions and scientific practices that range from mining to collecting and from fieldwork to autopsy. I have focused on the long history of links between geology and mining, between mineralogy and botany; on the connections between fieldwork, local knowledge, and the study of nature in a contextual sense, that informed both plant geography and geology with its attention to stratification. One of my reasons for this turn to the deeper, early modern past is linked to my thesis that the parameters of European traditions in natural science were set in the early-modern period – a thesis that goes beyond the scope of this chapter. Such long-term continuities in the history of the natural sciences have remained largely out of sight in part because the concept of the Scientific Revolution (itself a 20th-century historiographical invention) has conditioned many of us to think in terms of discontinuities.[104] Another reason is to avoid the pitfalls of teleology. Highlighting what we presently regard as relevant in Humboldt's work, in particular biogeography and comparative plant ecology, risks casting shadows over elements of Humboldt's work that are no longer regarded as relevant or correct and on people and practices that formed part of his relevant context. Those aspects and persons, however incorrect, curious, and outdated they may look to us now, were at the time an integral part of Humboldt's style of doing science. Finally, my attempt to contextualise Humboldt and link his American research as symbolised by his *Géographie des Plantes*, with a centuries-long European tradition of scientific practices is intended as a contribution to the cultural history of science. In the cultural approach, the heroic, European figure who almost singlehandedly invented new branches of science fades from view to allow room for collaborative processes, the role of assistants, informants, local knowledge, practice, women, the often laborious and circuitous process of developing ideas and 'making' knowledge, trial and error, the importance of the senses, and not least the relevance of place. It should need no emphasis that acknowledging the many ways in which Humboldt was anchored to longstanding traditions in no way takes away from his unique capacity or innovative way of fusing them and taking them in new directions. It might even be useful to see Humboldt's *reconnection* of scientific themes that, by his time, had become subjects of specialisation as, at the same time, both very modern and very early modern. And is it possible to find some Humboldtian *Wechselwirkung* also in Humboldt's typical worldwide comparisons and eye for the *transareal*, and yet find him bound to iconic locations and a strong, also stylistic, sense of place?

Notes

1 This chapter was written during COVID-19 lockdown in 2020. It would have been impossible without the help of many friends. I would like to especially thank Juan Pimentel and Peter Mason, who critically read all of it. Without the advice and thoughtful comments and suggestions of Alberto Gómez Gutiérrez, who also generously shared his vast Humboltian experience, work, and sources, this text would not have been the same. My title is inspired by Richard Cobb's *A Sense of Place* (London: Duckworth, 1975). Martin Rudwick's pathbreaking works on the history of geology are also a foundation for many of the themes I discuss in the following. These include *The Meaning of Fossils: Episodes in the History of Paleontology* (Chicago: University of Chicago Press, 1972), *Bursting the Limits of Time. The Reconstruction of Geohistory in the Age of Revolution* (Chicago: University of Chicago Press, 2005), and *Earth's Deep History: How it was Discovered and Why it Matters* (Chicago: University of Chicago Press, 2014).

2 'From their entrails Pichincha and Etna will no longer scatter glowing ashes and fiery rivers of sulphur and liquified stones through cultivated fields; vast chains of mountains, islands, and kingdoms will arise from their breast; and at their base will lie buried large cities. Who then will be bold enough to surmise our glory and our stony bones that lie so deep? Whoever will want to recognize them under the immense, alternating beds of marine remnants and weighty lava or there in the depths of the avaricious mines where alas the insatiable and wicked thirst for fatal gold only too often penetrates.' Translation by Peter Mason. The 9-page poem-essay was originally published on 1 October 1768 in *L'Europa letteraria* and republished in an Italian-English edition. For a discussion of Fortis and this poem, see Ferri, Sabrina, *Talking Ruins. Natural History and Philosophy of the Italian Enlightenment* (unpubl. diss. Stanford University, 2007).

3 Born Giovanni Battista Fortis; Alberto was his adopted name. On Fortis and North-Italian geology, see Luca Ciancio's essential *Autopsie della Terra: Illuminismo e geologia in Alberto Fortis (1741–1803)* (Florence: Olschki, 1995a), of which I have made abundant use in this essay; cf. Ciancio. 'Alberto Fortis e la pratica del viaggio naturalistico. Stile di ricerca e modalità di prova,' *Nuncius*, 10 (1995b), 617–644, and *Esploratori del tempo profondo. Scienza, storia e società nella cultura veneta* (Verona: QuiEdit, 2014). See also Toscano, Maria, *Alberto Fortis nel Regno di Napoli e il Naturalismo-Antiquario, 1783–1791* (Bari: Cacucci Editore, 2004) on Fortis in Naples.

4 For a summary of the Pompei-Herculaneum excavations, see Schnapp, Alain, *The Discovery of the Past. The Origins of Archaeology* (London: British Museum Press, 1996), 242–264.

5 Although no evidence has yet emerged of a personal meeting between Humboldt and Fortis, the latter was indeed known to Humboldt by 1798.

6 Ferri, *Talking Ruins*, 71–73. See also Ottaviani, Alessandro, *Stanze sul tempo. Sei variazioni tra rovine, fossili e volcani* (Rome: Storia e Letteratura, 2017); and dal Prete, Ivano, 'Being the World Eternal … The Age of the Earth in Renaissance Italy,' *Isis*, 105 (2014), 292–317 and 'The Ruins of the Earth. Learned Meteorology and Artisan Expertise in Fifteenth-Century Italian Landscapes,' *Nuncius*, 33 (2018), 415–441. See also Marchi, Gian Paolo, 'Una lettera di Goethe a Gian Giacomo Dionisi sui fossili di Bolca,' *Belfagor*, 59:3 (2004), 277–279 on antiquities and fossils in early modern collections.

7 Humboldt wrote: 'There are only three important and most peculiar works of mine: The Geography of Plants and the Physical Tableau of the Tropics connected with it, the theory of isothermal lines, and the observations on earth magnetism.' Humboldt, letter to Johann Georg von Cotta, Potsdam 31 October 1854 in Leitner, Ulrike and Eberhard Knobloch, eds., *Alexander von Humboldt und Cotta. Briefwechsel*, Beiträge zur Alexander-von-Humboldt-Forschung, 29 (Berlin: Akademie Verlag, 2009). See the inspiring article by Leitner, 'Alexander von Humboldt's Schriften-Anregungen und Reflexionen Goethe's,' in Jahn, Ilse and Andreas Kleinert, eds., *Das Allgemeine und das Einzelne. Johann Wolfgang von Goethe und Alexander von Humboldt im Gespräch* (Halle: Wissenschaftliche Verlagsgesellschaft, 2003), 127–149 on Humboldt's early interests in strata, plants, and geology. However important, I will not reference here the vast literature on the modes of visualisation in Humboldt's work.

8 The first published editions, both in French, are Humboldt, Alexander von, *Essai sur la Géographie des Plantes; accompagné d'un Tableau Physique des régions equinoxiales* (Paris: Levrault, 1807; first presented in Paris in 1805) and *Essai géognostique sur le gisement des roches dans les deux hémisphères* (Paris: G. Levrault, 1823), respectively. I have used the English translation of the latter for quotations, *A Geognostical Essay on the Superposition of Rocks in Both Hemispheres* (London: Longman & Co, 1823).

9 See Fiedler, Horst and Ulrike Leitner, *Alexander von Humboldts Schriften. Bibliographie der selbständig erschienenen Werke* (Berlin: Akademie Verlag, 2000), 239–240 and 344–347.

10 In using the term *style*, I follow Ciancio. On the spatial turn in the history of science, see Finnegan, Diarmid, 'The Spatial Turn: Geographical Approaches in the History of Science,' *Journal of the History of Biology*, 41 (2008), 369–388. On Humboldt's use of scientific instruments see Bourguet, Marie-Noëlle and Christian Licoppe, 'Voyages, mesures et instruments. Une nouvelle expérience du monde au Siècle des lumières,' *Annales. Histoire, Sciences Sociales*, 52:5 (1997), 1115–1151; and Dettelbach, Michael, 'The Face of Nature: Precise Measurement, Mapping, and Sensibility in the Work of Alexander von Humboldt,' *Studies in History and Philosophy of Biological and Biomedical Sciences*, 30:4 (1999), 473–504. On Humboldt, fieldwork, visualisation, and art, see Lubrich, Oliver, ed., *Alexander von Humboldt. Das graphische Gesamtwerk* (Darmstadt: Lambert Schneider-WBG, 2014).

11 Jackson, Stephen T., ed., *Alexander von Humboldt and Aimé Bonpland, Essay on the Geography of Plants* (Chicago, IL: University of Chicago Press, 2009), 61. See Gómez Gutiérrez, Alberto, 'Alexander von Humboldt y la cooperación transcontinental en la Geografía de las plantas: una nueva apreciación de la obra fitogeográfica de Francisco José de Caldas,' *HIN*, 17:33 (2016), 22–49, https://doi.org/10.18443/238 and *Humboldtiana*

neogranadina, 5 volumes (Bogotá, 2018) and contribution to the present volume. A key online resource in this field is Paessler, Ulrich, ed., 'Humboldt's Papers: Phytogeography and Life Sciences,' in Ottmar Ette, ed., *Edition Humboldt Digital* (Berlin: Brandenburg Academy of Sciences and Humanities, https://edition-humboldt.de/themen/biowissenschaften.xql?l=en).

12 See Humboldt, *A Geognostical Essay*, 8 and 71–72. And Bruhns, Carl (with R. Avé-Lallemant), *Alexander von Humboldt. Eine wissenschaftliche Biographie*, vol. I (Leipzig: Brockhaus, 1872), 171 on Humboldt's youthful enthusiasm for this topic.

13 As Humboldt put it:

> In this geognostical essay, as well as in my researches on the *isothermal lines*, on the *geography of plants*, and on the laws which have been observed in the *distribution of organic bodies*, I have endeavoured, at the same time that I presented the detail of the phenomena, to generalize the ideas respecting them, and to connect them with the great questions in natural philosophy.

> *A Geognostical Essay*, vi. See also Ette, Ottmar, 'Alexander von Humboldt: Wissenschaft im Feld. Transareale Wissenschaftsfelder in den Tropen,' *HIN*, 12:23 (2011), 9–25.

14 See Godlewska, Anne Marie Claire, *Geography Unbound: French Geographic Science from Cassini to Humboldt* (Chicago, IL: University of Chicago Press, 1999), 244–245. Cf. Leitner, *Alexander von Humboldt's Schriften*, 128. Humboldt's interest in basalt was certainly stimulated by Georg Foster, with whom Humboldt and his Dutch travel companion stayed about a week during their *naturhistorische Reise* in Germany of September to early November 1789. See on this Bruhns, *Alexander von Humboldt*, 89–93.

15 Humboldt, Alexander von, *Florae Fribergensis specimen plantas cryptogamicas praesertim subterraneas exhibens* (Berlin: Rottmann, 1793).

16 See Bratanek, F. Th., ed., *Goethe's Briefwechsel mit den Gebrüdern von Humboldt (1795–1832)* (Leipzig: Brockhaus, 1876), 307–308. Humboldt's essay seems to have never appeared. Cf. Humboldt's remark about his treatise on underground meteorology in his letter to Goethe of 16 July 1795, Bratanek, *Goethe's Briefwechsel*, 311. Cf. *Alexander von Humboldt's Schriften*, 130–131.

17 The literature on Humboldt and Goethe is extensive. See, for instance, Buttimer, 'Beyond Humboldtian Science,' and 'Renaissance and Re-membering Geography: Pioneering Ideas of Alexander von Humboldt, 1769–1859,' *South African Geographical Journal*, 85:2 (2003), 125–133.

18 See Hamm, E. P., 'Unpacking Goethe's Collections: The Public and the Private in Natural-Historical Collecting,' *The British Journal for the History of Science*, 34:3 (2001), 278–287. The term *suite* usually refers to a set of specimens from a single area or of a single kind. On fossil collecting in connection with Romantic natural philosophy, see Rupke, Nicolaas, 'The Study of Fossils in the Romantic Philosophy of History and Nature,' *History of Science*, 21 (1983), 389–413.

19 On Paris see Carlyle, Margaret, 'Collecting the World in Her Boudoir: Women and Scientific Amateurism in Eighteenth-Century Paris,' *Early Modern Women: An Interdisciplinary Journal*, 11:1 (2016), 149–161. Cf. Dietz, Bettina and Thomas Nutz, 'Collections Curieuses: The Aesthetics of Curiosity and Elite Lifestyle in Eighteenth-Century Paris,' *Eighteenth-Century Life*, 29:3 (2005), 44–75. On Portland, see Sloboda, Stacey, 'Displaying Materials: Porcelain and Natural History in the Duchess of Portland's Museum,' *Eighteenth-Century Studies*, 43:4 (2010), 455–472 and Cook, Alexandra, 'Botanical Exchanges: Jean-Jacques Rousseau and the Duchess of Portland,' *History of European Ideas*, 33 (2007), 142–156. For a survey of eighteenth-century French natural history, see Spary, Emma, *Utopia's Garden. French Natural History from Old Regime to Revolution* (Chicago, IL: Chicago University Press, 2000).

20 In 1769–1770, she followed courses at La Sapienza in astronomy and experimental physics and, during her Grand Tour (1794–1795), visited numerous collections and cabinets. In La Specola in Milan a geologist and mining inspector showed her the mineralogy

collection. In Florence, Felice Fontana of the anatomical cabinet explained his experiments with live electric eels kept in sea water, and showed her fossilised fish and minerals, machines, telescopes and magnets. In Naples (1795), she visited Sir William Hamilton's famous house and collection, learnt about Vesuvius's eruption of 1794, and saw the electric machine (an early kind of seismograph) of Duke Ascanio Filomarino della Torre as well as other instruments and displays linked to volcanism.

21 See Brianta, Donata, 'Education and Training in the Mining Industry, 1750-1860: European Models and the Italian Case,' *Annals of Science*, 57:3 (2000), 276–277; Engelhardt, Wolf von, 'Goethe und Alexander von Humboldt: Bau und Geschichte der Erde,' in Benno Parthier, ed., *Das Allgemeine und das Einzelne. Johann Wolfgang von Goethe und Alexander von Humboldt im Gespräch*, Acta Historica Leopoldina, 38 (Halle: Deutsche Akademie der Naturforscher Leopoldina, 2003), 21–31; and for a detailed discussion of Humboldt's early training and the importance of his mining education and work, Klein, Ursula, 'The Prussian Mining Official Alexander von Humboldt,' *Annals of Science*, 69:1 (2012), 27–68.

22 See Hamm, 'Unpacking Goethe's Collections' on Goethe, Werner's influence, and an interesting critique of the concept of neptunism in historiography.

23 For an inspiring essay on the role of two artificial volcanoes and basalt formations in the German court garden of Wörlitz on the debates of the period, see Umbach, Maiken, 'Visual Culture, Scientific Images and German Small-State Politics in the Late Enlightenment,' *Past & Present*, 158 (1998), 110–145.

24 Engelhardt, 'Goethe und Alexander von Humboldt,' 24 has argued that Werner's emphasis on precise observation led to the undermining of his own neptunism and to the later shift of many of his pupils towards vulcanism. Cf. Hamm, 'Unpacking Goethe's Collections,' 284–285.

25 Until the mid-eighteenth century (and in some regions much later), earth sciences and mining were not taught at universities or other educational institutions but were learnt mainly in practice, usually either in connection with mineral collections that needed experts for acquisitions or in direct relation with the mining industry. See Brianta, 'Education and Training in the Mining Industry, 1750-1860,' 267–300.

26 Klein, 'The Prussian Mining Official,' 33.

27 Hamm, 'Unpacking Goethe's Collections,' 286–288 where he refers here in particular to Werner 1774, and also argues that this style of mineralogy was rooted in a centuries-old tradition of hard-rock mining in Central Europe and Sweden. Cf. Klemun, Marianne, 'Classification and Experience, Rocks and Taste: 'Vulgar' Reasoning in the Earth Sciences,' *De Achttiende Eeuw*, 48:1–2 (2016), 113–126, on taste and smell in period mining expertise. On Humboldt's use of his own body to observe the effects of high altitude in the Andes, see Debarbieux, Bernard, 'The Various Figures of Mountains in Humboldt's Science and Rhetoric,' *Cybergeo: European Journal of Geography*, Epistémologie, Histoire de la Géographie, Didactique, document 618 (2012), 17.

28 Klein, 'The Prussian Mining Official,' 32.

29 Humboldt, *A Geognostical Essay*, 309.

30 As pointed out and quoted by Gómez Gutiérrez, *Humboldtiana neogranadina*, I, 2:185. This letter of 29 July 1822 was one of the two of that date from Humboldt to Bolívar.

31 Vaccari, Ezio, 'Mining and Knowledge of the Earth in Eighteenth-Century Italy,' *Annals of Science*, 57:2 (2000), 163–166. See Morello, Nicoletta, 'The Birth of Stratigraphy in Italy and Europe / La nascita della stratigrafia in Italia e in Europa,' in G. B. Vai and W. Cavazza, eds., *Four Centuries of the Word 'Geology.' Ulisse Aldrovandi 1603 in Bologna* (Bologna: Minerva Edizioni, 2003), 251–264.

32 See on the history of the *Tableau*, Gómez Gutiérrez in this volume.

33 On the complex history and associations, also pre-print, of the Chimborazo image, see Jackson, *Alexander von Humboldt and Aimé Bonpland*, but also Gómez Gutiérrez in the present volume; and Buttimer, 'Beyond Humboldtian Science,' and 'Renaissance.' For the Teide image with Von Buch's plant information, see Lubrich, *Alexander von Humboldt*, 130–131.

34 Ebach, Malte Christian, *Origins of Biogeography: The Role of Biological Classification in Early Plant and Animal Geography* (Dordrecht: Springer, 2015), 38, argues that both Buffon's and Humboldt's

> vegetation types or forms are governed by climate as well as morphology. However, unlike Buffon's Law these taxa have worldwide distributions, and do not require a common stem species or form. In other words, it was Humboldt and not Buffon, who had founded the first distributional law.

35 Ebach, *Origins of Biogeography*, 49–50, points out that Stromeyer's work appeared while Humboldt was in America and that Humboldt did not know it until much later. He also notes that in the years before and concurrent with Humboldt's own research similar works by Linnaeus (*Stationes Plantarum*, 1754, on the theme of plant habitats), Carl Willdenow, teacher of the young Humboldt (*Prodromus flora e Berolinensis*, 1787 and *Grundriss der Kräuterkunde*, 1792, on regional plant distributions), and Friedriech Stromeyer (*Commentatio Inauguralis Sistens Historiae Vegetabilium Geographiae Specimen*, 1800, an anthology of eighteenth-century plant geography) appeared. Further important works, especially on account of their innovative visual forms of reportage and rendering, were Auguste Pyramus de Candolle's first systematic, biogeographic map of France (*Carte Botanique de France*), commissioned for the third edition of the *Flore Française* by Lamarck and Candolle (1805); and Georg Wahlenberg's *De vegetatione et climate in Helvetia septentrionali* (1813), based on field work well before 1813 in the Alps, Tatra, and Lapponia, where he gives a comparative account of plant regions, and the lower and upper limits of various plants.

36 Debarbieux, 'The Various Figures,' 15. The French geologist-botanist L. Ramond de Carbonnières compared the Pyrenees and the Alps, and was deeply interested in stratification and the discussions about the age of limestone that centred on the theories of Dolomieu and Lapeyrouse.

37 Beretta, Marco, 'Linnaeans in Italy: The Case of Johann Jakob Ferber,' in Marco Beretta and Alessandro Tosi, eds., *Linnaeus in Italy: The Spread of a Revolution in Science* (Sagamore Beach: Science History publications/Watson Publishing, 2007), 97, points out that in 1739 Linnaeus was co-founder of the Royal Swedish Academy of Sciences, which had a utilitarian orientation, and in which mineralogists-geologists were the most important profession after physicians. See also Koerner, Lisbet, *Linnaeus: Nature and Nation* (Cambridge, MA: Harvard University Press, 1999).

38 Beretta, 'Linnaeans in Italy,' 95–96. Linnaeus proposed a division (in 1730s, *Pluto Svecicus*) into three classes: petrae (simple rocks), minerae (composite rocks), and fossils. He incorporated this classification in his *Systema Naturae*.

39 Beretta, 'Linnaeans in Italy,' 98.

40 Beretta, 'Linnaeans in Italy,' 92. Cf. Vaccari, 'Mining and Knowledge.'

41 Hodacs, Hanna, 'In the Field: Exploring Nature with Carolus Linnaeus,' *Endeavour*, 34:2 (2010), 47; she points out that Linnaeus, while teaching, generally focused more on special and single naturalia than on vegetation groups. Cf. Hodacs, 'Linnaeans Outdoors: The Transformative Role of Studying Nature 'on the Move' and Outside,' *The British Journal for the History of Science*, 44:2 (2011), 183–209, and 'Linnaean Scholars Out of Doors: So Much to Name, Learn and Profit from,' in Arthur MacGregor, ed., *Naturalists in the Field. Collecting, Recording and Preserving the Natural World from the Fifteenth to the Twenty-First Century* (Leiden: Brill, 2018), 240–257.

42 See Vaccari, 'Mining and Knowledge,' 169–170; Vaccari, 'The Classification of Mountains,' and 'Eighteenth-Century Classification of Mountains in the Alpine Region,' *International Geology Review*, 52:10–12 (2010), 1009–1020; Clementini, Daniela, *Luigi Ferdinando Marsili. Viaggio tra le scienze* (unpubl. PhD thesis, Bologna University, 2006–2007), 142–178; Sartori, Renzo, 'Luigi Ferdinando Marsili, Founding Father of Oceanography,' in G. B. Vai and W. Cavazza, eds., *Four centuries*; Franceschelli, Carlotta and Stefano Marabini, 'Luigi Ferdinando Marsili (1658–1730): A pioneer in geomorphological and archaeological surveying,' in G. B. Vai, W. Glen and E. Caldwell, *The Origins of*

Geology in Italy (Boulder Colorado: Geological Society of America, 2006), 129–139, and Olmi, Giuseppe, 'L'illustrazione naturalistica nelle opere di Luigi Ferdinando Marsigli,' in Giuseppe Olmi, Lucia Tongiorgi Tomasi and Attilio Zanca, eds., *Natura-Cultura. L'Interpretazione del Mondo Fisico nei Testi e nelle Immagini. Atti del Convegno Internationale di Studi. Mantova, 5-8 Ottobre 1996* (Florence: Olschki, 2000), 255–303. Marsili also developed a special fascination for fungi, which he investigated in the field both in his native Bologna, and in Hungary.

43 Olmi 2000, 'L'illustrazione naturalistica,' 263–264. Marsili left much of his collections, images, manuscripts, and books to the Senate of the University of Bologna.

44 See Olmi, 'L'illustrazione naturalistica,' 264. On the metaphors of the earth's body and anatomy etc., see Ciancio, *Autopsie della Terra*, 39, 49–55; and Vaccari, 'Mining and Knowledge,' 167.

45 On Marsili's place in the context of eighteenth-century mountain classifications and the Alps, see Vaccari, 'The Classification of Mountains,' and 'Eighteenth-Century Classification of Mountains in the Alpine Region,' 1009–1020. See Clementini, *Luigi Ferdinando Marsili*, 159–166, for Lake Garda.

46 Vaccari, 'The Classification of Mountains,' 172.

47 Vaccari, 'The Classification of Mountains,' 170. On the illustrations of naturalia in Marsili's work, see Olmi, 'L'illustrazione naturalistica.'

48 Olmi, 'L'illustrazione naturalistica,' 257, 268–270, also with further details about the making of both drawings in the field, the engravers and the illustrations for Marsili's publications. For the 'visual' Humboldt, see Lubrich, *Alexander von Humboldt*; Erdmann, Dominik and Oliver Lubrich, eds., *Alexander von Humboldt. Das Zeichnerische Werk* (Darmstadt: WBG, 2019), but also Buttimer, 'Beyond Humboldtian Science.'

49 Monte Baldo also plays a part in his observations concerning the parallel patterns of rocky strata in the mountains on either side of the same lake. See Vaccari, 'The Classification of Mountains,' 170, who refers to a manuscript text by Marsili, 'Note dei strati delle Cave de' Marmi, che sono in Molte Baldo' (University Library Bologna). See also Clementini, *Luigi Ferdinando Marsili*, 166.

50 Bolca is a village some 40 km from Verona and 30 km from Monte Baldo as the crow flies. The nearby basalt columns of San Giovanni Ilarione date from the Paleocene. The fish fossils of Bolca are found in limestones from the Eocene and Paleocene. A good introduction in English to their history is Roghi, Guido, Stefano Dominici, Luca Giusberti, Massimo Cerato, and Roberto Zorzin, 'Historical Outline,' in C. A. Papazzoni, L. Giusberti, G. Carnevale, G. Roghi, D. Bassi and R. Zorzin, eds., *The Bolca Fossil-Lagerstätten: A Window into the Eocene World*. [Rendiconti della Società Paleontologica Italiana, 4] (2014), 15–17.

51 Italian *lastra* = sheet or slab. Marsili's topographic map and description of his journey to Bolca were published in the second edition of Antonio Vallisneri's *De' corpi marini, che su' monti si trovano* (2nd edition, Venice: Domenico Lovisa, 1728), 141–150. See Guerra, Romano, 'Antonio Vallisneri e i fossili di Bolca,' *Studi e ricerche sui giacimenti terziari di Bolca, XVI - Miscellanea Paleontologica*, 13 (2015), 40.

52 The very earliest mention of the variety of Monte Baldo's medicinal herbs dates from 1277. See on its history Turri 1994, 38.

53 Although it is often argued that scientific fieldwork, especially in geology, began in the eighteenth century, systematic and repeated botanical exploration of particular regions and fieldwork as a teaching practice existed already by the 1550s to the 1570s. See Egmond, Florike, 'Into the Wild: Botanical Fieldwork in the Sixteenth Century,' in Arthur MacGregor, ed., *Naturalists in the Field. Collecting, Recording and Preserving the Natural World from the Fifteenth to the Twenty-First Century* (Leiden: Brill, 2018), 166–211.

54 See Ogilvie, Brian W., *The Science of Describing: Natural History in Renaissance Europe* (Chicago, IL: University of Chicago Press, 2006); Egmond, Florike, *The World of Carolus Clusius. Natural History in the Making, 1550-1610* (London: Pickering and Chatto, 2010) and *Eye for Detail: Images of Plants and Animals in Art and Science, 1500–1630* (London: Reaktion Books, 2017).

55 Quoted from Gessner, Conrad, *De raris et admirandis herbis, quæ sive quod noctu luce-ant, sive alias ob causas, Lunariæ nominantur, commentariolus* (Zurich: Gesnerus fratres, 1555), 46, translation by Peter Mason; also see Leu, Urs, *Conrad Gessner (1516–1565). Universalgelehrter und Naturforscher der Renaissance* (Zurich: Verlag Neue Zürcher Zeitung, 2016), 267–271. On Gessner and fieldwork, see Egmond, 'Into the Wild,' 177–182. On Gessner and palaeontology, fossils and mountain research, see Leu, Urs and Peter Opitz, eds., *Conrad Gessner (1516–1565). Die Renaissance der Wissenschaften/The Renaissance of Learning* (Berlin: De Gruyter-Oldenbourg, 2019).

56 They included Luca Ghini (1490–1566), founder of the Pisa hortus; the Bolognese natu-ralist Ulisse Aldrovandi (1522–1605); Andrea Cesalpino (1519–1603) who became direc-tor of the Pisan hortus in 1555; Luigi Anguillara (ca. 1512–1570), the first director of the Padua hortus; and the pharmacist Francesco Calzolari (1522–1609), owner of a famous *Kunst und Wunderkammer* in Verona.

57 Quoted from Calzolari, Francesco, *Il viaggio di monte Baldo* (Venice: Valgrisi, 1556), 7. My translation.

58 Pona, Giovann, *Monte Baldo descritto da Giouanni Pona veronese* (Venice: R. Meietti, 1617) is the fullest edition with some 250 pages account of the field trip. See for further discus-sion, Egmond, 'Into the Wild.'

59 See Rath, Ulrich von, 'The Function and Architecture of the Botanic Garden of the University of Montpellier (1593–1622),' in Zbigniew Mirek and Alicja Zemanek, eds., *Studies in Renaissance Botany*, Polish Botanical Studies, Guidebook Series 20 (Kraków: W. Szafer Institute of Botany, Polish Academy of Sciences, 1998), 87–112. On sixteenth-century notions of plant ecology, see Ubrizsy Savoia, Andrea, 'Environmental Approach in the Botany of the 16th Century,' in Mirek and Zemanek, *Studies in Renaissance Botany*, 73–86.

60 Richer de Belleval, Pierre, *Onomatologia, seu Nomenclatura Stirpium quae in Horto Regio Monspeliensi recens constructo coluntur* ([1598] reprinted in M. Broussonet, *Opuscules de Pierre Richer de Belleval*, Paris, 1785) and *Dessein touchant la recerche des plantes du Pays de Languedoc*, ([1605] reprinted in M. Broussonet, *Opuscules de Pierre Richer de Belleval*, Paris, 1785).

61 See Morello, Nicoletta, 'Agricola and the Birth of the Mineralogical Sciences in Italy in the Sixteenth Century,' in Vai et al., *The Origins of Geology*, 23–30; cf. Vaccari, 'Mining and Knowledge,' 164. For the links between practical and technical mining see Long, Pamela O., 'The Openness of Knowledge: An Ideal and Its Context in 16th-Century Writings on Mining and Metallurgy,' *Technology and Culture*, 32:2 (1991), 318–355.

62 On Mattioli, see Ferri, Sara, ed., *Pietro Andrea Mattioli, Siena 1501–Trento 1578: La vita, le opere: con l'identificazione delle piante* (Perugia: Quattroemme, 1997). On Mattioli and fieldwork, see Ciancio, Luca, 'Per questa via s'ascende a magior seggio.' Pietro Andrea Mattioli e le scienze mediche e naturali alla corte di Bernardo Cles, *Studi Trentini. Storia*, 94:1 (2015), 159–184, 163 and 171. Ciancio has identified 82 references to specific plant locations in the mountainous Trentino in Mattioli's first edition of 1544.

63 Mattioli, P. A, *Il Dioscoride dell'eccellente Dottor Medico M.P. Andrea Matthioli da Siena* (Venice: Valgrisi, 1550). In a slightly later edition (1553) Mattioli discusses these fossils in a section devoted to mineralogy. See on Hurtado and the fossils, Guerra, Romano, 'Don Diego Hurtado de Mendoza, primo collezionista di fossili di Bolca,' *Studi e ricerche sui giacimenti terziari di Bolca, XIV - miscellanea paleontologica*, 11 (2012), 59–83.

64 See the historical bibliography of Bolca by Guerra, Romano and Roberto Zorzin, 'Bibliografia e citazioni di Bolca. Opere dal 1550 al 1850 (primo contributo),' *Studi e ricerche sui giacimenti terziari di bolca, XV - miscellanea palcontologica*, 12 (2014), 43–100. On historical debates on earth history and fossils, see Rudwick, *The Meaning of Fossils, Bursting the Limits of Time*, and *Earth's Deep History*; for a recent discussion. Ciancio, Luca and Domenico Laurenza, 'Visual representation in earth sciences history after "The emergence"', *Nuncius*, 33 (2018), 397–414.

65 See Prete, Ivano dal, 'Echi Fracastoriani nelle cosmologie e nelle teorie della terra del settecento,' in Alessandro Pastore e Enrico Peruzzi, eds., *Girolamo Fracastoro: fra medicina, filosofia e scienze della natura* (Florence: Olschki, 2006), 280, and Prete, 'Being the World Eternal,' 295–298.

66 Naturally, Bolca and Monte Baldo were not the only privileged sites of knowledge in the Alps. On the special role of Mont Cenis (Savoie) in the development of French botanical and fieldwork traditions, see Pepy, Émilie-Anne, 'Montagne(s) des naturalistes: l'invention de territoires scientifiques, XVIe–XIXe siècle,' in A. M. Granet and S. Gal, eds., *Les territoires du risque* (Grenoble: PUG, 2015), 176. Shortly after his return from America, Humboldt travelled with Gay-Lussac (March–April 1805) via Mont Cenis to Genoa and eventually Naples. For their investigations on Mont Cenis, see Bourguet, Marie-Noëlle, *Le monde dans un carnet. Alexander von Humboldt en Italie* (Paris: Éditions du Felin, 2017), 111–123.

67 See Martini, Bartolomeo, *Catalogus Plantarum Montis Baldi* (Verona: Io. Bernus, 1707). The herbarium (in two volumes) is now in the library of the Botanical Garden of Padua (online at: https://phaidra.cab.unipd.it/detail/o:267292?mycoll=o:267330).

68 Letter Séguier to Linnaeus, 23 June 1757, Nimes to Uppsala, in Latin. Original consulted online (March 2020) https://www.alvin-portal.org/alvin/view.jsf?pid=alvin-record:226696. Cf. Turri, Eugenio, 'La montagne et les passions territoriales: l'exemple du Mont Baldo (Italie),' *Revue de géographie alpine*, 82:3 (1994), 39–40; and Stearn, William, 'Botanical Exploration to the Time of Linnaeus,' *Proceedings of the Linnean Society of London*, 169:3 (1958), 173–196.

69 Ciancio, Luca, '"Tuis impulsus consiliis" Antonio Turra, the Vicenza Academy of Agriculture and the Reception of Linnaeus' Thought in the Venetian "Terraferma" (1758–1797),' in Marco Beretta and Alessandro Tosi, eds., *Linnaeus in Italy. The Spread of a Revolution in Science* (Sagamore Beach: Science History publications/Watson Publishing, 2007), 178–180. Turra's letter appeared at the end of 1764 in Francesco Griselini's *Giornale d'Italia*. Turra's own botanical investigations had been stimulated by an encounter with Linnaeus's pupil Claes Ahlströmer in 1762.

70 See Ciancio, '"Tuis impulsus consiliis,"' 176, 186–187. See Vaccari, 'Linnaeus and Giovanni Arduino. Some Notes on a Difficult Reception in Mineralogy and Geology,' in Beretta and Tosi, *Linnaeus in Italy*, 189–198.

71 Ciancio, '"Tuis impulsus consiliis,"' 170.

72 See Marchi Marchi, 'Una lettera,' 2004;Sorbini, Lorenzo, *La collezione Baja di pesci e piante Fossili di Bolca* (Verona: Museo Civico di Storia Naturale Verona, 1983) and Guerra Guerra and Zorzin, 'Bibliografia e citazioni,' 2014, Guerra, 'Antonio Vallisneri,' 2015.

73 For many years Séguier was secretary and friend of Marquess Scipione Maffei. Séguier's manuscript, *Pétrifications du Véronois* (MS 90) and plates (MS 256) are held in the Municipal Library of Nîmes. On Séguier see Pugnière, François, 'De l'*Instrumentarium* au Muséum. Le cabinet de Jean-François Séguier (1703-1784),' *Liame. Histoire et histoire de l'art des époques moderne et contemporaine de l'Europe méditerranéenne et de ses périphéries*, 26 (2016), 1–20.

74 Vaccari, 'Mining and Knowledge,' 175.

75 See Cerato, Massimo, *Cerato. I pescatori del Tempo* (Verona: Grafica Alpone, 2011), 5–6 on the Ceratos. On continuities since the Renaissance and the importance of Italy to the study of trace fossils, see Baucon, Andrea, 'Italy, the Cradle of Ichnology: The Legacy of Aldrovandi and Leonardo,' *Studi trentini di scienze naturali. Acta geologica*, 83 (2008), 15–29.

76 Volta 1796 with its beautiful illustrations may be consulted online: https://www.e-rara.ch/zut/wihibe/content/titleinfo/14156939. See further Sorbini, *La collezione*, 11. There is a growing literature on early modern visual representations of fossils, a topic that goes beyond the present article; see Findlen, Paula, 'Projecting Nature: Agostino Scilla's Seventeenth-Century Fossil Drawings,' *Endeavour*, 42 (2018), 99–132, with many further references.

77 A second collection rebuilt by Gazzola suffered the same fate in 1806; a third one now forms the core of the magnificent fossil collection at the Verona Museum of Natural History. The confiscated items are still in the Parisian museum. See on the confiscations, Roghi, Guido, Stefano Dominici, Luca Giusberti, Massimo Cerato, and Roberto Zorzin, 'Chapter 2: Historical Outline,' in C. A. Papazzoni, L. Giusberti, G. Carnevale, G. Roghi, D. Bassi, and R. Zorzin, eds., *The Bolca Fossil-Lagerstätten: A window into the Eocene World* [Rendiconti della Società Paleontologica Italiana, 4] (2014), 5–17. [online at: http://paleoitalia.org/archives/rendiconti/83/vol-4-2014/]

'Historical Outline'; and Sorbini, *La collezione*, 11.

78 See Agassiz, Luis, *Recherches sur les poissons fossiles*, vol. I (Neuchatel: Petitpierre et Prince, 1833), 5. Mansfeld like Bolca was and still is today a world-famous fossil location.

79 Fortis noted similarities between Bolca fish and coral reef fish of the Pacific, which induced speculations about climate change in the long earth history of the Bolca region. See Ciancio 1995a, 245–248, and Gaudant 2014, who points out that Fortis (and a few others) could only note parallels with coral reef fish as an unintended consequence of James Cook's first voyage. Cook reached Tahiti in 1769, accompanied by the naturalists Joseph Banks and Daniel Solander (Linnaeus' pupil), who made information about tropical reef fish available for the first time to Europeans in the early 1780s. Humboldt's friend Georg Forster accompanied Cook on the latter's second journey (1872–1875), where he observed Pacific atolls and developed hypotheses about the rise of coral reefs.

80 See Ciancio, *Esploratori del tempo profondo*, 146–147, and the detailed study of this *Querelle* in Ciancio, *Autopsie della Terra*, 246–256. Testa's more conservative ideas, which reached back to the biblical Flood, originated in his study in Rome. He only knew the fossils through publications by others, and through one visit to the Veronese Museo Gazzola.

81 Fortis, Alberto, *Della Valle vulcanico-marina di Roncà* (Venice: Carlo Palese, 1778). For recent geological research and a history of this valley's exploration, see Zorzin, Roberto and Guido Roghi, 'Roncà, storia antica e recente del giacimento paleontologico. Gli scavi 2010-2012,' *Notizie di Archeologia del Veneto*, 1 (2012), 130–136.

82 Fortis was not the only geological expert in whom a relatively small-scale field study in the mountains of the Veneto generated far-ranging ideas. Speaking about Giovanni Arduino, Vaccari in 'The Classification of Mountains' (164) states:

the decisive turning point toward a broader reflection on the classification of rocks and mountains can clearly be linked to a trip that Arduino undertook at the end of October 1758 in the Agno Valley in the upper Vicentine area.

83 Ciancio, *Esploratori del tempo profondo*, 153–154.

84 See Marchi, 'Una lettera di Goethe,' 285–286. Goethe also owned Fortis' autograph.

85 Marchi, 'Una lettera di Goethe,' *passim*, 283, provides a list of connections between Goethe and Bolca. The gift came from Gian Giacomo Dionisi (1724–1808), canon of the Verona cathedral.

86 See Prescher, Hans, *Goethes Sammlungen zur Mineralogie, Geologie und Paläontologie: Katalog* (Berlin: Akademie-Verlag, 1978). Various Bolca items can be found under the heading 'Paleontological Collection' (118–156, nrs. 2185–2189); cf. 508–514 on the consignment to Goethe of various, non-Bolca, minerals and fossils in 1829 from Cristofori in Milan (a stones, minerals etc. dealer); cf. 260–262 (on Monte Baldo items).

87 The Veronese Dionisi collection of the 1780s was absorbed into the Gazzola collection, discussed above, confiscated by French troops together with the Canossa collection. See Marchi, 'Una lettera di Goethe,' 293.

88 Quoted from Bratanek, *Goethe's Briefwechsel*, 311; letter written from Bayreuth. Humboldt used similar words in the letter of 28 July 1795 from Triest to Freiesleben, quoted in Bruhns, *Alexander von Humboldt*, 169. Humboldt travelled via Innsbruck, Trent, and Treviso to Venice, and then turned westward, via Verona, Padua and the Euganean hills just south of it (another famous area of volcanic origin with a long history of geological interest) to Vicenza, Parma, Mantua, Genua and eventually Milan, before going on to the Swiss Alps. For more detail, see https://edition-humboldt.de/chronologie/index.xql?&offset=201.

89 Bruhns, *Alexander von Humboldt*, 169.

90 Von Buch's standard work on the Canary Islands (1825 German edition; 1836 in French) has an extensive discussion of their volcanic origins. On Humboldt and the Canaries, see Peter Mason's contribution in the present volume.

91 Much of this fieldwork-based research was published by Von Buch between 1802 and 1809.

92 See Bourguet, *Le monde dans un carnet*, 124–128. Buch had arrived in Rome (July 1805) during Humboldt's stay there, and they travelled together to Naples. Baroness Elisa von der Recke (1754–1833) was a well-known writer, traveller, and acquaintance of Humboldt since his youth in Berlin. Just a year earlier (September 1804) she visited Monte Baldo (referring to its botanical riches) and Bolca. See Recke, Elisa von der, *Tagebuch einer Reise durch einen Theil Deutschlands und durch Italien in den Jahren 1804 bis 1806*, vol. I (Berlin: Böttiger, 1815), 138–139, 144.

93 Engelhardt, 'Goethe und Alexander von Humboldt,' 25–26, argues he remained an explicit Neptunist until well after his return to Europe and until his reading of Von Buch's 1809 publication on the volcanoes of the Auvergne.

94 Humboldt, Alexander von, *Briefe Alexander's von Humboldt an seinen Bruder Wilhelm, herausgegeben von der Familie Von Humboldt* (Stuttgart: Cotta, 1880), 217–223. On basalt and its symbolism, see Umbach, 'Visual Culture,' 130–138.

95 Goethe remained a Neptunist: on his complex reactions to Humboldt's shift towards vulcanism, see Engelhardt, 'Goethe und Alexander von Humboldt,' 25–31. But see Hamm, 'Unpacking Goethe's Collections,' 285 on the complexities of the contrast itself.

96 Humboldt 1880, 94–95 (letter to Wilhelm, 11 October 1822, from Verona). Von Buch's unparalleled field experience spanned not only much of the Alpine zone, but also further parts of France and Italy as well as Scandinavia (1806), the Canaries (1815), Scotland, and Ireland.

97 Humboldt, *Briefe Alexander's*, 94–95 (letter to Wilhelm, 11 October 1822, from Verona). Humboldt showed Von Buch printed but as yet unpublished parts of his *Geognosy* in Verona; Humboldt was clearly pleased when Von Buch showed himself to be in agreement. For Von Buch, too, the conversations and trip were important. In 1824, he published a group of letters to various naturalists on the stratification and geognosy of South Tirol with special, comparative attention to limestone and fish fossils, the Val di Fassa, and the latest finds near Predazzo. This 1824 edition includes a long letter to Humboldt (4 February 1823) on 'Le Tableau Géologique du Tyrol méridional.'

98 Humboldt, *Briefe Alexander's*, 98.

99 Humboldt, *A Geognostical Essay*, 398; cf. 50, 369. Humboldt reflected on the possible origin of the Bolca fish as partly pelagic and partly fluviatile.

100 Humboldt, *Briefe Alexander's*, 93–95. They also visited the quarries of S. Ambrogio in Valpolicella (northwest of Verona) where the red-pink marble of Verona is quarried; it is a fossiliferous limestone.

101 The geology museum in Predazzo shows evidence of Humboldt's visit (30 September 1822) on http://www.predazzoblog.it/inagurazione-museo-geologico-delle-dolomiti-di-predazzo/.

102 Marzari Pencati also published on the mountain vegetation and the origins of basalt formations. In 1786, Goethe stayed in the family palazzo in Vicenza built by Palladio, according to a commemorative plaque. On the Alps and eighteenth-century mountain classifications, see Vaccari, 'Eighteenth-century Classification of Mountains.'

103 Humboldt, *Briefe Alexander's*, 93; and Humboldt, *A Geognostical Essay*, 338–342, with a lengthy discussion of the problem and many references to Von Buch. As Dal Piaz, Giorgio Vittorio, 'The Birth and Development of Geological Sciences in the Veneto,' *Rendiconti Lincei. Scienze Fisiche e Naturali*, 25 (2014), 421, discusses, Marzari Pencati's findings, published in 1819 and 1820, also conflicted with certain ideas of Von Buch. This seems to have led to a personal clash between Marzari Pencati and Von Buch, who apparently stated that even if Pencati was right official science would believe him and not a provincial nobody. If this is true, then it might illustrate a growing rift between official science and practice-based, local expertise.

104 That thesis also underlies Egmond, *Eye for Detail*.

3

SIX DAYS ON TENERIFE

The Making of Humboldt's Tropical Antique

Peter Mason

On April 28, 1831, Charles Darwin wrote to his sister Caroline:

> All the while I am writing now my head is running about the Tropics: in the morning I go and gaze at Palm trees in the hot-house and come home and read Humboldt: my enthusiasm is so great that I cannot hardly sit still on my chair. … I never will be easy till I see the peak of Teneriffe *[sic]* and the great Dragon tree; sandy, dazzling, plains, and gloomy silent forest are alternately uppermost in my mind.[1]

But Darwin was in for a disappointment. He was denied access to the peak that he saw towering in the sky 'twice as high as I should have dreamed of looking for it'[2] because the consul in Santa Cruz insisted on a 12-day quarantine to guard against the risk of cholera. The captain of the *Beagle*, Robert Fitzroy, refused to wait that long and did not land in the Atlantic until they reached the island of St Jago in the Cape Verde group.[3]

The words 'I never will be easy till I see the peak of Teneriffe and the great Dragon tree' could just as well have come from the pen of Alexander von Humboldt. He regarded his six-day stay on Tenerife, which enabled him to study both the Pico del Teide and the dragon tree, as 'the most enjoyable days of my life.' We shall see how they came to serve as icons of the island in his writings, but first it is necessary to offer a brief sketch of these two protagonists.[4]

The Pico del Teide and the Dragon Tree

With its height of 3,718 metres above sea level, the active volcano Teide on Tenerife is the highest mountain in Spanish territory [1]. Although Christopher Columbus never landed on the island, en route from La Gomera to Gran Canaria in August

DOI: 10.4324/9781003231479-4

1492 he was able to witness the eruption of the Boca Cangrejo volcano on the north-western flank of Tenerife.[5] Two further eruptions followed at the beginning of the eighteenth century that destroyed the port of Garachico on the northern coast of the island, and only a year before Humboldt's arrival the Pico Viejo of the Teide volcano was in a state of eruption for three months.

On June 21, 1799, Humboldt, Bonpland and their guides set out from La Orotava on the north coast of Tenerife, the regular starting point for those who wanted to make the ascent of the Pico del Teide. During his initial stay in Santa Cruz on the south side of the island, he had already noted that 'nobody in Santa Cruz had ever climbed to the summit of the mountain.'[6] He hoped that the local guides in La Orotava would be more experienced, although he regretted that, being obliged to follow guides, he would only see what had already been seen and described by previous travellers. In fact, the route and the difficulties that Humboldt encountered, both physical and human, are virtually identical to those met with a few years earlier by the companions of Lord Macartney, the first British ambassador to China, who had attempted the ascent in unfavourable conditions during their stay on Tenerife from October 21 to 27, 1792.[7] Still, Humboldt was not out to discover; he was out to take more precise measurements than any of his predecessors had done, even though the instruments that he and Bonpland had with them had been originally acquired to take measurements in Algeria and Tunisia in the wake of the Napoleonic conquests in North Africa.[8]

After spending a dismal night at about 3,000 metres above sea level, the next morning they pressed onwards and upwards. Humboldt did not have a good word for the guides:

> The laziness and bad temper of our guides made this ascent more difficult. They were despairingly phlegmatic. The night before they had tried to convince us not to pass beyond the limit of the rocks. Every ten minutes they would lie down to rest; they threw away pieces of obsidian and pumice-stone that we had carefully collected.[9]

He concluded that they had never visited the summit of the volcano before. The descent was difficult too: thoroughly frozen from the icy winds blowing on the peak, Humboldt found the temperature of 22.5 °C suffocating when they had dropped to 3,000 metres, and his anger at the guides was rekindled when he discovered that they had drunk all the malmsey wine and broken the water jugs.[10] Still, it should not be thought that Humboldt was always negative about his local guides. When the *Pizarro*, under the Spanish captain Luis de Artajo's command, had mistaken the island of La Graciosa for Lanzarote, it was a fisherman who got over his initial fright at the sight of the unfamiliar boat and sailors and set them right.[11]

A less physically and mentally demanding experience of the natural wonders of Tenerife was Humboldt's encounter with the dragon tree. This tree, actually a shrub because the trunk is hollow, is endemic to the Azores, the Canary Islands, Madeira, and Cape Verde Islands; attempts to cultivate it in northern Europe were generally unsuccessful because of the cooler climate there. The species that Humboldt saw was the *Dracaena draco*,[12] 'whose trunks are often rightly compared to snakes' bodies.'[13] They can be extremely long-lived: the present-day dragon tree in Icod de los Vinos, a small village located above Garachico in north-western Tenerife, is mentioned in deeds of land from 1503. When Humboldt visited the island, the oldest dragon tree[14] was the one he saw in La Orotava in the gardens of the Irish trader Don Juan Cólogan de Franchi,[15] Marqués del Sauzal, where it stood opposite a pre-conquest palm tree:

> Even though we knew about Franqui's dragon tree from previous travellers, its enormous thickness amazed us. We were told that this tree, mentioned in several ancient documents, served as a boundary mark and already in the fifteenth century was as enormous as it is today. We calculated the height to be about 50 to 60 feet; its circumference a little above its roots measured 45 feet. The trunk is divided into many branches, which rise up in the form of a chandelier and end in tufts of leaves similar to the Mexican yucca.[16]

Two engravings by Simon Cattoir from around 1770 show the dragon tree in the garden. The caption in four languages states that by then the trunk of the tree had a circumference of 51 feet and that a table placed in the branches could accommodate ten guests.[17]

After a fire completely destroyed the Cólogan de Franchi residence in La Orotava in 1745, the then owner Juan Bautista Domingo de Franchy y Benítez de Lugo decided to build an ostentatious new residential complex. Boasting an unparalleled 54-metre balcony-corridor that looked out seaward over the garden with its artificial lakes and mazes and the family estate, it was all intended to demonstrate its owner's familiarity with contemporary French and Italian styles and impress the cosmopolitans who visited the island.[18] The building, which was destroyed in another fire in 1905, occupied the site of what is now the Casa de los Marqueses de El Sauzal in La Orotava.

Hospitality and company of the kind offered by the Cólogan family were familiar and congenial to Alexander von Humboldt, accustomed as he was to move in society circles. Moreover, Bernardo Cólogan, who was only a couple of years younger than Humboldt, had visited the Pico Viejo del Teide a few days after the eruption in June 1798 and shared his unpublished notes on that ascent with Humboldt during the latter's stay on Tenerife.[19] This youthful scientific and literary member of the Cólogan family offered his guests the same hospitality that his predecessors had offered to Cook, Banks and Lord Macartney. As bearers of a royal passport and the latest news from Europe, Humboldt and Bonpland were assured of a warm welcome.[20] Needless to say, they were delighted to mingle with people with a taste for literature and music.[21]

Another member of the Canarian cultural élite was the Scottish wine merchant Archibald Little. He arrived on Tenerife, where his uncle had been British consul, in 1774. He was responsible for the laying out of the pleasure garden in Puerto de la Cruz which is still in existence today as Sitio Litre. On June 23, 1799, Humboldt and Bonpland were invited to 'a country party in Little's garden.'[22] This idyllic setting attracted many foreign visitors to the island, and its dragon tree was immortalised by the artist Marianne North during her stay on Tenerife in 1875.[23]

Little's wealth and easy-going nature made him popular with a number of influential figures on the island, including José Tomás de Armiaga y Navarro, the military commander in Santa Cruz. Humboldt met with a lukewarm reception from Armiaga at first.[24] This attitude changed after the commander had read the letter of recommendation that Humboldt carried from the director of the Real Gabinete de Historia Natural in Madrid, the Canary-born scientist and translator of Buffon, José Clavijo y Fajardo (1726–1806),[25] whose nephew assisted Humboldt and Bonpland in boarding the *Pizarro* in La Coruña.[26] Now, Armiaga's nephew was sent to invite the travellers to his home, where the commander apologised for his initial lack of hospitality and made ample amends.[27]

The example of Armiaga in particular demonstrates that access to the exclusive social circles in which Humboldt moved was limited. It was an élite world in which letters of recommendation played an important part. In turn, the social aloofness of the people with whom Humboldt mixed suggests that their knowledge of the day-to-day situation on the island was also limited. As local informants, they nevertheless enjoy a higher status in Humboldt's writings than, for example, the anonymous 'old Canarians' who told him that camels become very high-spirited when they drink the first winter rains; Humboldt's use of the German word *Sage* (hearsay) for this knowledge immediately qualifies it as untrustworthy.[28]

Somewhere between the anonymous fisherman and old Canarians who offered Humboldt local knowledge and the exalted social circles of the wealthy Canarian landowners was the Salcedo family, who resided next to the customs building in Santa Cruz. Among the fellow passengers on board the *Pizarro* were two Canarians, whom Humboldt refers to as Salcedo and Eduardo. Nothing more is known of this Eduardo, but we obtain a better picture of Salcedo, whom Humboldt characterises as 'very amiable' (*sehr liebenswürdig*).[29] Humboldt encounters Juan Manuel de Salcedo, the father of Humboldt's fellow passenger, in the street. He describes this lieutenant, who had spent part of his youth in France, as wise, cultured and calm. Salcedo's wife had been very concerned during the absence of her son Francisco and for two months had spent large sums on masses in a convent on the Pico del Teide praying for his safe return, while her elder son Emanuel scanned the deck of the *Pizarro* with a telescope for signs of his younger brother.[30] Humboldt does not mention the Salcedos in the published *Relation historique*, but contacts like these were probably a treasure of useful information and further contacts for him, on whose reports he would construct – others would say *invent* – a new identity of the islands for contemporary Europeans.

Alexander von Humboldt: Portrait of the Scientist as a Young Artist

From the moment of setting sail for America, Humboldt seems to have kept a keen eye out for subjects that would make a good artistic composition. Already when the *Pizarro* was manoeuvring clumsily and dangerously to get out of the channel from

La Coruña to the open sea and many of the passengers were seasick, he noted in his diary that the negress (*Negerin*) on board, stretched out on a bunk in what he calls a 'very oriental' (*sehr orientalisch*) pose with bared breasts while her two-year-old child played nearby, formed a 'very picturesque' (*sehr malerisch*) scene.[31] The following day the sight of the setting of Mars and the Moon at 22 hours, 'reflected in the turbulent water,' was likewise 'very picturesque' (*sehr malerisch*).[32]

Humboldt dedicated more than 10% of the almost 2000 pages of the *Relation historique* to a description of Tenerife. To be sure, many of these pages consist of digressions on ocean currents, the visibility of mountain peaks from a distance, and arid statistics. The images of Tenerife in Humboldt's published work, however, are few and indeed are exclusively devoted to the dragon tree and to the Pico del Teide. Two of these are diagrams: the 'Tableau physique des Iles Canaries. Géographie des Plantes du Pic de Ténériffe,' from the *Atlas géographique et physique des régions équinoxiales du Nouveau Continent*, shows the distribution of the plants of Tenerife in five strata, each corresponding to a different altitude.[33] The other, from the first volume of the *Relation historique*, explains how to calculate the height of the volcano from an observation point in the garden of Franchy in La Orotava or, more precisely, from the dragon tree in that garden. There could be no more direct link between volcano and dragon tree.[34]

Humboldt reproduced two images of the dragon tree at La Orotava [2][3].[35] The first represents it in its original splendour. The picture is attributed to the French marine painter Pierre Ozanne and was probably made in 1776 during one of the expeditions of Jean-Charles de Borda to measure the height of the Pico del Teide. Humboldt scrupulously records his source: 'Dessiné par Marchais d'après un esquisse de M. d'Ozonne.'[36] The second, which shows the damage caused by the strong winds of 1819, is based on a drawing by the English artist J.J. Williams. Humboldt declares that he found the drawing (perhaps a copy) among the papers of the travel diary of Jean-Charles de Borda. He published it in the form of an engraving made by a certain Mercier in an article dedicated to the 'Dragonier d'Orotava' that appeared in the periodical *La Belgique Horticole* in 1852, seven years before his death.[37]

Besides these second-hand images, Humboldt also produced his own sketches. Before reaching Tenerife, as they criss-crossed between La Graciosa and Lanzarote, Humboldt took advantage of the different viewpoints to produce several sketches of the coastal contours of the islands,[38] and on Tenerife, he sketched the view of the interior of the crater of the Pico del Teide that served as the basis for the engraving LIV in the *Vues des cordillères* [4].[39] Humboldt's description of that sight slides gently from geological characterisation to what that geological formation tells us about processes that took place long ago. At the same time, he does not fail to mention 'the picturesque beauties offered to those who keenly feel the splendours of nature.'[40] In this way, as Marie-Noëlle Bourguet has written, 'he converts sensory experience into a piece of field data to be taken into account, being simultaneously a direct access to the world and a first stage in the elaboration of knowledge.'[41]

Nevertheless, it would be mistaken to take Humboldt's words and the engraving in *Vues des cordillères* as a direct record of his impression on the spot. Printed between 1810 and 1813, the engraving is taken from a drawing by Wilhelm Friedrich Gmelin in Rome in 1805 that in turn is based on Humboldt's original sketch from 1799. Both *Vues des cordillères* and the *Relation historique* postdate the events they describe by a considerable number of years, and the images they contain often passed through various hands before publication.

Nor does the manuscript diary of the voyage offer us direct experience, for although no doubt written at the end of his stay on Tenerife and/or during the voyage from Spain to America, the date of penning by no means corresponds on a one-to-one basis to the date of the events it describes but is instead retrospective. Moreover, that section of the first volume of the diary dealing with Tenerife breaks off at the point where Humboldt is being introduced to the local authorities in Santa Cruz, before his encounter with either the dragon tree of La Orotava or the ascent of Teide.[42] As for the notes on Tenerife that appear in other volumes of the diary, they do not relate to his personal experience but consist mainly of extracts from the writings of earlier travellers.[43] Five pages of Volume IX are written in French and in a clearer hand than his usual scrawl, suggesting that they form part of his publication project, and much of the information they contain was later to appear in the Canarian pages of the *Relation historique*.[44] They consist of a brief account of the approach to Santa Cruz and comparisons of the visibility of the Pico del Teide with Antisana and other volcanoes in Ecuador. This in itself would already indicate that they postdate the events described by some years; indeed, Humboldt explicitly states that he made these notes in May 1803 in Mexico and in May 1804 in Philadelphia. Otherwise, they consist mainly of snippets of information about the Guanches, other islands, the history of eruptions and other diverse material, largely culled from his reading of José de Viera y Clavijo, whose *Diccionario de Historia Natural de las Islas Canarias* was completed in 1799,[45] and other writers during his travels. I have been unable to find any mention of the dragon tree in these five pages.

The Shaping of Humboldt's Aesthetic Sensibility

Since on the route from the Iberian Peninsula to America a stop on the Canaries or one of the other island groups in the eastern Atlantic was obligatory to take on supplies, many travellers and the modern editors and historians of their travel accounts have taken the passage through the Canaries for granted and passed over it in silence. One of the first to grace them with a more than passing mention was the young Milanese Girolamo Benzoni, who embarked in 1541 from San Lúcar de Barrameda for Gran Canaria and La Palma before securing a passage from La Palma to the Caribbean. While mention of the Canaries was limited to a brief account of his journey in the first edition of his extremely popular *Historia del Mondo Nuovo* in 1565, the second edition published in Venice in 1572 included a seven-page appendix, 'Breve Discorso di alcune cose notabile delle Isole di Canaria,' which included a woodcut of a remarkable tree that grew on the island of El Hierro.[46]

This notable neglect of the importance of the Canaries in shaping first impressions of America is all the more surprising given Columbus's frequent, and frequently cited, comparisons. Already on the day of his first encounter with Caribs on the island of Guanahaní, Columbus wrote: 'They paint themselves a dark colour, and they are of the same colour as the Canarians, neither black nor white.'[47] Two months later, the Pico del Teide came in handy to describe the view from a port, situated between the island of San Tomás and the Cape of Caribata: 'From that port could be seen a vast and cultivated valley …; and no doubt there are mountains there higher than the Canarian island of Tenerife, which is considered to be one of the highest in existence.'[48] Echoing Columbus, Alexander von Humboldt would likewise use the Canaries as a mental template with which to gauge his experience of the New World.

Columbus was in the back of Humboldt's mind if not on the tip of his pen throughout the Atlantic crossing. He was well aware that 'every traveller who writes his adventures begins by describing Madeira and Tenerife, though the natural history of these islands remains quite unknown.'[49] Columbus himself had not landed on Tenerife during his first voyage when his landings were confined to Gran Canaria and La Gomera,[50] where he took on water, firewood, meat and other supplies before setting sail into the Atlantic.

When Humboldt caught his first glimpse of moving lights on the coast of what he thought was the Canarian island of Lanzarote:

> During the voyage we had been reading the ancient Spanish navigators, and those moving lights reminded us of Pedro Gutiérrez, Queen Isabel's page, who saw similar lights on Guanahani island on the memorable night the New World was discovered.[51]

After leaving Tenerife, Humboldt claimed that he 'followed the same route as Columbus had taken on his first voyage out to the Antilles,' but the evidence suggests that the route varied somewhat.[52] The great masses of floating seaweeds encountered to the north of the Cape Verde islands called to Humboldt's mind and pen the banks of weeds that Columbus had compared to the great meadows of the Canary Islands, although the Prussian did not confuse the two.[53] In this respect at least, he differed little from Columbus, who had set sail from the Iberian mainland with his head filled with images drawn from his readings of Pliny the Elder, Marco Polo, Sir John Mandeville, Pierre d'Ailly and others.[54]

Another element of Humboldt's mental baggage may be traced to his own travels with the naturalist Georg Forster through the Netherlands to England and France in 1790. The two friends had arrived in the French capital in the first week of July, in time to witness the preparations for the celebration of the first anniversary of the storming of the Bastille; their departure on July 6 prevented them from observing the actual celebration on the 14th. These travels with Forster, Humboldt noted in his diary, had strengthened his desire to visit Tenerife, a dream since early youth.[55] He also recorded there that during his time on board the *Resolution* during Cook's

second circumnavigation of the world, Forster had spent a month on Tahiti and
the other Society Isles in 1773. Forster had assured Humboldt that the few days
that he had spent on Tenerife had been of equal importance to his whole existence
on Tahiti. A marginal note by Humboldt corrects this mistake: Forster had been
speaking of Madeira.[56] Perhaps we find a reflection of Forster's comparison between
Tenerife/Madeira and Tahiti in the dystopian passage that concludes the chapter on
Tenerife in the *Relation historique*:

> A short time after the discovery of America, when Spain was at the zenith
> of her glory, the gentle character of the Guanches was the fashionable topic,
> just as in our times we praise the Arcadian innocence of the Tahitians. In
> both these pictures the colouring is more vivid than true. When nations are
> mentally exhausted and see the seeds of depravity in their refinements, the
> idea that in some distant region infant societies enjoy pure and perpetual
> happiness pleases them.[57]

Humboldt echoes Forster's words when he writes 'we had stayed at Tenerife for
a few days only, yet we left the island feeling we had lived there for a long time.'[58]
In the corresponding diary entry, he goes even further: 'For six days we were on
Tenerife, in Santa Cruz, La Laguna, the port of Orotava and the Pico del Teide –
the most enjoyable days of my life, highlights.'[59] And in a letter to his brother, he
confessed that the departure from the island had left him 'on the verge of tears.'[60]

The travels with Forster in 1790 left another lasting impact on Humboldt, for in
that year, they would have been able to see *Landscape, Ruins and Figures* by William
Hodges at the Royal Academy in London. Indeed, the title *Landscape, Ruins and
Figures* would have been an apt title for the work that we know as *Vues des cordillères,
et monumens des peuples indigènes de l'Amérique*. On board the *Resolution*, Forster
had shared the company of Hodges, whose paintings executed during the voyage
earned him the title of 'pioneer of English *plein-air* painting.'[61] Some of his paintings of scenes witnessed during the voyage must surely have been finished after his
return to England when he was contracted by the Admiralty to work up some of his
drawings and sketches into a series of epic paintings to commemorate the second
voyage. Several of these were exhibited at the Royal Academy in 1776 and 1777.[62]
In the case of our remarkable painting, we see a fusion of the exotic with the
antique, of the remote in space with the remote in time, as the artist lifts the dorsal
view of a seated female nude from one of the paintings of Tahiti to inhabit a space
dominated by the ruinous Roman monument known as the Temple of Minerva
Medica.[63] If the scene of disporting nymphs in an idyllic setting recalls the classical
picturesque landscapes of Richard Wilson, whose studio Hodges had entered in
1758, the so-called Temple seems rather to reflect the dramatic ruins of Giovanni
Battista Piranesi, who had indeed included a representation of the 'Temple' as one
of the plates in the first volume of his *Le Antichità Romane*.[64]

Between 1780 and 1783 Hodges was travelling again, this time in India on behalf
of the East India Company. The result was a mass of drawings, paintings, prints,

an architectural treatise, a stage design of Calcutta, and a travel narrative.[65] These absences from England and particularly the London art scene did not make it easy for Hodges to exhibit his work in venues like the Royal Academy, but he sought to compensate by appealing to alternative exhibition spaces such as John Boydell's Shakespeare Gallery and the European Museum.[66] Forster and Humboldt saw some of Hodges's Indian production in the collection of his patron Warren Hastings, the controversial first Governor General of Bengal, where he organised military protection for Hodges and commissioned several oil paintings.[67] Ninety sketches and various oil paintings of Indian subjects formed part of Hastings's collection until its sale in 1797 and would thus have been viewable in 1790 too.[68] These pictures of exotic scenery impressed upon the young Humboldt the value of landscape painting to supplement the message conveyed in the printed word.[69] Moreover, their presentation of the complexity of Anglo-Indian relations in a distant land may have prepared him for the ambiguity of lush tropical landscape and inhuman slavery that he was to find in America.

Years later, in his *Kosmos* lectures, Humboldt looked back on the early sense impressions that had pointed him in the direction of the systematic study of nature, singling out Georg Forster's descriptions of the South Sea islands, paintings by Hodges of the banks of the Ganges that he had seen in the London residence of Warren Hastings, and 'a colossal dragon tree in an old tower in the botanical garden in Berlin.'[70] By this time, however, while paying tribute to Hodges and the seventeenth-century Dutch artists who had been to Brazil, he considered the new popular phenomenon of panoramas and dioramas a means of immersing the viewer even more deeply into the exotic world of the tropics.[71]

In the *Kosmos* lectures, he returned to another theme from the same years: *Lebenskraft* or *vis vitalis*. He had already tackled it in 1795 in a short allegorical fable that he wrote for publication in Friedrich Schiller's monthly literary journal *Die Horen*. In this fiction, an exposition of the vital force of nature is put into the mouth of the philosopher Epicharmus when called on to explain the meaning of a Rhodian painting that had stood for a hundred years in Syracuse without anyone being able to come up with an adequate explanation of it.[72] It is symptomatic of Humboldt's interests at this time that he should have couched his thinking about the concept of nature as a living organism using the allegorical device of a *painting*.[73]

Eight years after the travels of Forster and Humboldt to the Netherlands, England and France, Humboldt returned to Paris, where another encounter with the work of an artist was to further shape his aesthetic sensibility. Humboldt met the acclaimed French artist François Gérard during the drawing classes that he was taking under the supervision of Marie-Éléonore Godefroid, whom he affectionately called 'ma bienfaitrice' in a later letter to Gérard.[74] From July 19 to October 6, 1798, Gérard exhibited three works in the Paris Salon, including a painting on the popular theme of *Cupid and Psyche*,[75] the centre of attention at the show.[76] While Humboldt had spent only a few days in the French capital in 1790, in 1798 he was in Paris from late April to October, giving him ample time to view that work, either in the Salon or in the artist's studio. This must have been the moment when he made an outline drawing of the two mythological figures.[77] At the Salon he could also have seen two paintings by Hubert Robert,[78] Keeper of the King's Paintings, who exhibited at the Salon from 1767 to 1802.[79] As a young student in Rome, Robert had been accommodated for a time in the residence of the Order of Malta on the Aventine Hill, where Piranesi's design of the piazza and gardens probably had an influence on 'Robert des ruines,' a nickname he earned from his penchant for Piranesi-inspired interiors or demolition scenes.

François Gérard had achieved a major success at the Salon in 1795 with his painting *Bélisaire*.[80] The meeting with Humboldt three years later marked the prelude to an enduring friendship, as is demonstrated by the various portraits that Gérard made of his friend both before and after the American expedition [5]. Humboldt attended Gérard's social gatherings in his Parisian home, and the correspondence between the two men was to last over a period of decades. Under the watchful eye of his mentor, the sensibility that had been awakened by the encounter with the work of William Hodges was augmented with the technical ability of a draughtsman that is evident in the *plein-air* sketches that Humboldt made himself.

It was during Humboldt's activities in Gérard's atelier in Paris that he came into contact with and befriended the young Carl Wilhelm von Steuben from Thüringen, who painted a full-length portrait of Humboldt (now lost) perched on an outcrop of basaltic rock in 1812.[81] Given the significance of basalt in the debate between the Vulcanists and their opponents, the choice of this rock formation on which to place the Prussian scientist is no arbitrary one.[82]

The Tropical Antique

It would be tempting to suppose that the volcano and the exotic vegetation of Tenerife prepared Alexander von Humboldt for the task of interpreting the physical and natural world of America. Indeed, the division of the vegetation zones of Teide along climatic lines and their visualisation in a cross section of the volcano is a technique that he was to apply to the volcanoes of South America too. But his remarks on the dragon tree of Tenerife repeatedly emphasise that, although he was fascinated by the strange appearance of the tree, what impressed Humboldt most was its *antiquity*. Absent from the diary of his stay on Tenerife, it is described in two phrases in the letter to his brother: 'In the town of Orotava there is a dragon tree (Dracaena Draco) with a circumference of 45 feet. Four hundred years ago, in the time of the Guanches, it was as thick as it is today.'[83] The account published seventeen years later in the *Relation historique* repeats the same information, and the 1852 article follows the same pattern, like a mantra. What links the two phenomena, the volcano and the dragon tree, in his mind is the time depth they offer. Like the North American artist Frederic Church (especially during his second trip to Ecuador in 1857),[84] the view of the interior of a volcano afforded a glimpse into the deep history of the earth's formation. As for the dragon tree, its thickness was a symbol of temporal profundity. It was a living antique. Although Humboldt never met the British ambassador to the Neapolitan court, William Hamilton, who died in 1803, the two men shared a common interest in volcanoes and antiquities.[85]

These are the same two elements that we find in the famous frontispiece designed for the *Voyage de Humboldt et Bonpland* by Alexander's friend François Gérard [6].[86] Gérard had no difficulty with the iconography of the classical figures of Minerva and Mercury, representing knowledge and trade, respectively, who raise the defeated Aztec warrior from his tomb, but he required Humboldt's assistance for the American material. For this purpose, Humboldt lent him a copy of his *Vues des cordillères* to derive the details of the engraving from the almost seventy 70 plates depicting archaeological sites, Mexican codices and volcanic landscapes.[87] The Ecuadorian volcano Chimborazo that dominates Gérard's frontispiece is based on two of those plates,[88] which, in turn, derived from sketches done by Humboldt on the spot. Although it is well known that Humboldt did not manage to reach the peak at 6263 metres, he did arrive at an elevation of almost 6000 metres, which is how Goethe represented him climbing the flank of the volcano in his 1807 'Esquisse des principales hauteurs des deux continens.' Humboldt must have felt that with his ascent Chimborazo had become *his* volcano.[89] Hence, Chimborazo appears as an iconic image inextricably associated with Humboldt in the portrait painted in Mexico by Rafael Ximeno y Planes in 1803, in the painting of Humboldt and Bonpland in the Andes by Friedrich Georg Weitsch in 1810, and the portrait of Humboldt as an old man, painted by Julius Schrader some months before the Prussian's death in 1859.[90] Neither Ximeno y Planes, Weitsch nor Schrader, it should be noted, ever travelled to the Andes.

As for the antiquities, the *monumens des peuples indigènes* complement the *vues des cordillères*; although Chimborazo is located in the central Andes, the dress and attributes of the native personification of America and the two monuments situated in the background are Mexican. Although these monuments are a fairly faithful representation of archaeological reality, they have been merged in a unified landscape with Chimborazo to evoke a generic pre-Columbian world being saved by reason, juxtaposed side by side like the exotic objects united within the space of a cabinet of curiosities.

There is a parallel to our frontispiece in other graphic works of François Gérard. The French painter was born in Rome, where his father was working for the French ambassador, and grew up in a cultural milieu in which the question of the superiority of Greek art and architecture as advocated by Winckelmann was questioned by those who, like Piranesi, championed the merits of Roman architecture and engineering. In 1794, Gérard designed an illustration for the sixth book of Vergil's *Aeneid* depicting the journey of Aeneas to the underworld and his reunion with his father Anchises, who guides him through the world of the dead.[91] In a certain sense, the *monumens des peuples indigènes* that interested Humboldt so much represent a world of the dead, the lost pre-Columbian civilisations. The upturned bust in the frontispiece of *Vues des cordillères* corresponds to the basalt figure of a 'Buste d'une prêtresse aztèque' represented in the first two plates in that work. In his discussion of the bust, Humboldt widens the scope of his inquiry from the Old

World of classical antiquity to the even Older World of pharaonic Egypt, as he draws a comparison with Egyptian and Greek sculpture and takes what he regards as the absence of hands on the bust to be a sign of 'l'enfance de l'art.'[92] This interpretation of Mexican art as representing a particular historical phase, that of infancy, corresponds to the Humboldtian vision of American nature, in which both the landscape and the 'indigenous monuments' appear as the results or traces of an activity remote in time. They are all antiquities.

Humboldt's ongoing invention of what may be called 'the tropical antique' was one of the factors that prompted him to travel to Italy after his return from America. Ever since his ascent of the Pico del Teide in Tenerife, he had been plagued with doubts regarding the effects of volcanic fire. He hoped that the ascent of Vesuvius would help to resolve these doubts.[93] He left the French capital in March 1805 for Italy, where he was to spend three and a half months, one in Naples and the remaining period in Rome. In Naples, he was accompanied by his old friend from their student days together in the Freiberg School of Mines, the geologist Leopold von Buch, with whom he had travelled through the Austrian Alps in 1797. For cultured company he could count on support from the salon of poets and artists under the aegis of Baroness Elisa von der Recke, who spent the summers in Naples. He was taking tea with her on the evening of 12 August when Vesuvius suddenly began to erupt. They rushed to the balcony to view the spectacle before Humboldt and some male companions dashed off to observe the eruption from a more proximate vantage point.[94]

But Rome exerted a strong pull too. In 1803 Alexander had admired the vast collection of plaster casts of acknowledged masterpieces of classical sculpture assembled in the Academia de San Carlos in Mexico City under the aegis of Charles III of Spain.[95] He considered them to be 'a much finer and more complete collection of casts than is to be found in any part of Germany.'[96] If Naples could satisfy his curiosity about volcanoes, it was the Eternal City that would help to throw light on the other topic that had fascinated him ever since his vision of the dragon tree on Tenerife, antiquities, and he was no doubt eager to view the original sculptures *in situ*. Personal reasons also guided his steps towards Rome, as his brother Wilhelm was then the Prussian ambassador to the Vatican. What is more, in Rome Alexander could count on the congenial company of writers, intellectuals and other members of the cultural élite as he had done on Tenerife. Among those who frequented the salon of Wilhelm von Humboldt and his wife, Caroline, were Elise von der Recke during the winter months and the Danish consul general, archaeologist and antiquary Georg Zoëga, who had settled permanently in Rome in 1783. He and Alexander became close friends, and on their promenades through the city, Zoëga acted as the perfect guide to the antiquities of Rome. Zoëga, or rather his daughter Laura, was to introduce the young Jean-Auguste-Dominique Ingres to archaeological circles in Rome after his arrival in October 1806,[97] and many years later, after Ingres had been appointed director of the Académie de France in Rome, he wrote to Alexander von Humboldt to thank him for his constant support of the Néo-Grec painter.[98] As for Mexican antiquities, though the death of the young novice from Honduras, José Lino Fábrega, in 1797 prevented Humboldt from meeting him in person, Fábrega's pioneering commentary on the Codex Borgia and his studies of the two Vatican codices were to prove valuable aids to Humboldt in his task of interpretation.

Invention and Publication

In the opening sentence of his *Historia del Mondo Nuovo*, Girolamo Benzoni gave as one of the motives for his departure for the New World the natural curiosity of a 22-year-old to see the world. When embarking from Sanlúcar de Barrameda in 1541, he can hardly have foreseen that his journey would take him through various parts of the Caribbean, Central America and Peru before ending up in Quito 25 years later. Alexander von Humboldt was apparently motivated by a similar desire to see the world, without concerning himself unduly about which direction his journey might take. Constitutionally unfit for the life of a Prussian civil servant which his family expected him to adopt, the death of his mother in 1796 left him free to set his sights on a destination, but the turbulent international political events of those years left him with limited choices. He and Aimé Bonpland hoped to be able to sail from Marseille to North Africa and then travel overland to reach Egypt, but no ship was to be found. So the fact that his present reputation is based mainly on the five years of travel in America is the result of a particular concatenation of circumstances. At this point in time, his urge was to travel, no matter where.[99]

He will have been aware that an American destination entailed passing through and stopping in the Canary Islands, but the six days on Tenerife were exactly that: a short stop on the way to what had serendipitously become his main destination, the American continent. Nevertheless, his emphasis on the antiquity of both the Pico del Teide and the dragon tree of La Orotava suggests that he was already looking back in time in the search for origins. And once on American soil, his encounters not only with tropical vegetation but also with monuments of tropical antiquity were to point the way to a project that would eventually assume the form of the *Vues des Cordillères*. A further step in the chain was his Italian sojourn in 1806, where he was able to study volcanic phenomena and Mexican codices first-hand.

It thus appears that Humboldt was inventing his project as he went along, accumulating and filtering experiences that would later form the basis of more synthetic studies. Even if the diary entries that Humboldt made during the American journey do not reflect experiences on a direct, day-to-day basis, they are nevertheless the closest that we can get to those experiences themselves.

Clearly the distance from the events increases considerably with the publication of the *Relation historique*, beginning in 1814. The same is true of the account of the Italian journey of Humboldt's friend Goethe between 1786 and 1788, which was not written until a quarter of a century after the journey itself. Not only does the published account combine events that actually took place in different years for the sake of providing a smooth narrative, but it also reinterprets those events in the light of the changes in Goethe's literary strategy that had taken place in the intervening years.[100] Comparison of Charles Darwin's *Beagle* diary from 1831 to 1836 with his narrative of the voyage of the *Beagle* published in 1839 reveals examples not only of similar disparities between the written journal and the published narrative, but even within the entries in the written journal itself.[101] In short, Humboldt's practice in this respect is in line with that of his near-contemporaries in relation to the genre of the travel narrative.

The picture of Humboldt's project that emerges from the *Relation historique* (and from the *Vues des cordillères*) is thus a synthetic one. Looking back, it weaves experiences together as though they formed part of a preconceived, forward-looking project, whereas, in fact, that is the result of the narrative strategy. On this reading, the six days on Tenerife form the prelude, the narrative of the five years on the American continent is the main work, and the Italian journey becomes a coda to the whole enterprise.[102] At this point the invention of Humboldt's project shades into the Humboldtian project of dealing with his intellectual legacy.

Notes

1 Darwin Correspondence Project, 'Letter No. 98,' https://www.darwinproject.ac.uk/letter/DCP-LETT-98.xml, accessed on 28 October 2019.
2 R.D. Keynes, ed., *Charles Darwin's Beagle Diary* (Cambridge: Cambridge University Press, 1988), 19.
3 Janet Browne, *Charles Darwin Voyaging* (London: Pimlico, 1996), 177–183.
4 For more on the iconography of Teide and the Drago, see Peter Mason, *Before Disenchantment. Images of Exotic Animals and Plants in the Early Modern World* (London: Reaktion, 2009) and *El drago en el Jardín del Edén. Las Islas Canarias en la circulación transatlántica de imágenes en el mundo ibérico, siglos XVI y XVII* (Madrid: Iberoamericana Editorial Vervuert, 2018).
5 Cristóbal Colón, *Textos y documentos completos*, prólogo y notas de Consuelo Varela, 2nd ed. (Madrid: Alianza Editorial, 1984), 18; J.C. Carracedo, et al., 'La erupción que Cristobal Colón vio en La Isla de Tenerife (Islas Canarias),' *Geogaceta* 41 (2007), 39–42.
6 Alexander von Humboldt, *Personal Narrative of a Journey to the Equinoctial Regions of the New Continent*, abridged and translated by Jason Wilson (London: Penguin, 1995), 26; Alexander von Humboldt, *Voyage aux régions équinoxiales du nouveau continent. Tome Premier* (Paris: J. Smith, 1816), 220.
7 Sir George Staunton, *An Historical Account of the Embassy to the Emperor of China* (London: John Stockdale, 1797), 52–67, which includes a report of the more successful ascent by the British surveyor of Madeira, William Johnstone, during a previous summer.
8 Irina Podgorny and Wolfgang Schäffner, "La intención de observar abre los ojos.' Narraciones, datos y medios técnicos en las empresas humboldtianas del siglo XIX,' *Prisma, Revista de historia intelectual* 4 (2000), 217–227, here 221.
9 Humboldt, *Personal Narrative*, 31; Humboldt, *Voyage*, 275.
10 Humboldt, *Personal Narrative*, 37; Humboldt, *Voyage*, 313.
11 Humboldt, *Personal Narrative*, 23; Humboldt, *Voyage*, 179; Alexander von Humboldt, *Voyage d'Espagne aux Canaries et à Cumaná Obs. astron. De Juin à Oct. 1799*, Tagebuch 1, 10r, https://edition-humboldt.de/reisetagebuecher/detail.xql?id=H0016412, accessed on 5 March 2020.
12 The species *Dracaena tamaranae* is endemic to Gran Canaria, where Humboldt did not land.
13 Humboldt, *Personal Narrative*, 28; Humboldt, *Voyage*, 236.
14 The tree subsequently suffered damage from storms and was finally destroyed by a hurricane in 1867, leaving the dragon tree of Icod de los Vinos as the oldest surviving specimen.

15 Various spellings are found of the name of this descendant of the Genoese noble merchant Antonio de Franchi Luzardo.

16 Humboldt, *Personal Narrative*, 29; Humboldt, *Voyage*, 249–250.

17 The print was inserted in some editions of Staunton's account of Lord Macartney's journey from London to serve as the first British ambassador to China in 1793; see Carlos Gaviño de Franchy, 'La estampa en Canarias. Desde los comienzos del reinado de Felipe V hasta la subida al trono de Isabel II,' in María de los Reyes Hernández Socorro, Gerardo Fuentes Pérez and Carlos Gaviño de Franchy, eds., *Historia cultural del arte en Canarias V. El despertar de la cultura en la época contemporánea* (Santa Cruz de Tenerife and Las Palmas de Gran Canaria: Gobierno de Canarias, 2008), 247–267, here 263–264.

18 David Martín López, 'La casa de Franchy de La Orotava: megalomanía y estética arquitectónica en Canarias (1745-1908),' *Coloquio de historia Canario-Americana* XVII (2006), 1325–1362.

19 Marcos Guimera Peraza, 'Bernardo Cólogan y Fallon (1772-1814),' *Anuario de Estudios Atlánticos* 25 (1979), 307–355, here 323. The manuscript, titled 'Noticias del volcán de la montaña de Chahorra,' is reproduced in Juan Tous Meliá, *La medida del Teide. Historia: descripciones, erupciones y cartografía* (San Cristobal de La Laguna, 2015), 202–206.

20 Alexander von Humboldt in a letter to his brother Wilhelm, 23 June 1799. In *Briefe Alexander's von Humboldt an seinen Bruder Wilhelm* (Stuttgart: Verlag der J.G. Cotta'schen Buchhandlung, 1880), 8.

21 Humboldt, *Personal Narrative*, 37; Humboldt, *Voyage*, 317.

22 Humboldt, *Personal Narrative*, 38; Humboldt, *Voyage*, 317.

23 Michelle Payne, *Marianne North. A Very Intrepid Painter* (Kew: Royal Botanic Gardens, 2011), 30–31.

24 This initial coolness is glossed over in Humboldt, *Personal Narrative*, 26; Humboldt, *Voyage*, 218.

25 Humboldt, *Voyage d'Espagne*, Tagebuch I, 18v. José Clavijo y Fajardo was a cousin of the Canarian naturalist José de Viera y Clavijo, who included a description of the dragon tree in the 1799 autograph manuscript of his *Diccionario de historia natural de las islas Canarias* (Archivo Catedral de Canarias). See *Viera y Clavijo. De Isla en Continente*, exh. cat. (Madrid: Biblioteca Nacional de España, 2019), 275.

26 Humboldt, *Personal Narrative*, 18; Humboldt, *Voyage*, 93.

27 Humboldt, *Voyage d'Espagne*, Tagebuch I, 66r.

28 Humboldt, *Voyage d'Espagne*, Tagebuch I, 18v.

29 Humboldt, *Voyage d'Espagne*, Tagebuch I, 2v. Compare Alexander von Humboldt in a letter to his brother Wilhelm, 23 June 1799. In *Briefe Alexander's von Humboldt an seinen Bruder*, 7.

30 Humboldt, *Voyage d'Espagne*, Tagebuch I, 19v and 66r.

31 Humboldt, *Voyage d'Espagne*, Tagebuch I, 3r.

32 Humboldt, *Voyage d'Espagne*, Tagebuch I, 3v.

33 It was drawn in 1817 by [Pierre-Antoine] Marchais and Louis Aubert, and engraved by Louis Coutant, compiling Humboldt's 1799 observations and Leopold von Buch's and Christen Smith's 1815 observations. See *Alexander von Humboldt. Das graphische Gesamtwerk* (Darmstadt: Lambert Schneider Verlag, 2015), 130–131. The *Atlas géographique et physique des régions équinoxiales du Nouveau Continent* forms volume 17 of the *Voyages aux régions équinoxiales*. An undated sketch among Humboldt's papers shows a cross section of the Pico del Teide but is confined to a description of its geological composition without references to vegetation. See *Alexander von Humboldt. Das zeichnerische Werk*, Dominik Erdmann and Oliver Lubrich, eds. (Darmstadt: Wissenschaftliche Buchgesellschaft WBG, 2019), 213.

34 *Das graphische Gesamtwerk*, 55.

35 José Barrios García, 'La imagen del drago de La Orotava (Tenerife) en la literatura y el arte. Apuntes para un catálogo cronológico (1770–1878),' *Coloquio de Historia Canario-Americana* XIX (2012), 748–758.

36 Alexander von Humbolt, *Vues des Cordillères, et des monumens des peuples indigènes de l'Amérique* (Paris: J. Smith, 1810), Planche 69; *Das graphische Gesamtwerk*, 125. The two volumes of the *Vues des Cordillères* form volumes 15 and 16 of the *Voyages aux régions équinoxiales*. Pierre-Antoine Marchais, a protégé of Hubert Robert, was a landscape painter; his first entry to the Salon de Paris was in 1793 with two paintings, one of which was a historical landscape with an episode from the legend of Bélisaire, now in the musée des augustins, Toulouse.

37 Alexander von Humboldt, 'Le dragonier d'Orotava,' *La Belgique Horticole* 2 (1852), 79–86; *Das graphische Gesamtwerk*, 742.

38 Humboldt, *Voyage d'Espagne*, Tagebuch I, 9v. The sketches are in the Staatsbibliothek Preußischer Kulturbesitz, Nachl. Alexander von Humboldt, gr. Kasten 8, nos 1a, 15v–17v. See *Das zeichnerische Werk*, 209–212.

39 Humboldt, *Personal Narrative*, 34; Humboldt, *Voyage*, 288; *Das graphische Gesamtwerk*, 110.

40 Humboldt, *Personal Narrative*, 34; Humboldt, *Voyage*, 296.

41 Marie-Noëlle Bourguet, 'El mundo visto desde lo alto del Teide: Alexander von Humboldt en Tenerife,' José Montesinos, Javier Ordóñez and Sergio Toledo, eds., *Ciencia y Romanticismo* (La Orotava: Fundación Canaria Orotava de Historia de la Ciencia, 2003), 279–302, here 288.

42 Humboldt, *Voyage d'Espagne*, Tagebuch I, 66r.

43 Marie-Noëlle Bourguet, *Le monde dans un carnet. Alexander von Humboldt en Italie (1805)* (Paris: Éditions du félin, 2017), 286, referring to Tagebuch 3 and 9.

44 Alexander von Humboldt, *Varia*, Tagebuch IX, 9r–11r.

45 Cristóbal Corrales and Dolores Corbella, 'Diatopismos léxicos en la obra de Viera y Clavijo,' in *Viera y Clavijo. De Isla en Continente*, 253–281.

46 Girolamo Benzoni, *La historia del mondo nuovo di M. Girolamo Benzoni milanese. La qual tratta dell'isole & mari nuovamente ritrovati & delle nuove città da lui proprio vedute, per acqua & per terra in quattordeci anni. Nuovamente ristampata & illustrata con la giunta d'alcune cose notabile dell'isole di Canaria* (Venice: Pietro y Francesco Tini, 1572), 176r–179v.

47 Colón, *Textos y documentos completos*, 30–31.

48 Colón, *Textos y documentos completos*, 88.

49 Humboldt, *Personal Narrative*, 26; Humboldt, *Voyage*, 217.

50 Colón, *Textos y documentos completos*, 18–19.

51 Humboldt, *Personal Narrative*, 22, Humboldt, *Voyage*, 167.

52 Humboldt, *Personal Narrative*, 39; *Voyage aux régions équinoxiales du nouveau continent*, 3.

53 Humboldt, *Personal Narrative*, 40; Humboldt, *Voyage*, v. 2, 11.

54 Valerie I.J. Flint, *The Imaginative Landscape of Christopher Columbus* (Princeton: Princeton University Press, 1992); Peter Mason, *Deconstructing America. Representations of the Other* (New York and London: Routledge, 1990).

55 Humboldt, *Voyage d'Espagne*, Tagebuch I, 15r.

56 Humboldt, *Voyage d'Espagne*, Tagebuch I, 15r. See George Forster, *A Voyage Round the World*, Nicholas Thomas and Oliver Berghof, eds., v. I (Honolulu: University of Hawaii Press, 2000), 21–30. Forster passed the Canarian islands of La Palma and El Hierro, but not Tenerife.

57 Humboldt, *Personal Narrative*, 38, Humboldt, *Voyage*, 425. Cf. Antonio Sánchez Jiménez, '¿Leyenda Negra o lascasianismo?: la polémica del Nuevo Mundo y la reescritura de la historia de *Los guanches de Tenerife*,' in Yolanda Rodríguez Pérez and Antonio Sánchez Jiménez, eds., *La leyenda negra en el crisol de la comedia. El teatro del Siglo de Oro frente a los estereotipos antihispánicos* (Madrid: Iberoamericana/Vervuert, 2016), 89–100.

58 Humboldt, *Personal Narrative*, 39; Humboldt, *Voyage*, v. 2, 2.

59 Humboldt, *Voyage d'Espagne*, Tagebuch I, 15v.

60 Alexander von Humboldt in a letter to his brother Wilhelm, 23 June 1799. In *Briefe Alexander's von Humboldt an seinen Bruder*, 10.

61 A. Bernard Smith, *Imagining the Pacific. In the Wake of the Cook Voyages* (New Haven: Yale University Press, 1992), 123.

62 Peter Mason, *The Colossal from Ancient Greece to Giacometti* (London: Reaktion, 2013), 82.
63 Geoff Quilley and John Bonehill, eds., *William Hodges 1744-1797. The Art of Exploration*, exh. cat. (New Haven, CT and London: National Maritime Museum Greenwich and Yale University Press, 2004), cat. no. 76. This architectural structure from the fourth century C.E. was a pavilion that formed part of a luxurious residential complex. See Fabiola Fraioli, 'Regione V. Esquiliae,' in Andrea Carandini, ed., *Atlante di Roma Antica* (Milan: Electa, 2012), v. 1, 323–336, here 336.
64 The first edition was published in Rome in 1756, followed by a second in 1784: see Luigi Ficacci, ed., *Piranesi. The Complete Etchings* (Cologne: Taschen, 2000), n. 165.
65 Quilley and Bonehill, *William Hodges*, 138.
66 Quilley and Bonehill, *William Hodges*, cat. no. 11.
67 Quilley and Bonehill, *William Hodges*, cat. no. 36.
68 *A View of the West Side of the Fortress of Chunargarh on the Ganges* (cat. no 55), and *View of a Mosque at Rajmahal* (cat. no. 49) may have formed part of the Hastings collection, and *View of the City of Rajmahal* (cat. 48) certainly did.
69 Katherine Emma Manthorne, *Tropical Renaissance. North American Artists Exploring Latin America 1839-1879* (Washington, DC: Smithsonian Institution Press, 1989), 29.
70 *Kosmos* (J.G. Gotta'scher Verlag, Stuttgart and Augsburg, 1845–1858), II, 5. The reference is puzzling: Hans Walter Lack, director of the Botanic Garden and Botanical Museum in Berlin, writes:

> I have never heard or read about a colossal specimen of *Dracaena draco* cultivated during Humboldt's youth (i.e., before his departure for South America) 'in the old tower of the botanical garden in Berlin.' In the fourth edition of his *Species Plantarum*, Willdenow notes 'v.s.' [vidi siccum] and the respective specimens in his herbarium have no reference to the botanical garden.
> (personal communication, 4 December 2019)

71 Joachim Rees, *Die verzeichnete Fremde. Formen und Funktionen des Zeichnens im Kontext europäischer Forschungsreisen 1770-1830* (Paderborn: Wilhelm Fink, 2015), 396.
72 Alexander von Humboldt, 'Die Lebenskraft oder der rhodische Genius, eine Erzählung,' *Die Horen*, 5, Stück (1795), 90–96. Reprinted in Alexander von Humboldt, *Ansichten der Natur*, 3rd ed. (Stuttgart and Tübingen: J.B. Gotta, 1849), Band II, 297–308.
73 For Humboldt's thoughts about *Lebenskraft* see, with extensive bibliography, Alicia Lubowski-Jahn, 'A Comparative Analysis of the Landscape Aesthetics of Alexander von Humboldt and John Ruskin,' *British Journal of Aesthetics* 51:3 (July 2011), 321–333.
74 Baron [Henri-Alexandre] Gérard, ed., *Lettres adressées au baron François Gérard*, 3rd ed. (Paris: A. Quantin, 1888), v. 2, 54. Godefroid was to become Gérard's invaluable assistant from 1812 on.
75 Psyche was also the subject of a drawing by Denis-Antoine Chaudet and a painting by J.J. Lagrenée. See *Explication des ouvrages de peinture et dessins, sculptures, architecture et gravure exposés au Muséum central des Arts, Imprimerie des Sciences et Arts* (Paris, 1798), 15 and 38. In Rome, Antonio Canova was working on a sculpture of Cupid and Psyche during the same period, which was eventually shown in the Paris Salon in 1808: see Giuseppe Pavanello, *Canova. Eterna Bellezza*, exh. cat. (Milan: Museo di Roma, Silvana Editoriale, 2019), cat. no. 136.
76 Thomas Crow, *Emulation. Making Artists for Revolutionary France* (New Haven, CT: Yale University Press, 1995), 211–214.
77 *Das zeichnerische Werk*, 65.
78 These represented an ancient building in use as a public bath, and the entrance to an ancient villa: *Explication des ouvrages*, 61.
79 On Hubert Robert see Guillaume Faroult and Cathérine Voiriot, eds., *Hubert Robert 1733-1808. Un peintre visionnaire*, exh. cat. Musée du Louvre (Paris: Somogy Éditions d'Art, 2016).

80 Crow, *Emulation*, 205ff.; Florike Egmond and Peter Mason, 'A Horse Called Belisarius,' *History Workshop* 47 (1999), 240–252.

81 Alexander von Humboldt in a letter to his sister-in-law Caroline, 19 August 1813. In *Briefe Alexander's von Humboldt an seinen Bruder*, 218–219. The painting is known from an engraving published in *Katalog zur Ausstellung deutscher Kunst aus der Zeit von 1775-1875 in der Königlichen Nationalgalerie Berlin* (Munich: Verlag F. Bruckmann, 1906).

82 Maiken Umbach, 'Visual Culture, Scientific Images and German Small-State Politics in the Late Enlightenment,' *Past and Present* 158 (1998), 110–145; Juan Pimentel, *Testigos del mundo. Ciencia, literatura y viajes en la Ilustración* (Madrid: Marcial Pons, 2003), 206–207.

83 Alexander von Humboldt in a letter to his brother Wilhelm, 23 June 1799. In *Briefe Alexander's von Humboldt an seinen Bruder*, 10.

84 Manthorne, *Tropical Renaissance*, 77–80.

85 Ian Jenkins and Kim Sloan, eds., *Vases and Volcanoes. Sir William Hamilton and His Collection*, exh. cat. (London: British Museum Press, 1996).

86 *Das graphische Gesamtwerk*, 127. The frontispiece was published in the *Atlas géographique et physique des régions équinoxiales du Nouveau Continent* that was incorporated in volume 18 of the *Voyages aux régions équinoxiales*. See Helga von Kügelgen-Kropfinger, 'El frontispicio de François Gérard para la obra de viaje de Humboldt y Bonpland,' *Jahrbuch für Geschichte von Staat, Wirtschaft und Gesellschaft Lateinamerikas*, Band 20 (1983), 575–616.

87 They include the Mexican temple of Mitla, which Humboldt singled out for the elegance of its ornaments, see Benjamin Keen, *The Aztec Image in Western Thought* (New Brunswick: Rutgers University Press, 1971), 333. Besides the fundamental article on Gérard's frontispiece by Helga von Kügelgen-Kropfinger, see Peter Mason, *The Lives of Images* (London: Reaktion, 2001), 140–147.

88 *Vues des cordillères….*, Planches XVI and XXV.

89 Pimentel, *Testigos del mundo*, 179–210 and the chapter by Pimentel in this volume.

90 Halina Nelken, *Alexander von Humboldt. Bildnisse und Künstler. Eine dokomuntierte Ikonographie* (Berlin: Reimer, 1980).

91 Compare the artist's *Bélisaire* of 1795, in which the blind Belisarius leads his young guide, who has been bitten by a serpent.

92 *Vues des cordillères*, 54; cf. Alexander von Humboldt, *Political Essay on the Kingdom of New Spain*, trans. John Black, v. 1 (New York: I. Riley, 1811), 160. The eminent archaeologist of Rome Ennius Quirinus Visconti (1751–1818) contested his interpretation in a letter sent to Humboldt from Paris on December 12, 1812, and published at the conclusion of the *Vues*, 299–304.

93 Bourguet, *Le monde dans un carnet*, 138ff.

94 Bourguet, *Le monde dans un carnet*, 126–127.

95 The project was conceived by the Bohemian *pintor de Cámara* of Charles III, Anton Raphael Mengs, during his stay in Italy from 1769 to 1774, see Almudena Negrete Plano, ed., *Anton Raphael Mengs y la Antigüedad*, exh. cat. (Madrid: Real Academia de Bellas Artes de San Fernando, 2013).

96 Humboldt, *Political Essay*, 159. Cf. Ana Leticia Carpizo González, 'Museo Nacional de San Carlos, genealogía de una colección,' in Giovanni Capitelli and Stefano Cracolici, eds., *Roma en México / México en Roma. Las academias de arte entre Europa y el Nuevo Mundo 1843-1867*, exh. cat. (Mexico City and Rome: Museo Nacional de San Carlos and Campisano Editore, 2018), 143–147.

97 P. Picard-Cajan, 'Capturer l'antique: Ingres et le monde archéologique romain,' in P. Picard-Cajan, ed., *L'illusion grecque. Ingres et l'antiquité*, exh. cat. musée Ingres (Arles: Montauban, 2006), 41–50.

98 Ingres to Humboldt, 17 July 1842. *Nachlass Alexander von Humboldt*, gr. Kasten 2, Mappe 3, Nr. 110.

99 Andrea Wulf, *The Invention of Nature. The Adventures of Alexander von Humboldt, the Lost Hero of Science* (London: John Murray, 2015), 39–48.

100 On this shift from the *Sturm und Drang* of *Werther* to the classical voice of his rewritten *Iphigénie auf Tauris* see Peter Mason, *The Ways of the World. European Representations of Other Cultures from Homer to Sade* (Canon Pyon: Sean Kingston Publishing, 2015), 208–216.

101 R.D. Keynes, 'Introduction,' *Charles Darwin's Beagle Diary*, xi–xxiv. Darwin's *Journal and remarks. 1832–1836* was first published as volume three of the *Narrative of the Surveying Voyages of His Majesty's Ships Adventure and Beagle between the Years 1826 and 1836, Describing Their Examination of the Southern Shores of South America, and the Beagle's Circumnavigation of the Globe*. On the contributions of the various authors to this three-volume work, see Marta Penhos, *Paisaje con figuras. La invención de Tierra del Fuego a bordo del Beagle (1826–1836)* (Buenos Aires: Ampersand, 2018).

102 When the elderly Humboldt began to organise his notebooks and diaries in numbered volumes, he included the Italian journey of 1806 in the same volume as the diaries of the American travels. See Bourguet, *Le monde dans un carnet*, 14–15.

4

CALDAS AND HUMBOLDT IN THE ANDES

Who Invented Biogeography?[1]

Alberto Gómez Gutiérrez

In a recent study, Stephen T. Jackson has argued that the Creole biogeographer Francisco José de Caldas was a significant influence on Alexander von Humboldt's 'decision to use the Andes as the ideal region to illustrate his ideas.'[2] In this chapter, I argue further that, in the Andes, both Caldas and Humboldt participated in the founding of the new science of biogeography.

The coincidence of scientific findings has been studied under the concept of 'multiple discoveries,' insofar as they have been reported by independent, contemporary, or non-contemporary researchers.[3] This concept contrasts with the 'heroic theory' that has sought to associate specific scientific findings to the mind of a single individual. Two emblematic cases of multiple, and eventually simultaneous, discoveries are that of Isaac Newton (1643–1727) and Gottfried Leibniz (1646–1716) on infinitesimal calculus, and that of Charles Darwin (1809–1882) and Alfred Russel Wallace (1823–1913) on evolution by natural selection. The degree of public recognition of this simultaneity and protagonism has varied, but it is evident that the popular imaginary preferably registers individual biunivocal relations between the discoveries and their postulants.

In the field of biogeography, or the study of the distributions of organisms in space and time,[4] the most representative individual referent has been Alexander von Humboldt (1769–1859). He is widely considered to be the pioneer or 'inventor' of this theory followed, in chronological order according to their respective dates of birth and publication, by Friedrich Stromeyer (1776–1835) with *Commentatio inauguralis sistens historiae vegetablium geographiae specimen* (1800); Augustin Pyramus de Candolle (1778–1841) with his *Essai élémentaire de géographie botanique* (1820); Hewett C. Watson (1804–1881) with his four-volume work, *Cybele Britannica: Or British Plants and Their Geographical Relations*

DOI: 10.4324/9781003231479-5

(1847–1859); and Alfred Russel Wallace (1823–1913) with *The Geographical Distribution of Animals* (1876).[5]

The work that positioned Alexander von Humboldt as a pioneer of biogeography was published between 1805 and 1807 in Paris under the title *Essai sur la géographie des plantes*. The Prussian acknowledges in his preface some decisive preliminary moments of his discovery associated with Georg Forster (1754–1794), to whom he would have communicated 'the first sketch of a Geography of Plants' since 1790,[6] followed by Augustin P. de Candolle and Louis-François Ramond de Carbonnières (1755–1827), who provided him with 'interesting material on the geography of the plants of the High Alps [and] on the flora of the Pyrenees.'[7] Humboldt also cites his friend and master Carl Ludwig Willdenow (1765–1812) in this regard. According to Humboldt, French scholars would have provided him with some measurements and corrections to different scales in the first printed version of the 'Tableau physique' published in Europe,[8] as he wrote referring to Baron Ramond de Carbonnières:

> But perhaps no one is better qualified to work successfully on the geography of Europe's alpine plants than Mr. Ramond, who for so many years has climbed the highest peaks of the Pyrenees and who combines the ability to unite geognostic, botanical and mathematical knowledge with philosophical observations of nature.[9]

Alexander von Humboldt leaves out in this explicit account the works of Abbé Jean-Louis Giraud-Soulavie (1751–1813), who had already published a biogeographic profile reporting variations in climate and vegetation according to altitude in his seven-volume work titled *Histoire naturelle de la France méridionale* (1780–1784), with emphasis on olive trees, vines, and chestnut trees, in ascending barometric order, all below the limits of the 'great alpine trees'[10] (Figure 4.1). In North America, Humboldt cites the precedent of Benjamin Smith Barton (1766–1815), a physician and traveller 'who works ceaselessly in zoology, botany, and Indian linguistics, [and] is currently engaged in these same research for the temperate regions of the United States.'[11] Notably, there are absolutely no references to either Central or South American preliminary works on biogeography.

Let us return for a moment to Humboldt's written acknowledgement of his preliminary communication to Georg Forster of his 'first sketch' as early as 1790. This acknowledgement requires two clarifications. First, Forster, his first peer in the domain of plant geography (according to Humboldt's own report), was the son of Johann Reinhold Forster (1729–1798), a Prussian Lutheran naturalist and pastor whom Georg had accompanied on James Cook's second voyage (1728–1779) between 1772 and 1775. Second, on their return to England, Forster (the father) published his *Observations Made during a Voyage Round the World, on Physical Geography, Natural History and Ethic Philosophy* (1778), in which he included one

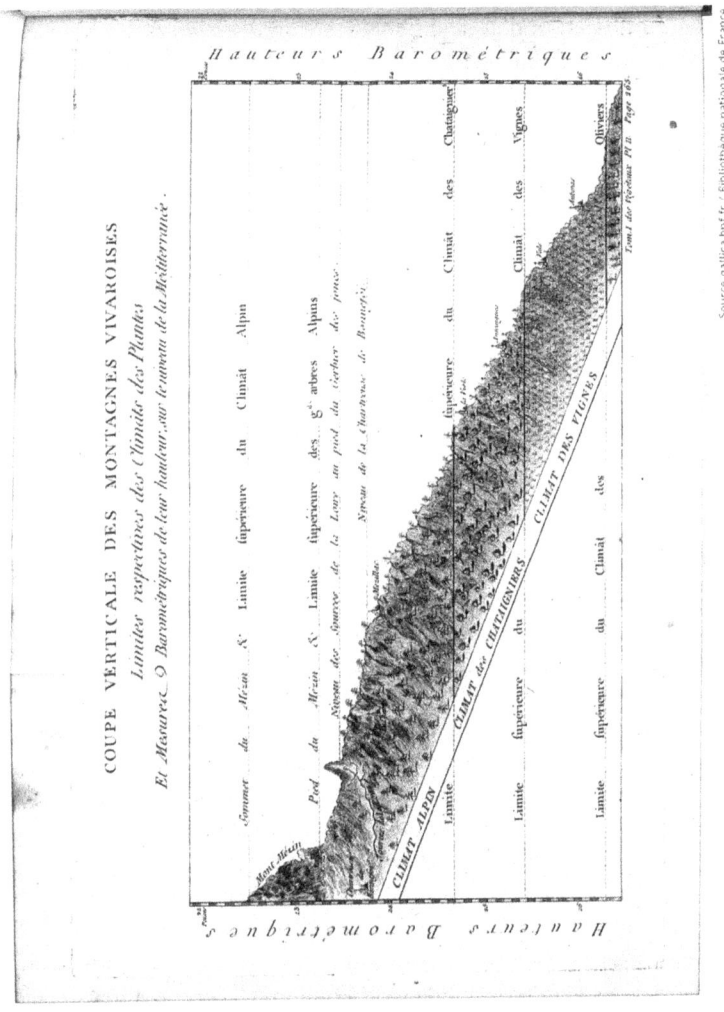

FIGURE 4.1 Jean-Louis Giraud-Soulavie, 'Coupe verticale des montagnes vivaroises. Limites respectives des climâts des plantes et mesures barométriques de leur hauteur, sur le niveau de la Méditerranée,' in Buttimer, A. (2012, 28): https://journals.openedition.org/cybergeo/docannexe/image/25478/img-6.jpg

of the first systematic representations of diverse biotic regions. His observations supported the view of Georges Louis Leclerc, Count of Buffon (1707–1778) that is now referred to as Buffon's law, which states that isolated regions contain plants and animals that vary according to their distance with respect to a hypothetical original node. Thus, the 'first sketch of a geography of plants' reported in 1790 by Alexander von Humboldt to his friend Georg Forster, could have been a Forsterian inspiration. In fact, Reinhold Forster himself complemented Buffon's law with a novel concept, noting the decreasing trend of plant diversity from the equator to the poles as a function of temperature variation according to latitude. All this was already stated and published since 1778 by the father of his youthful friend and peer.[12]

This was not the only precedent. One should also consider the lessons learned from Willdenow, one of Humboldt's main botanical contacts and mentors in Berlin before (and after) his American trip. Humboldt explicitly notes in his Preface to the *Essai sur la géographie des plantes*, 'I consulted Mr. Willdenow's other classics. It was important to compare the phenomena of equatorial vegetation with those of our European soil.'[13] In addition to the Forsters and Willdenow, the influence of Johann Wolfgang von Goethe (1749–1832) on the aesthetic representation of Humboldt's biogeographical ideas should be considered. Indeed, Goethe, whom Humboldt visited at least twice before travelling to America, had already published two reference works on 'Naturphilosophie' and botany: *Einfache Nachahmung der Natur* (1789) and *Die Metamorphose der Pflanzen* (1790). In these as well as in his personal conversations with the Humboldt brothers and Schiller in Jena, Goethe insisted on the importance of direct, intimate, and sensitive observation of natural phenomena, connecting reason and emotion, poetry, and aesthetics, in what has been defined as 'delicate empiricism: observation, reflection, association.'[14]

Humboldt's debt to Goethe in this domain is clear. Indeed, Goethe, upon receiving the printed copy of *Ideen zu eine Geographie der Pflanzen* (1807) that Humboldt himself had dedicated to him, produced and reciprocated to its author an alternate version of the 'Naturgemälde' or 'Tableau physique' centred on the Chimborazo, comparing the geography of Europe with that of America (Figures 4.2 and 4.3). Goethe wrote:

> I have read through the volume several times with great attention, and I have begun, even without the promised cross-sectional diagram, to imagine a landscape myself where, at a scale of 4000 toises (approx. 8 000 m) to a page, the heights of the European and American mountains are sketched side by side; the snowlines and the vegetation are also sketched. I enclose a copy of this sketch, partly for fun, partly seriously, and I ask you to make corrections on it with feather pen and with colours if you like, and to make some notes on the page and return it to me as soon as possible.[15]

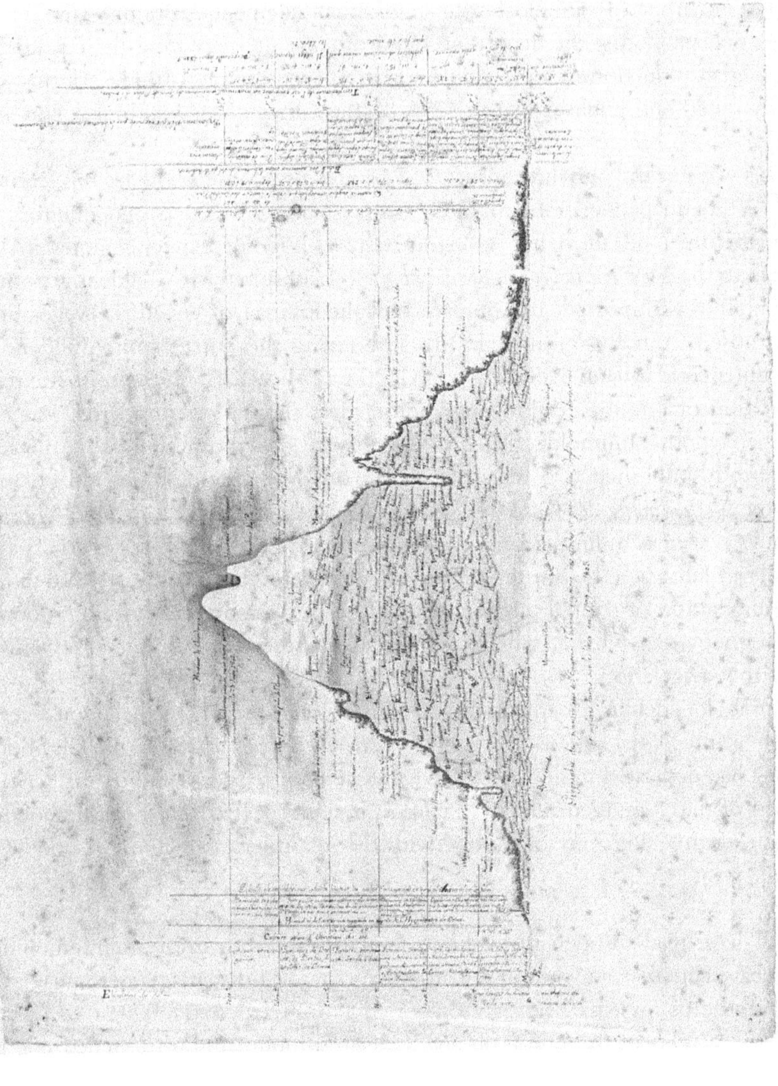

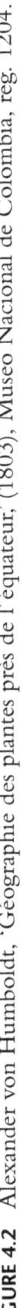

FIGURE 4.2 Alexander von Humboldt, 'Géographie des plantes près de l'équateur,' (1803), Museo Nacional de Colombia, reg. 1204.

FIGURE 4.3 Johann Wolfgang von Goethe, 'Tableau comparatif des altitudes de l'Ancien et du Nouveau Monde,' (1807), in Buttimer, A. (2012, 28): https://journals.openedition.org/cybergeo/docannexe/image/25478/img-6.jpg.

Years later, in a letter to Friedrich Schiller's sister-in-law, Caroline von Wolzogen (1763–1847), Humboldt clearly and concisely recorded Goethe's influence on his perception and representation of nature with these words:

> In the jungles of the Amazon, as in the peaks of the Andes, I had the feeling that life itself infiltrated rocks, plants, and animals, as well as the expansive chest of humanity, as animated from pole to pole by a single spirit. *Everywhere I deeply felt how powerfully these relationships forged in Jena now influenced me, and thanks to Goethe's perspective on nature, I acquired virtually new organs of perception.*[16]

Another German harbinger, fundamental to Humboldtian biogeography, is Zimmermann with his *Specimen zoologiae geographicae quadrupedum* (1777), also explicitly mentioned by Humboldt in his 1807 *Essai*:

> Mr Zimmerman's classic work indicates the homeland of the animals, according to the difference of the heights they inhabit. It would be interesting to fix in a profile the different heights to which these rise under the same latitude.[17]

In the preliminary handwritten draft of the *Essai sur la géographie des plantes* sent by Humboldt to José Celestino Mutis (1732–1808) from Guayaquil in February 1803, while travelling from Lima to Acapulco in the middle of his American journey through colonised and decolonised territories of the Spanish and British empires,[18] Zimmermann's work already appears as a key reference in zoogeography: 'This scale of animals is a fragment of a map similar to the one that represents the vegetables of the Andes and is part of a work that will complete the excellent *Geographia animalium*, published by Mr. Zimmerman.'[19] Mutis was the director of the Real Expedición Botánica del Nuevo Reino de Granada since 1783 and, eventually, according to the Sweedish taxonomist Carl Linnaeus (1707–1778), the best-known European naturalist established in South America.[20]

Following a reference by Francisco José de Caldas (1768–1816) included in his *Memoria sobre el plan de un viaje de Quito a la América septentrional* sent to José Celestino Mutis between April 21 and May 6, 1802, it can be inferred that Humboldt had already personally referred him to Zimmermann's work on zoogeography, since Caldas states, in the 'Zoology' section of his *Memoria*:

> Mr Baron has told me that a wise man, whose name I do not have in mind, has begun to work on this matter: his observations, his ideas would perfect ours, and our works would be appreciated as the first of this species in America.[21]

This explicit reference to a biogeographic work focused on the distribution of animals clearly reveals that Caldas dealt with Humboldt on this issue during their cohabitation in Quito. Nevertheless, the self-taught Neogranadian naturalist, mathematician, and geographer who upon his appointment to Mutis's Real Expedición Botánica met Humboldt in the first semester of 1802,[22] is conspicuously absent from all of Humboldt's explicit references to previous or simultaneous biogeographic works. For some reason, Humboldt relied on Mutis as a peer by sending him his first manuscript on and graphic design of the geography of plants, brushing aside the self-taught Creole. Thus, in the process of elaborating his biogeographical theory, whose first drafted manuscript was made in February 1803 while waiting in Guayaquil for the departure of the ship that would carry Humboldt to Acapulco, Caldas was excluded. Humboldt himself referred to this moment in his published *Essai sur la géographie des plantes* in these terms:

> I drew this *Tableau* for the first time in the port of Guayaquil, in February 1803, when I was returning from Lima by the South Sea, and when I was preparing to sail to Acapulco. I sent a copy of this first sketch to Mr. Mutis in Santa Fe de Bogotá, who honours me with a particular kindness. No one was in a better position than he to pronounce on the accuracy of my observations, and to extend them with his own observations obtained over 40 years of journeys in the Kingdom of New Granada. ... *No other botanist had more opportunities to make important observations on the geography of the plants, since during the collection of these he always made altimetric measurements and has climbed many times the high peaks of the mountain ranges ...*[23]

In the original manuscript of this work, which is written in Bonpland's hand with several corrections by Humboldt in the margins,[24] and sent to the Parisian editors of the *Essai* in 1805, Mutis was further praised but then finally downgraded and downplayed. After stating that

> No one was in a better position than him to judge the accuracy of my observations, and to extend them with his own observations obtained over the space of 40 years of journeys in the kingdom of the New Granada,

Humboldt proceeded to add a most significant sentence that he later crossed out and removed from the definitive version published in Paris and Tübingen: 'This great botanist who, despite his distance from Europe, followed the progress of our physics [. . . *combines the vast knowledge of a naturalist with that of a distinguished geometer*]. . .[25] In print, only botany should be associated with Mutis, not 'distinguished geometry' or geography. For Humboldt, Mutis was not to be exalted as one of the accepted harbingers in the new field of biogeography. The 'geography' of plants would be introduced as an original Prussian affair, in which only a few safe harbingers, carefully chosen in Europe, would be accommodated. Nor is there any mention of Caldas in print in this regard despite the fact that, according to Caldas, verbally Humboldt did so. As stated by Caldas:

> So we came to Ibarra, ate with [Humboldt, who] turned to me and said: 'I have seen your precious works in astronomy and geography. They have been taught to me in Popayán. I have seen corresponding heights taken with such precision that the biggest difference is no more than four seconds.'[26]

Humboldt dedicated his first manuscript on the geography of plants to Mutis, which was published in Spanish by Caldas on page 121 of the April 23 edition of the *Semanario del Nuevo Reyno de Granada* (1809): 'dedicated with the deepest gratitude to the illustrious Patriarch of Botanists, Don José Celestino Mutis, by Federico Alexandro Baron Humboldt.' But this inscription would be ephemeral, and the 'feelings of the deepest recognition' would vanish amid the evening mists of his transatlantic voyage. The first printed edition of the *Essai* in France in 1807 was dedicated to the botanists Antoine-Laurent de Jussieu (1748–1836) and René Desfontaines (1750–1833), and its first German edition to Johann Wolfgang von Goethe. It should be pointed out, however, that Humboldt and Bonpland later dedicated the first volume of their *Plantes équinoxiales* to Mutis in 1808.[27] Again, it was a recognition of his work on plants, not plant geography.

In an article published in 1960, Spanish geographer Pablo Vila i Dinarès (1881–1980) critiques Humboldt's additional disregard of Caldas, noting that 'the Baron must have realised that the Creole was on the way to establish the existing relationships between plants, temperature, and altitude, which did not cease to surprise him.' 'Both,' says Vila, 'were found in the course of the same geobotanical studies.'[28] In short, Humboldt and Caldas found themselves in an explicit, very eloquent silence. A particularly paradoxical element noted by Vila is that Humboldt wrote a letter from Lima on November 25, 1802, to French astronomer and mathematician Jean-Baptiste Joseph Delambre (1749–1822). This letter was only six months after saying goodbye to Caldas after having worked with him side by side in Quito. In the letter he refers to 'Mutis, [the] president of the [Audiencia de Quito,] Montúfar, [but] not to Caldas.' He writes that

> there is no vegetable from which we cannot indicate the rock it inhabits and the height in toises up to which it rises; to such an extent that the geography of the plants will have in our manuscripts very exact data.[29]

It seems very likely that, as may be deduced from the Humboldt–Caldas conversation on Zimmerman's zoogeography, that in the days of their cohabitation in the first semester of 1802, as Joaquín Acosta put it in his footnote to Caldas's *Memoria sobre el plan de un viaje* in Paris in 1849, Caldas had already begun to 'fertilize it in his own way.'[30]

In this regard, three preliminary paragraphs in the 'Botany' section of Caldas's *Memoria* must be considered. These passages are critical to support the degree of originality of Caldas's own practical and theoretical elaborations before the eventual hybridisation of his ideas with those of the Prussian traveller. The first passage reads as follows:

> I have always seen with annoyance a map in which only the names of miserable villages are read. It is better to see in it the place, the homeland of a plant, of a mineral, of a kind of animal, of a thermal spring, etc., than that pile of barbaric names that we can barely pronounce. How beautiful, how

interesting would it be to include a botanical map of the kingdom in the *Flora of Bogotá*, ... in which, suppressing so many dark villages, so many streams of no consideration, plants useful to the arts, to commerce, to health would be instead substituted [!] What a pleasure to see at a glance the homeland of cocoa, tea, nutmeg, almond, quinine, etc.

The second passage of interest follows:

I have worked a lot on this genre, and the most precious fruit I have produced is a certain habit of seeing, measuring, and designing countries with ease. ... *I have told Mr. Baron nothing about these materials, with the exception of Timaná's map*, which is one of my first essays. One of the things that I have noticed in the geographical works of this wise man is that he mixes what is true with what is doubtful, and that eager to embrace everything, he designs next to a piece worthy of D´Anville, another by simple relations of ignorant people.

The third passage reads as follows:

The barometer is held at the limit of snow to 16 inches, in the sea to 28, or near: the difference is 12 inches. *Wouldn't it be new, and at the same time beautiful, to divide into 12 zones of an inch in the barometer of width each, all the part of the earth that is able to vegetate? Wouldn't it be new to assign to each plant its limits, and in a laconic and exact way say: inhabits the first zone, inhabits from the third to the fifth, and so on? I have projected some barometric-botanical levellings similar to those that Mr. Baron de Humboldt has constructed with the only object of giving idea of the diverse heights of the land.* I divide them into twelve zones which will not be equal in width, because the higher zones will gradually increase in elevation, and I place in each one the plants that vegetate in it. If one grows into two, three, or more, it is put into the lower and into the last, and this announces that it prospers in the middle. *This idea touches me, I think it is new and worthy to be tried.*[31]

Caldas, it seems, had 'told Mr. Baron nothing about these materials' (with the exception of the map of Timaná) and had 'projected barometric-botanical levellings similar to those that Mr. Baron de Humboldt has built with the sole purpose of giving an idea of the various heights of the terrain.'

That is, as Caldas noted, *only barometric levellings are found in Humboldt* in the first semester of 1802. There is no iconography related to the geography of plants in the Prussian manuscripts before their first encounters, even though Humboldt included various observations on plant geography in his diaries from 1799 on. This fact can be substantiated by comparing Humboldt's first American geographical levellings as depicted in his 1802 publication, 'Barometrische Nivellirung der Gegend zwischen Cartagena und Santa Fé,'[32] with the handwritten 'Tableau physique' sent to Mutis via Caldas in February 1803. Caldas, on the other hand, had designed and applied a model that he considered 'new' and original, based on his first phytogeographic transect and shared with Mutis in the first semester of 1802. This model included a biogeographical division 'into 12 zones of one inch in the barometer of width each, [of] all the part of the earth that is able to vegetate.' Caldas would perfect his concept with further collections and observations in the second semester of 1802, and finally draw it on paper in the second semester of 1803, following the principles defined by Caldas in his *Memoria sobre el plan de un viaje*.

Imbabura: A Formal Scenario for Caldas's Biogeography

Humboldt's claim to pioneering biogeography is based for the most part on the referred French and German editions of the *Essai sur la géographie des plantes* [*Ideen zu einer Geographie der Pflanzen*] published in 1807. These editions are printed as upgraded products of Humboldt's 1803 manuscript, sent to Mutis from Guayaquil and accompanied by the 'Tableau physique des régions equinoxiales.' Similarly, Caldasian biogeography can be traced-back to one of his original although undated manuscript drawings, entitled 'Levelling of 30 species of plants, on the western view of Imbabura, a mountain near Ibarra, by F.J. de Caldas.' This pictogram is conserved in the archives of the Royal Botanical Garden in Madrid (Figure 4.4).[33] To fix the date of this phytogeographic profile, I present documentary evidence indicating that it was conceived at least one year before its physical execution in the second semester of 1803. In addition, I consider the preliminary developments to Caldas' handwritten 'levelling.' Reviewing what has been previously reported in secondary sources that have addressed this issue,[34] I discuss the successive documentary sources written by Caldas between 1802 and 1809. These sources support the notion of the parallel construction or invention, by Humboldt and Caldas, of what is today called 'biogeography.'

First, in the first semester of 1802 Caldas sent to Mutis his *Memoria sobre el plan de un viaje de Quito a la América septentrional*. In this plan he clearly proposes a research project that he still considered 'new' despite having cohabited with Alexander von Humboldt and Aimé Bonpland for nearly six months in Quito.[35] Second, in a letter to Mutis from Ibarra dated August 8, 1802, Caldas refers to a scientific principle that he defines as 'laws of level.' Caldas wrote:

> Vegetation, which makes my first object, does not keep *the laws of level that I have constantly observed on all the high hills that I know and have climbed.* ... I have astronomically fixed in latitude all the places of my transit, I have drawn the map of the country that I have crossed, I have sketched the views of Cayambe, Cotacache and Imbabura, I have traced the levelling of my path, the mercury heights at all the main points, and, in short, further things that I will communicate to you from Quito. The first is to determine geometrically the perpendicular height of several points on the slopes of Imbabura, to verify in them my observations of the barometer, and to examine if the formulas of Schevrbuch (sic)[36] and Tralles[37] agree also in the torrid zone and great elevations as in the temperate zone and at medium heights.

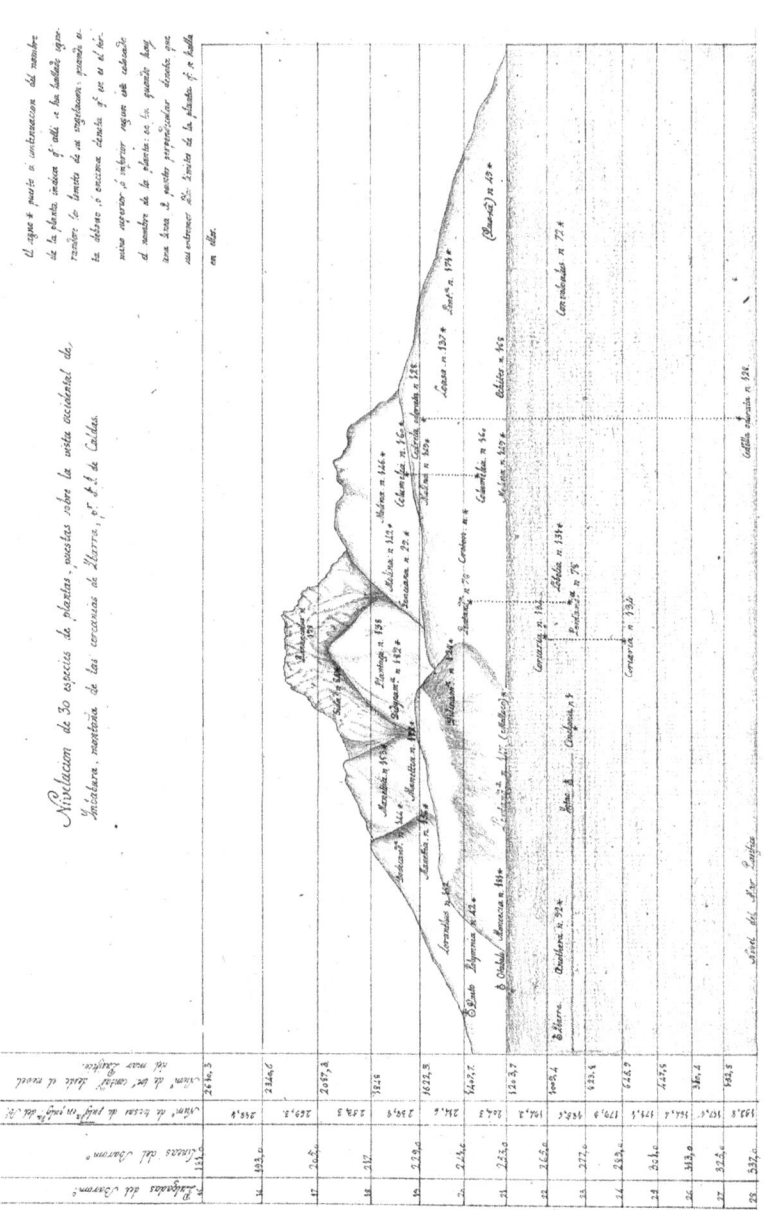

FIGURE 4.4 Francisco José de Caldas. Nivelación de 30 especies de plantas, sobre la vista occidental de Imbabura, montaña de las cercanías de Ibarra. Archivo del Real Jardín Botánico de Madrid, AJB, División III, M00529.

> *I believe that ithat no one has thought of subjecting their formulas to examination south of the [equatorial] line.* I regret not having at hand the works of these sages to direct my operations.[38]

Third, in a following letter, sent from Ibarra on September 23, 1802, Caldas writes to Mutis in these terms:

> On Imbabura, a mountain of which I have so much to tell you, I have found a *Syngenesia polygamy*. ... I have measured the base of the extinct volcano of Imbabura, on whose slopes is this village in the surroundings of Ibarra at one thousand eight hundred and fifty-five [Spanish] yards, by a network of triangles. I have made a chart of the country and the plane of the volcano, from which I have taken four views of the cardinal points. I have given preference to this mountain because nothing is known of it until now. The academics and Mr. Baron despised it absolutely. I have climbed twice this dreadful hill. ... I am going to copy here an excerpt from my diaries, and if you care for me, I think it will make you shudder.[39]

Fourth, having 'so much to tell' about Imbabura, Caldas included the details of this stretch in his travel diary[40] with corresponding botanical collections and barometric measurements. Caldas also included notes on how one of the three indigenous guides who accompanied him, and whose name was Salvador Chuquin, had saved his life:

> On the 6th we climbed Imbabura. ... I came down with many plants, and I spent the [following] days 7, 8, 9, 10, etc., in designing, skeletoning, dissecating and describing them. ... The 14th of September of [1]802 was destined for a trip that touched me so much and filled me with enthusiasm. Armed with my barometer, thermometer, octant and compass we left, don José Valentin and myself, with many Indians and experienced mountain boys.... What was left of the day I used to design and describe the plants I had removed from the crater. *As [this crater] is one inch of the Barometer below the lowest term of vegetation, but it already has some plants, and many mosses at the bottom.* On the 16th we descended the south face of the volcano ..., and on the 17th we returned to the village of Ibarra loaded with plants. The remaining days were spent drawing and describing plants.[41]

Fifth, a month and a half later, on November 7, 1802, Caldas wrote to Mutis from Otavalo (an Indian town south of Ibarra) in these terms:

> I will give you an account of my activities since I came down from the terrifying heights of Imbabura. *I have levelled all the surrounding areas of Ibarra*, and I have found by the fruit of my labours that all are higher than the plane on which this village is located. ... *In botany I have worked without limits.* I would never finish [this letter] if I told you all I have collected in this genre ...[42]

Sixth, and still in Otavalo ten days later, Caldas wrote to Humboldt a highly significant six-plus-page letter with details of his measurements and collecting:

> Mr. Bonpland had told me that he wished to return to Ibarra because many plants had escaped him; I witnessed the speed with which Your Lordship passed through these countries; Ulloa, Bouguer and De la Condamine barely mention Imbabura, Lake Mojanda and Cuicocha; in short, I felt I was entering an almost virgin country, and I left Quito at the end of last July. My activities have included measuring with uncommon care in the *ejido* [commons] of Ibarra a perfect map levelled at 795 toises. ... I took angles of height with the quarter circle that Your Lordship was kind enough to leave in my hands, and I calculated its height. On this basis I have formed a system of triangles, and I have undertaken a scrupulous topography of the whole country that I have traversed or will traverse... *Your Lordship can add a prodigious number of angles with the compass, and more astronomical determinations of latitude and you will have what I have done in terms of geography.*[43]

Seventh, in this same letter of November 1802, Caldas includes a reference to further geographical and botanical work in Cotacache:

> I have climbed Cotacache on the northface to the snowline and from there up until it is absolutely inaccessible. I verified my observation of the barometer and boiling water and collected a considerable number of plants. On this northface there is not the smallest vestige of eruption. I have barometrically determined many points which I would refer to Your Lordship now if I did not have other more interesting things to say. ... It is true that botany is my first occupation today, because Mr. Mutis wanted it so, and the plan of my work in this field is immense. As I don't have Humboldt's lights, nor Bonpland's, I could not leave any plant in the field, I had to describe them all, to dissect them all and to draw those that are not in my miserable books. ... In January, I will make my first remission of at least a thousand plant skeletons. Judging by my books and by the genres I copied from Willdenow, Gmelin, Schreber, and the compendium of the *Flora of Peru*, I have much new material, which I have been sending by mail to my generous benefactor. I don't know how it escaped me to take from these authors the characters of the genre *Dichondra* that doesn't even have [my] Palau: I pray Your Lordship to copy it and send it to me. I believe that in that [city] there will be good books of botany, if you find a Schreber, I ...[44]

This letter addressed to Humboldt is critical evidence, for it reveals that Caldas had communicated to him a series of geographical and botanical findings and works, although not explicitly associated with one another. Unfortunately, this letter is truncated in the transcription of Bateman and Arias,[45] as well as in the transcription

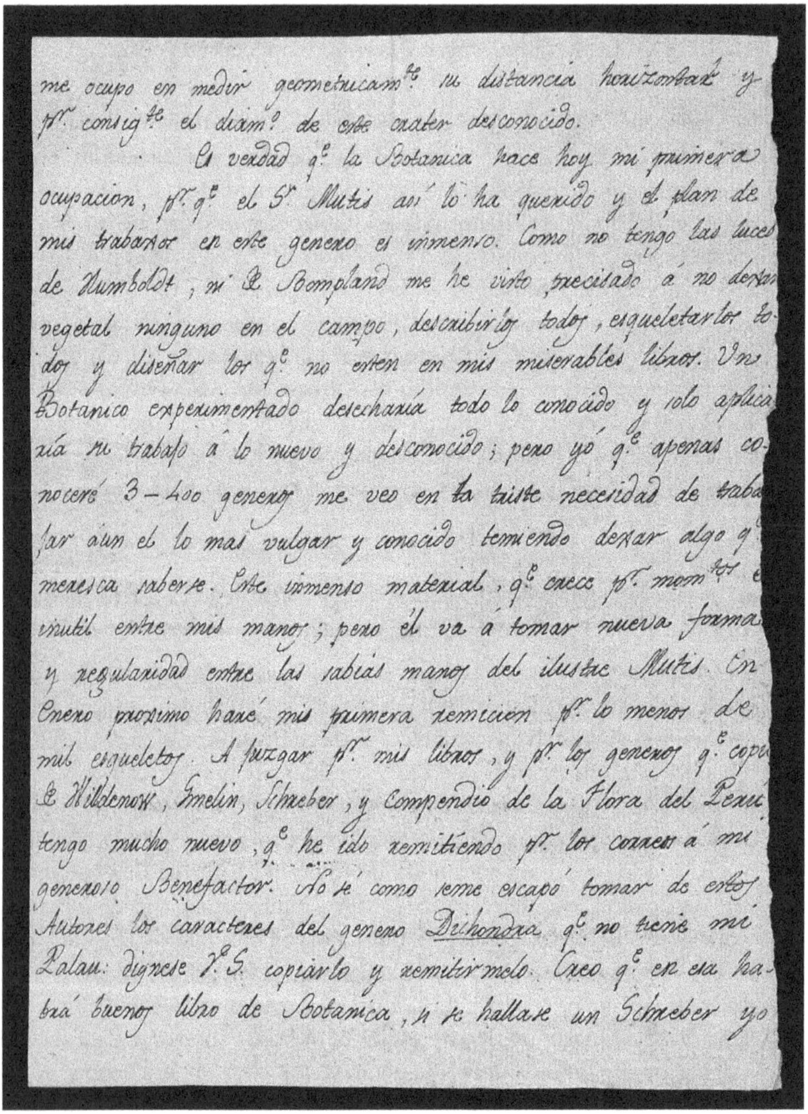
me ocupo en medir geometricamte su distancia horizontal y
pr consigte el diamo de este crater desconocido.
Es verdad qe la Botanica hace hoy mi primera
ocupacion, pr qe el Sr Mutis asi lo ha querido y el plan de
mis trabajos en este genero es inmenso. Como no tengo las luces
de Humboldt, ni de Bompland me he visto precisado á no dexar
vegetal ninguno en el campo, describirlos todos, esqueletarlos todos y diseñar los qe no esten en mis miserables libros. Un
Botanico experimentado desecharía todo lo conocido y solo aplica
ría su trabajo á lo nuevo y desconocido; pero yó qe apenas co-
noceré 3–400 generos me veo en la triste necesidad de traba
jar aun el lo mas vulgar y conocido temiendo dexar algo qe
merezca saberse. Este inmenso material, qe crece pr momtos e
inutil entre mis manos; pero él va á tomar nueva forma
y regularidad entre las sabias manos del ilustre Mutis. En
Enero proximo haré mi primera remicion pr lo menos de
mil esqueletos. A juzgar pr mis libros, y pr los generos qe copio
de Willdenow, Gmelin, Schreber, y Compendio de la Flora del Peru
tengo mucho nuevo, qe he ido remitiendo pr los correos á mi
generoso Benefactor. No sé como se me escapó tomar de estos
Autores los caracteres del genero Dichondra qe no tiene mi
Palau: dignese V. S. copiarlo y remitirmelo. Creo qe en esa ha-
brá buenos libros de Botanica, y si se hallase un Schreber yo

FIGURE 4.5 Francisco José de Caldas, Letter to Alexander von Humboldt (1802), Tagebücher der Amerikanischen Reise VIIbb et VIIc, folium 475, Staatsbibliothek, Berlin, http://digital.staatsbibliothek-berlin.de/werkansicht?PPN=PPN779884310&PHYSID=PHYS_0927&view=overview-toc&DMDID=DMDLOG_0001.

of Ulrike Moheit,[46] and, more definitively, in Humboldt's archive held at the Staatsbibliothek in Berlin, where it can be noted that the final folium of the Caldas letter was ripped out (Figure 4.5).[47] This letter is also critical for its timing, as it was received by Humboldt before his first elaboration of a 'Tableau physique.' Indeed, five months later, on April 21, 1803, Caldas wrote again to Mutis:

Mr. Baron de Humboldt, who left Guayaquil two months ago, sent to the hands of Mr. Marqués de Selva Alegre a tin tube, which contained *a report on the geography of the plants. I don't know why he kept it in his possession for such a long time, and didn't give it to me for remission by my hand, according to the wishes of the Baron himself. I have held it for fifteen days to make a copy, and I am sending it now accompanied by a trifle of mine, almost in the same genre, which I hope you will receive with kindness.* The collection of plants that Humboldt has sent to you in my care has still not arrived from Guayaquil. I hope they arrive in time so that I may remit them to you together with mine.[48]

The 'trifle' referred to here by Caldas consisted of a *Report on the Distribution of Plants Cultivated in the Vicinity of the Equator,* including a four folia profile divided into nine altimetric zones in which eight different plants appear distributed with their maximum and minimum heights (Figure 4.6). Based on the dates reported by Caldas in this multiple profile finished and dated April 6, 1803, his work and considerations on plant distribution had begun in 1796, that is, at least five years before his personal encounter with Humboldt in early 1802. Later, in this same letter to Mutis, Caldas specifically refers to his botanical observations on the Imbabura volcano: 'In the following I will send to you the description and drawings of some plants that have seemed particularly interesting to me, with a *Report on Imbabura*.'[49]

But Imbabura would only reappear in the correspondence of Caldas five years and five months later, on September 30, 1808, two weeks after Mutis had died in Santafé. The reference appeared in a letter addressed to the Secretary of the Viceroyalty of New Granada and the Judge Commissioner for Matters of the Royal Botanical Expedition. There, the Imbabura volcano was represented once again in the context of the 'Profiles of the Andes' drawings made by Caldas with phytogeographic inscriptions in very original, coloured rectangles in the lower section of this pictogram, indicating geographic zones of distribution of different plant species, both in altitude and in latitude (Figures 4.7 and 4.8). Caldas notes at length here the following critical points:

The informal reports of Baron de Humboldt, and some works that I had sent to Mr. Mutis, made me known to this botanist. In 1802 he added me to his Expedition with the most flattering hopes and expressions, as I can prove with his correspondence. ... With this task I began my field excursions in July 1802. I left Quito and went to Ibarra and Otavalo; I travelled through these two regions; I made the map supported by astronomical and geodesic observations; I measured the mountains of Cotacache, Mojanda and Imbabura; I entered the crater of this last volcano, and, above all, I collected all the plants presented to me, described them, and drew them by my own hand. This is where I began to collect the materials for my great work, which must be entitled *Geography of the Plants of the Viceroyalty of Santafé*: an immense, complicated, and original work, a work that demands deep knowledge in geography, astronomy, meteors, and above all in the barometer and its measurements.

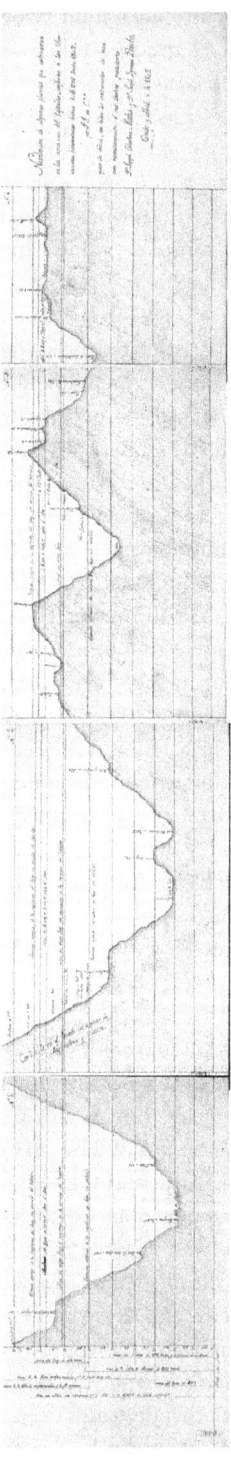

FIGURE 4.6 Francisco José de Caldas, 'Nivelación de algunas plantas que cultivamos en las cercanías del Equador conforme á las Observaciones barométricas hechas desde 1796 hasta 1802 por F[rancisco] J[osé] de C[aldas].' (1803), Real Jardín Botánico de Madrid, CSIC, Archivo, AJB, Div. III, M00519–M00522.

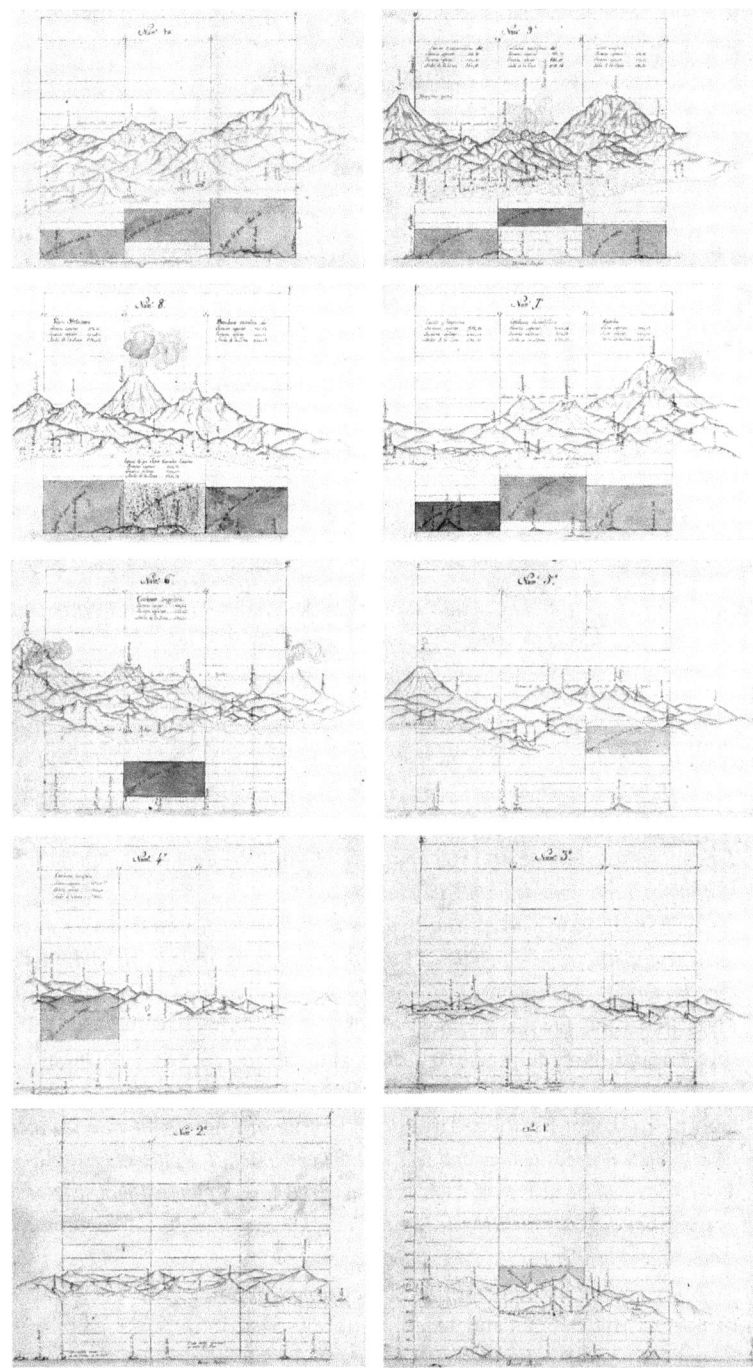

FIGURE 4.7 Francisco José de Caldas, Profiles of the Andes, [1807–1808], in Mauricio
Nieto, *La obra cartográfica original de Francisco José de Caldas* (Bogotá:
Universidad de los Andes, 2006), 2.1.

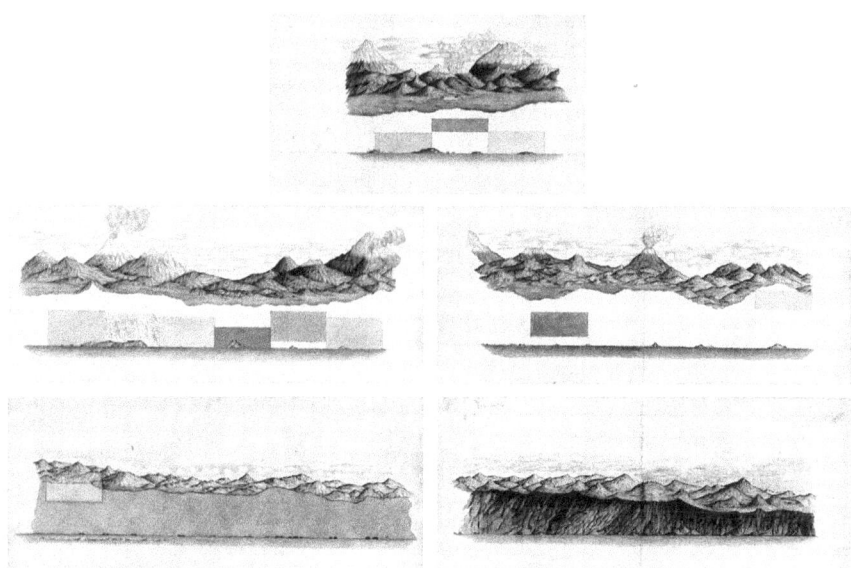

FIGURE 4.8 Francisco José de Caldas, 'Profiles of the Andes,' [1808–1809], in Mauricio Nieto, *La obra cartográfica original de Francisco José de Caldas* (Bogotá: Universidad de los Andes, 2006), 2.12.

Hence the care of perfecting this instrument, hence my inquiries and perhaps discoveries, hence the fact that I carried it on my back to all these places and pointed out with my hand all the points where each plant grows. Among Mutis' manuscripts there must be a *Report of the Levelling of the Plants that are Cultivated in the Vicinity of the Equator, which I made in 1802, fruit of my trip from Popayán to Quito in 1801, and which I sent and dedicated to Mutis.*[50]

This small work is like a very light essay, which I have subsequently undertaken with new journeys, new books, and new knowledge. In it you will find original and very important observations on the cultivation of wheat and other crops. How my ideas about this favourite object of my investigations have widened! If I find support and have the necessary time, the nation will see a *Botanical Chart of the Kingdom*; it will see all the Andes in profiles from 4½ degrees of southern latitude to 9½ of northern latitude; *it will see at what height each plant is born, what climate it needs to live, and which is the one in which it thrives best. All Mutis's dependents will be able to confirm that this general and philosophical way of looking at vegetation has not been learned in his home, where he never thought of leaving the common and beaten path …*

After these field operations, I began to organize my botanical works, to correct my designs and to continue my observations. … The President, Baron Carondelet, entertained a thousand doubts about the goodness, length, and expenses of Malbucho's road. Two ignorant commissioners were in contradiction, and it was said that these regions were rich in productions and quinas. Mutis's commission on this point, and the commission I deserved from Carondelet to reconnoiter these countries, obliged me to leave Quito in June 1803. I entered these burning and unhealthy solitudes; I undertook imponderable works; I drew a choreographic [*sic*] chart of these forests; I fixed the astronomical position of many points; I described the course of [the rivers] Mira, Bogotá, Santiago, Cayapas; I probed the port, and established its position forever at 1° 29′ boreal latitude; I collected and designed a respectable herbarium; I advanced my work on the *Geography of Plants*; I formed a barometric profile from the Ocean to the eternal snows of Imbabura. …

I climbed down into the forests of Yntac in search of the quinas. I picked up the letter, collected many plants, and found my first species of *Cinchona* and I brought out rich materials for the *Geography of Plants*. … Astronomy has formed the basis of these many determinations in latitude. *The works of the barometer, water boiling, geography of plants, profiles, charts, etc., I have held them up to this capital city with the same interest and activity that I began them.* To all this must be added the numerous collections of ectypes or impressions of living plants on paper with the help of the portable press that I carried everywhere. I keep this large collection in my possession, and I can make it manifest. …

The summary of all my works made from 1802 to the end of 1805, is reduced to a respectable herbarium of five to six thousand skeletons dissected in the midst of the anguish and speed of a trip; two volumes of descriptions, many designs of the most remarkable plants made by my own hand, because not even a painter wanted to give me; seeds, barks of the useful ones, some minerals, the necessary material to form the geographical chart of the Viceroyalty, the necessary ones for the botanical chart, for the zoographic chart, the profiles of the Andes in more than 9°; the geometric height of the most famous mountains; more than 1500 heights of the different peoples and mountains barometrically deduced; a prodigious number of meteorological observations; two volumes of astronomical and magnetic observations, some animals and birds. With this material contained in 16 boxes I introduced myself to Mutis. I put everything in his hands, I consecrated everything to his glory, with a generosity and a disinterest that he did not deign to correspond.[51]

A final documentary source, published by Caldas in the first semester of 1809, may be enlisted here in support of the thesis of the simultaneous development of biogeography in the minds of Caldas and Humboldt in the northern Andes of South America. These are Caldas' own 'Preface' and 'Notes' to the 1803 Spanish translation of Humboldt's *Essay on the Geography of Plants*, which he included in issues 16 through 25 of the original edition of the *Semanario del Nuevo Reyno de Granada* (1809). In all, these 'Notes' occupy more than thirty pages in the original edition of the *Semanario*. One passage in his 'Preface' allows us to contrast Caldas's work with that of Humboldt in the field of phytogeography:

We who have travelled within the viceroyalty by order and at the expenses of the Royal Botanical Expedition of Santa Fe and D[on] José Ignacio Pombo; we who *have visited many places that are common to us and Humboldt; in a word, we who have closely followed in the footsteps of this illustrious traveller, with the same objects and with the Geography of Plants in hand, are authorized to warn the public about what we have noticed about this interesting production* made by the voluntary martyr of galvanism.

It is not the pruritus of writing; it is not the foolish vanity of exaggerating the carelessness of great men that obliges us to make these notes. It is *the love of truth, the desire to illustrate some points of physics and natural history of our countries, that are the reasons that move us. While respecting the lights, the vast knowledge, and the great talents of this extraordinary traveller, we respect more the truth.*[52]

In addition to this paragraph, there is perhaps a more forceful one that demonstrates the originality of the work of Caldas as compared to the work of Humboldt on this matter. In this passage, Caldas reveals that he has mapped all the plants he has examined:

It has been many years since we gathered materials and observations for a work entitled Fitografía del Ecuador [Phytography of Ecuador], working on a broader plan and perhaps more useful to trade, agriculture and plant medicine. Like Humboldt, the quina has attracted all our attention. Going up and down the Andes in every sense, from 4° 30' of southern latitude to 5° 25' of northern latitude, we have been able to fix irrevocably not only the genus *Cinchona* as [Humboldt] has, but also those of all the species that constitute [this latitude]. The plants that we cultivate, those that serve in the arts and to restore our health, are those that have deserved our preference.

Humboldt limits himself to heights, and we, after establishing the precise terms to which each species under the equator is reduced, dare to point out the latitude as far as its existence extends and, so to speak, to fix the tropics of all the plants that we have subjected to our examination. We establish general principles and laws on the geography of vegetation, and we believe that we have made a step forward in this science, which, by Humboldt's confession, is still in its cradle.

Despite these efforts to perfect our *Phytography* we still have many observations to verify, and a trip to the Andes of Quindío. If circumstances [and] my fortune allow it, and if I complete my knowledge in this important branch of botany, I will present them to the public as a testimony of the love I profess for my country and my fellow citizens.[53]

Chimborazo: A Formal Scenario for Humboldt's Andean Biogeography

To close this review of what may be called Caldasian biogeography, I will refer to the first two graphic representations of Chimborazo in Humboldt's printed work as well as to later, post-American elaborations made in Europe. Neither the subsequent evolution nor the contents of the manuscript or published editions of the *Essai sur la géographie des plantes* between 1805 and 1807, in its French and German editions (the latter under the title *Ideen zu eine Geographie der Pflanzen* in 1807), will be dealt with at this point. Nevertheless, it should be noted that these two editions should be contrasted in detail with the Spanish translation, made by Jorge Tadeo Lozano, of the *princeps* manuscript that Humboldt sent to Mutis from Guayaquil in February 1803, as well as with its printed version in *Semanario del Nuevo Reyno de Granada* in 1809.

The very first post-Ecuadorian evidence of a graphic representation of the bio-geography of Chimborazo by Humboldt (Figure 4.9) appears to be the one referred as 'dessiné par A. de Humboldt à Mexico 1803, par F. [*sic*] Marchais à Paris en 1824.' This original description clarifies several points. First, it suggests that the primary source of the 1824 engraving was a drawing produced by Humboldt in Mexico in 1803, that is, a few months *after* the one he had sent to José Celestino Mutis from Guayaquil in February of that same year. Second, it suggests that Humboldt's draw-ing was reworked by 'F' [*sic*] Marchais, an artist eventually related to Pierre-Antoine Marchais (1763–1859), a French landscape painter and cartographer associated with several graphic representations in Humboldt's opus. Third, it suggests that this tran-scription by 'F' Marchais of Humboldt's original drawing made in Mexico was made in Paris in 1824.[54] Fourth, it suggests that the engraving was later published in Paris in 1825, based on a drawing of 1824, and included in the *Atlas géographique et physique des régions équinoxiales du Nouveau Continent*, whose first 'livraison' in 1814 had been produced by Friedrich Schoell in that same city. Fifth, it suggests that unlike the 'Tableau physique' or the 'Naturgemälde' illustrations in the French and German editions of the *Geography of Plants* from 1807, this 1803 representa-tion of Chimborazo printed in 1824 only showed a scale of heights in toises (and not in meters and toises as in the 'Tableau physique' and the 'Naturgemälde'), and none of the lateral biophysical scales of the original 'Tableau physique' illustrating Humboldt's *Geography of Plants*. Sixth, it suggests that this 1803/1824 pictograph includes a tight relation (one might say a *potpourri*) of plants, with only a few selected temperatures and geographical references at different altitudes, beside an essentially geophysical, and not botanical, depiction, although the subtitle of the engraving refers to an eventual 'esquisse' of its geobotanical component in the Andes of Quito between 0° 20′ boreal and 4° 12′ southern latitudes. Furthermore, while Humboldt conceived the idea for this drawing in 1803, the plants included in the 1824 illus-tration were taken from the first four volumes of the *Nova Genera*, as these botanical data had been corrected and improved by Carl S. Kunth (1788–1850). Finally, it is clear that this profile is related to the attempt to climb Chimborazo, as can be deducted from its title 'Journey to the Summit of Chimborazo, attempted on June 23, 1802, by Alexander von Humboldt, Aimé Bonpland and Carlos Montúfar,' and in the descriptive sentences inserted at an approximate height of 2700 toises and at a measured height of 3016 toises, in the area corresponding to the snowy section of the volcano, which read 'Travelers began to bleed in their eyes, lips and gums,' and that a 'crevice prevented travelers from reaching the top.'[55]

The second biogeographical plate to be considered here centred on the Chimborazo volcano in the first years of the nineteenth century (Figure 4.10). It corresponds to the first 'Tableau physique' engraved by Louis Bouquet (1765–1814)

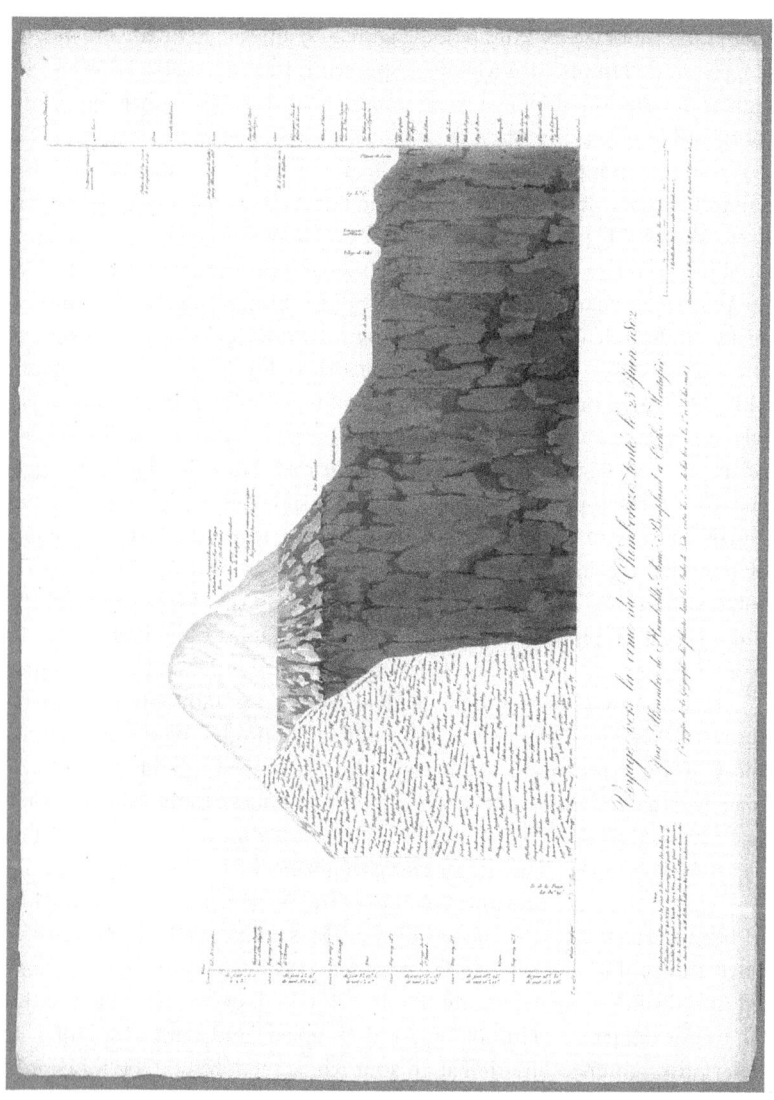

FIGURE 4.9 Alexander von Humboldt 'Voyage vers la cime du Chimborazo, tenté le 24 juin 1802,' Geographie des plantes dans les Andes de Quito … (1831), https://www.davidrumsey.com/luna/servlet/detail/RUMSEY~8~1~292857~90064432.

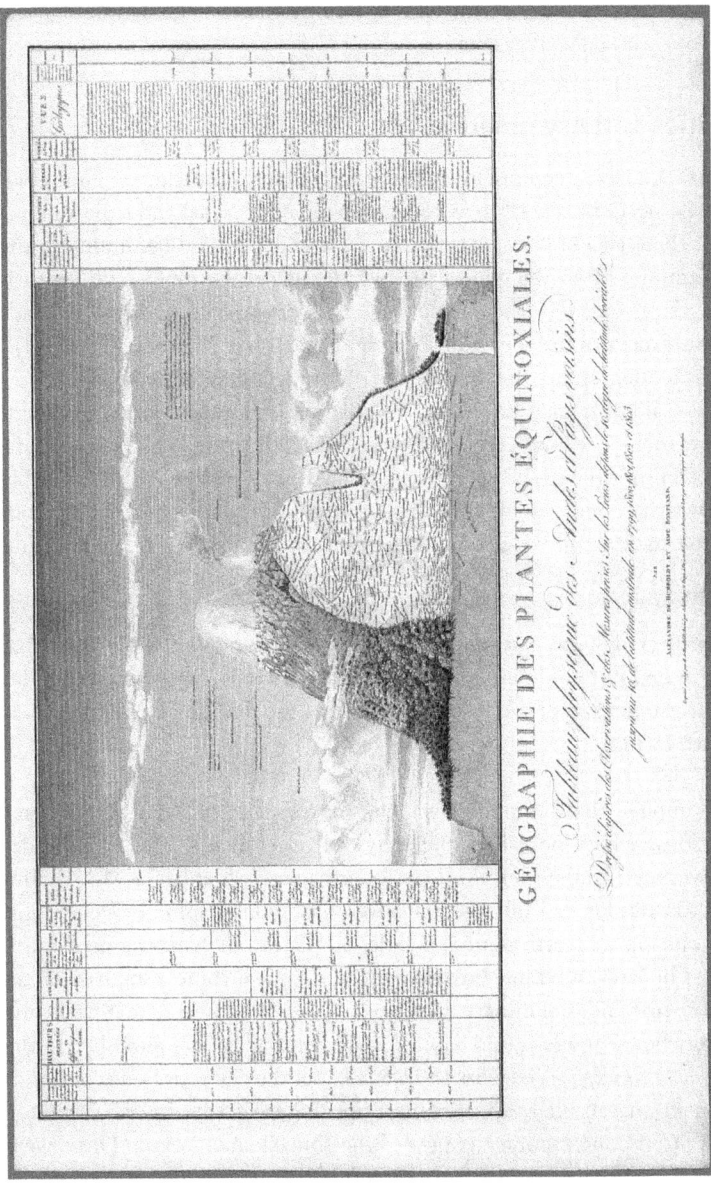

FIGURE 4.10 Alexander von Humboldt, 'Géographie des plantes équinoxiales. Tableau physique des Andes et pays voisins. Dressé d'après des observations & des mesures prises sur les lieux depuis de 10e degré de latitude boréale jusqu'au 10e degré de latitude australe en 1799, 1800, 1801, 1802 y 1803 …' (1807), in Gómez Gutiérrez, A. *Humboldtiana neogranadina* (2018, v. 2, 251). https://bibliotecanacional. gov.co/es-co/colecciones/biblioteca-digital/humboldtiana/Documents/index.html#book/254.

from an original drawing made by Humboldt of which there is no trace today other than that conserved in Bogotá (Figure 4.2). That drawing was redrawn by Lorenz Adolf Schoenberger (1768–1847), a German artist, and by Pierre-Jean François Turpin (1775–1840), who was considered one of the greatest botanical illustrators of the Napoleonic era in France, and who also collaborated with Humboldt and Bonpland in *Plantæ æquinoctiales* (1808).

Synchronicities and Asymmetries

Up to this point we have been discussing the findings and postulates of Humboldt and Caldas in the period from 1801 to 1803. But the conceptual and experimental coincidence of these two savants takes place over a wider period beginning in the 1790s and continuing up to the end of the first decade of the nineteenth century, when Caldas got involved in the revolution of independence in New Granada which led to his execution by the Spanish army on October 29, 1816 (Table 4.1).

As for the iconographic dimension of the biogeographic elaborations of Humboldt and Caldas, it may be postulated that these began as land 'levellings.' Humboldt drew such 'levellings' in New Granada in 1801, prior to his first encounter with Caldas. His profile of the land from Cartagena to Santa Fé is most probably the infography that, once shared, unleashed the counterpoint, stimulating both men to proceed to draw phyto-geographical profiles. Caldas was explicit in this intention which, as we noted earlier, he believed was original:

> I have projected some barometric-botanical levellings similar to those that Mr. Baron de Humboldt has constructed with the only object of giving idea of the diverse heights of the land. ... This idea touches me, I think it is new and worth trying.[56]

Caldas further implied that Humboldt was the source of his inspiration in subsequent elaborations, which he called, precisely, 'Nivelaciones,'[57] after Humboldt's 'Nivellement barométrique du terrain depuis Carthagene à S[an]ta Fe' (1801). But it is clear that Humboldt had not yet conceived any iconographic biogeography. His 'Profil du chemin de Carthagene des Indes au plateau de Santa Fé de Bogotá' (Figure 4.11), as he later titled his 'Nivellement,' does not include a single plant, as can be seen in its first and preliminary engraving, titled 'Barometrische Nivellirung der Gegend zwischen Cartagena und Santa Fé.' This engraving was published in the *Allgemeine geographische Ephemeriden* in 1802 while still travelling in South America and is based on an unavailable original manuscript (Figure 4.12).

Two handwritten contemporary copies of the 'Nivellement' exist. These were reportedly elaborated between July and August 1801 in Santafé de Bogotá by two enlightened relatives of Caldas, Antonio Arboleda (1770–1825) and Santiago Arroyo (1773–1845). The first is handwritten in French (Figure 4.13), the second translated to Spanish (Figure 4.14). Caldas commented on these copies, first to Santiago Arroyo on October 6, 1801, noting, 'The profile of the barometric

TABLE 4.1 Chronology of biogeographical works of Alexander von Humboldt and Francisco José de Caldas, 1790–1851

Date	Alexander von Humboldt	Francisco José de Caldas
1790	Travels to England with Georg Forster	Measurements and registers
1793	*Flora Fribergensis*	
1796–1801	Measurements and registers	
1801	*Nivellement barométrique de Carthagène-Sta Fe*	- *Medición de la altura de Guadalupe* - *Discurso sobre el calendario rural del Nuevo Reino*
1802	Measurements and registers	- *Memoria sobre un plan de viaje* - Measurements and registers - *Nivelación de 7 especies: Santafé – Quito*
1803	- *Tableau physique des Andes* (manuscript) - *Géographie des plantes* (manuscript)	- *Nivelación de 30 especies en el Imbabura* - *Nivelación de oro y platino*
1804–1805	Oral presentations - Institut de France (Paris)	- *Determinación del término de la vegetación (13 especies) en las cercanías del ecuador* - *Nivelación de las quinas*
1807	- *Essai sur la géographie des plantes* - *Tableau physique des Andes / Naturgämalde* - *Ideen zu eine Geographie der Pflanzen*	Materiales para la Memoria sobre la quina
1808	*Plantæ Æquinoctiales*	
1809		- *Comments on Humboldt's Geografía de las plantas* - *Perfiles fitogeográficos Quito-Santafé*
1816		Caldas is shot by the Spanish Army
1817	*De distributione geographica plantarum*	
1851	*Physicalischer Atlas H[einrich] Berghaus*	

levelling that you have had the kindness to delineate and send, has given me a big idea of the knowledge, the exactitude and extension of the plans of the Baron.'[58] Then, to Antonio Arboleda in the same month, Caldas wrote:

Arroyo writes to me and talks to me about the Baron, he tells me that he left on August 9 for [Popayán], he includes a profile of a barometric levelling from Cartagena to Santafé with expression of latitudes, elevation of the land, nature of minerals, and other very curious things.[59]

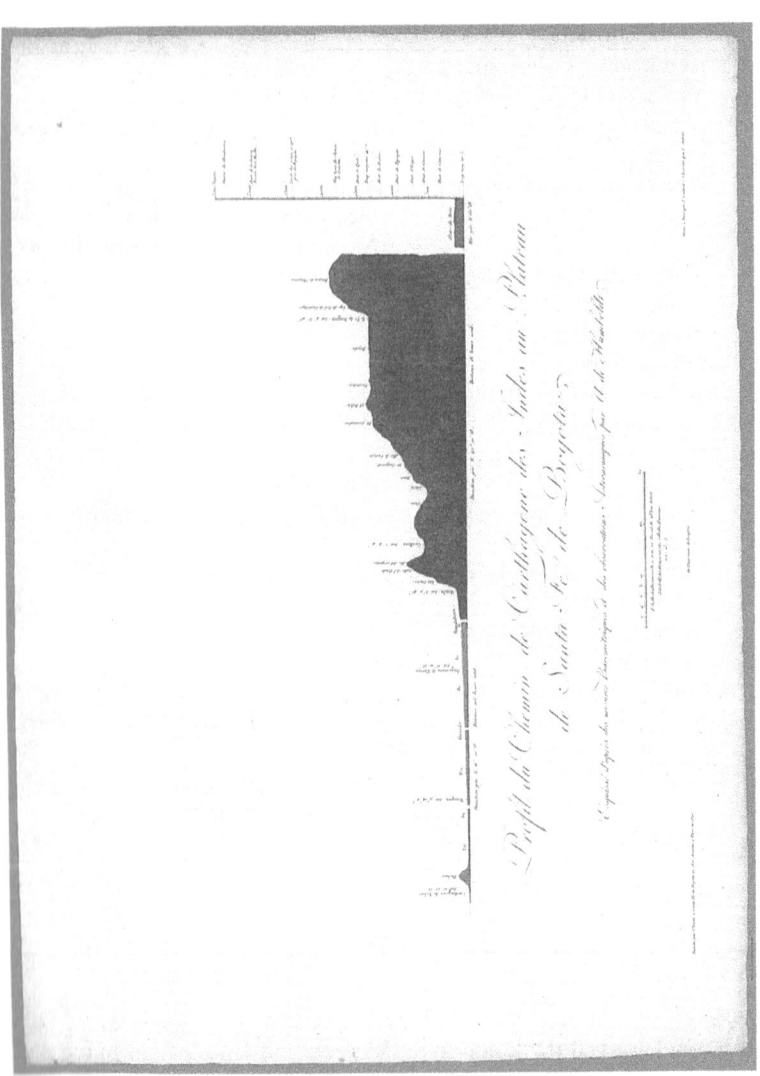

FIGURE 4.11 Alexander von Humboldt, 'Profil du chemin de Carthagène des Indes au plateau de Santa Fe de Bogotá (1814), https://www.davidrumsey.com/luna/servlet/detail/RUMSEY~8~1~292854~90064435:VI–Profil-du-chemin-de-Carthagene–

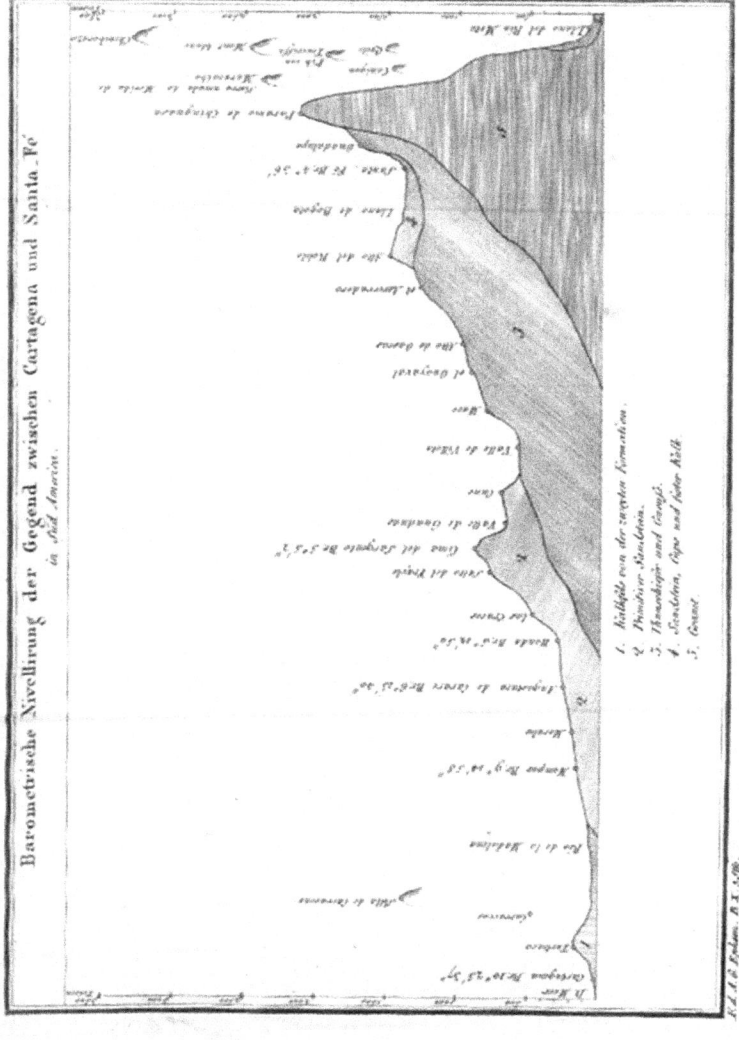

FIGURE 4.12 Alexander von Humboldt, 'Barometrische Nivellirung der Gegend zwischen Cartagena und Santa Fé' (1802), https://zs.thulb.uni-jena.de/rsc/viewer/jportal_derivate_00207057/Allg_geogr_Ephemeriden_1802_10_%200291a.tif

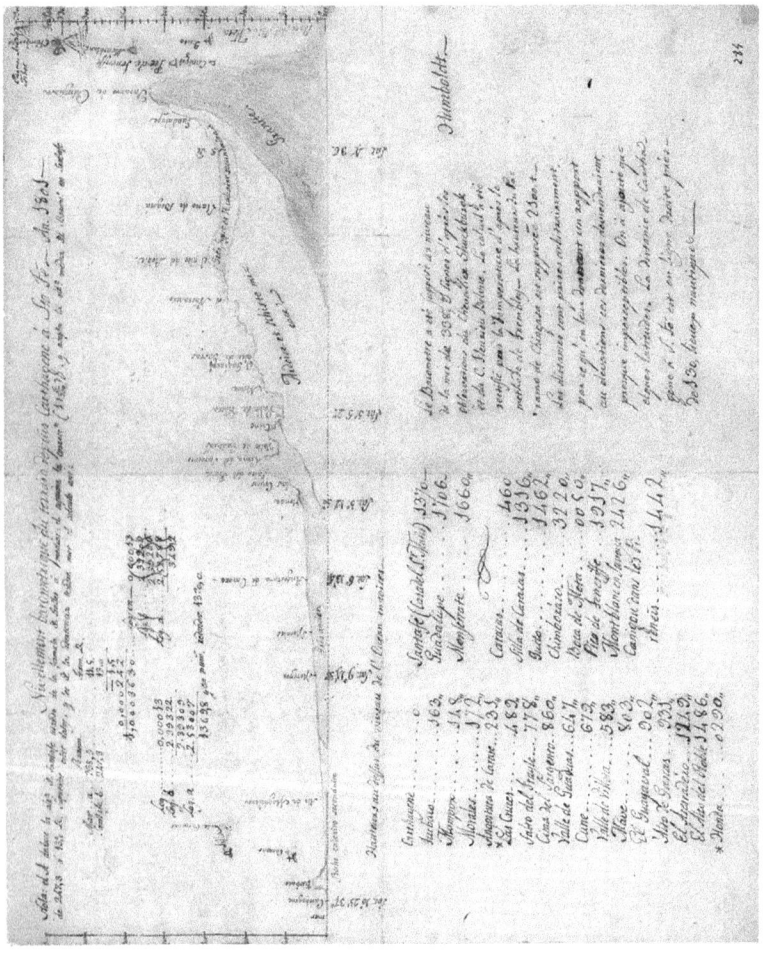

FIGURE 4.13 Alexander von Humboldt, 'Nivellement barométrique du terrain depuis Carthagene á S[an]ta Fe,' [Anonymous] (1801), Archivo Cartográfico y de Estudios Geográficos del Centro Geográfico del Ejército, Ministerio de Defensa de España, Ar.J-T.7-C.1_10Bis.

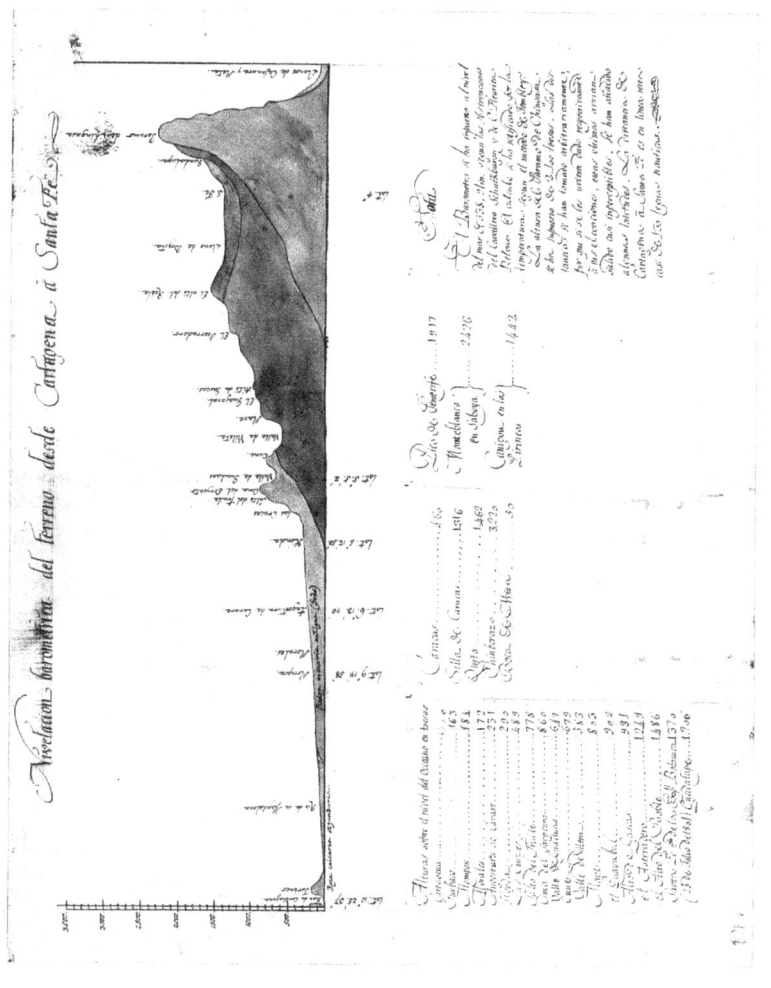

FIGURE 4.14 Alexander von Humboldt, 'Nivellement barométrique du terrain depuis Carthagène á S[an]ta Fe,' [Anonymous] (1801), Archivo Cartográfico y de Estudios Geográficos del Centro Geográfico del Ejército, Ministerio de Defensa de España, Ar.J–T.7–C.1_10.

In addition to thanking his Payanese informants, Caldas worked on these copies according to his own perception, experience, and knowledge, as can be deduced from the critical notes on the measurement of the altitude of Santafé, which he included under the title of one of the copies:

> The author deduces the elevation of Santafé using an incomplete Tralles formula: He suppresses the [correction] (1 + 5/200) and adopts the mean elevation of the barometer in Santafé of 247,3 at 13°,0 R[éaumur]. Assuming these data and those of the warning, we will have the calculation as follows. …[60]

According to Caldas, the altitude of Santafé should be exactly 1370.0 toises. The date of this note is unknown, but it reveals, once again, the critical spirit of Caldas. Further evidence of this spirit is found in a letter to Santiago Arroyo sent two weeks later and dated October 21, 1801:

> I have reviewed in the profile, or barometric levelling, that the latitudes expressed … do not keep proportion; you must observe either the cut by the meridian or by any other direction inclined to it; you notice that from Cartagena to Mompós there is a difference in latitude of 1° 10′ 39″, and from Mompós to the Angostura de Carare, 3° 1′ 18″, and despite the difference expressed between Cartagena and Mompós, it seems almost to triple that between Mompós and Angostura. I observed the same deformity between this point and Honda, the valley of Villeta and Santafé; consider this critique and let´s say something about this [when we meet in] Quito.[61]

In the same domain of geography and then biogeography, Caldas had published in the second semester of 1801 in a local journal in Santafé, his 'Medición de la verdadera altura de Guadalupe,' or Measurement of the true altitude of Guadalupe, which Humboldt would have had the opportunity to read while staying at the house of José Celestino Mutis, since it was published in July of that same year. Another publication by an enlightened member of Caldas' circle[62] appeared in the same journal (*Correo Curioso*, issue 35, October 1801) after Humboldt had left Santafé for Popayán. However, there is no evidence that this second 1801 Neogranadian paper was read by the Prussian. In the contemporary anonymous publication 'Discurso sobre el calendario rural del Nuevo Reyno,' or Discourse on the rural calendar of the New Kingdom, Caldas' several biogeographical determinants were also referenced:

> Sugar cane, which occurs in hot lands, requires a light, crumbly and vegetable soil. … The 'turmas' or potatoes (*Solanum tuberosum*), which make up the main crops of the cold lands. … Wheat is grown in cold and some temperate lands.'[63]

It seems clear from the evidence that Caldas and his Neogranadian peers were at this early date well prepared to interact with Humboldt in the Andean making of the field of biogeography. But Caldas didn't, as he shyly retracted from any open controversy with the Prussian naturalist until he published his Spanish translation of Humboldt's *Geography of Plants* in the *Semanario del Nuevo Reyno de Granada* in 1809, almost eight years later. The *Semanario*, a judicious weekly publication then under his direction, with fifty-two issues in 1808, fifty-two issues in 1809, and a series of *Memorias* or Reports in 1810, was a timely but discrete journal that circulated mainly in New Granada. As historian Mauricio Nieto has noted, it

> was published during a decisive period in the political history of Spain and its colonies; and it is significant that in the midst of the crisis of the Spanish empire, there is a periodical that aims to disseminate, among a group of Creoles, knowledge considered useful for the good government and prosperity of the Kingdom of New Granada.[64]

Although the journal was significant, it is clear that Humboldt possessed better means, both financial and academic, to peddle his theories in high centres of science and society: Paris and Berlin. This decisive asymmetry in the circulation and means of Caldasian and Humboldtian knowledge production can be visually metaphorised by two portraits painted in the early years of the nineteenth century. On the one hand, we have the famous 1.26 × 0.92 m oil canvas of Humboldt painted in 1806 by Friederich Georg Weitsch (1758–1828), depicting a bountiful scenario of natural plants and facing the Orinoco River, with a barometer, collected specimens and botanical journal in hand (Figure 4.15). This portrait adorns the cover of many books on Humboldt. Caldas's portrait, on the other hand, was painted in miniature (7.5 × 6.5 cm) in Quito by an anonymous artist on an undetermined date circa 1803 (Figure 4.16).

It was ten years after the shooting of Caldas by the Spanish army that Humboldt finally referenced his works on plant geography, albeit only in a preliminary prospect for a book that would never be published. In this prospect, Humboldt submerged (should we say, drowned?) Caldas in a long list of fifty-six naturalists that had worked in the new field he had supposedly pioneered (Figure 4.17): 'In the last 15 years, [the following botanists] have either addressed questions relative to this science, or else furnished materials that would enlarge its limits.'[65] But there is a curious (voluntary or involuntary?) mistake in this belated acknowledgement. Being the very precise measurer and quantifier that he was, Humboldt was surely keenly aware that Caldas had worked in botanical barometry since at least early 1802. In effect, Caldas's work in this area was 24 years, not '15,' prior to 1826.

FIGURE 4.15   Friederich Georg Weitsch (1806), 1.26 × 0.92 m. Alexander von Humboldt, https://es.m.wikipedia.org/wiki/Archivo:Alexandre_humboldt.jpg.

FIGURE 4.16   Anonymous (c. 1803), 7.5 × 6.5 cm. Francisco José de Caldas, Museo Casa de la Independencia, Casa del Florero, reg. 3355.

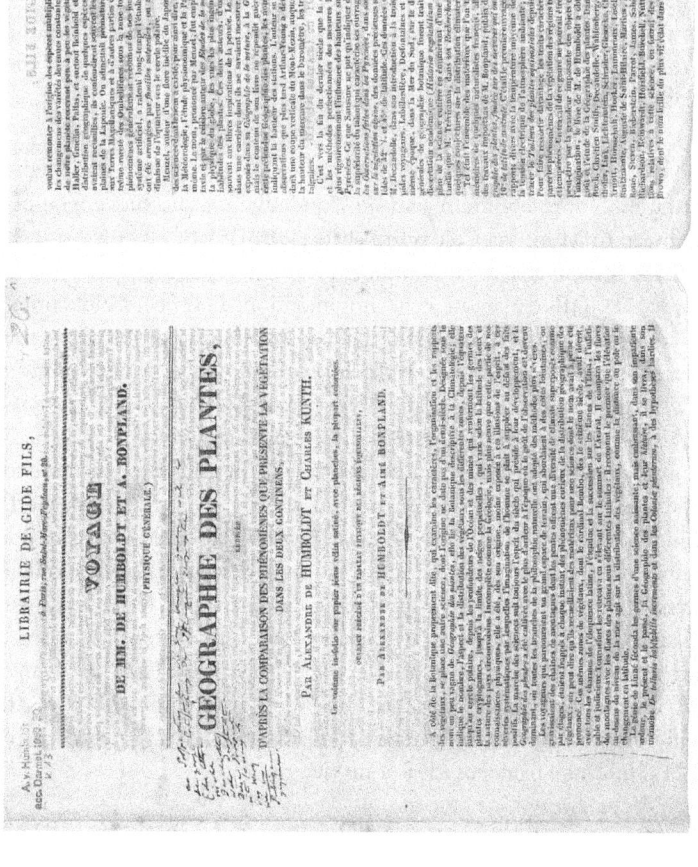

FIGURE 4.17 Alexander von Humboldt and Carl Kunth, 'Géographie des plantes redigée d'après la comparaison des phénomènes que présente la végétation dans les deux continens – Prospect (1826),' Staatsbibliothek zu Berlin Preußischer Kulturbesitz, Nachl. Alexander von Humboldt, gr. Kasten 13, Nr. 26, Bl. 1–2 [Communicated by Ulrich Paessler], in https://digital.staatsbibliothek-berlin.de/werkansicht?PPN=PPN838243452&PHYSID=PHYS_0001

Conclusions

In conclusion, our contentions and queries in this chapter may be summarised as follows. First, Caldas clearly registered his novel idea of 'levelling' plants (an idea that he considered original while cohabitating with Humboldt and Bonpland in Quito) in letters to Mutis in the months of April and May 1802, and then in his 'Memoria sobre el plan de un viaje proyectado de Quito a la América septentrional.' Second, Caldas reported to Mutis the measurement of altitudes at the Imbabura volcano taken before August 8, 1802. Third, Caldas reported to Mutis his plant collections in the foothills of Imbabura on September 23, 1802. Fourth, Caldas reported to Humboldt from Otavalo on November 17, 1802, the results of his geographic and botanical works in the northern districts of Quito. Fifth, in a letter dated in Lima on November 25, 1802, Humboldt reported to the French astronomer and mathematician Jean-Baptiste Joseph Delambre, that

> there is no vegetable of which we cannot indicate the rock it inhabits and the height in toises to which it rises; to such an extent that the geography of the plants *will have* in our manuscripts very exact data.[66]

In light of these five points, we should ask: Was Caldas's report from Otavalo the stimulus that led Humboldt to configure and subsequently remit his biogeographic profile of Chimborazo to Mutis via Caldas in February 1803?

In Caldas's report of activities to the authorities of the viceroyalty of Santafé dated September 30, 1808 (five years after the aforementioned events), Caldas notes that, in the second semester of 1803, he had made 'a barometric profile from the Ocean to the eternal snows of Imbabura.'[67] With precision, the date of execution of the profile of Imbabura can be fixed after November 21, 1803, since the village of Yntac visited by Caldas on that date is duly georeferenced in it. Although the phytogeographic drawing of Imbabura would have been elaborated at the end of 1803 or at the beginning of 1804 when he had already received in his hands Humboldt's Chimborazo sketch in April 1803, it is clear that Caldas's Imbabura profile is an original and independent product of the activities that he made explicit in his 'Memoria sobre el plan de un viaje' in the second quarter of 1802, and then specifically on August 8 of that same year in a letter to Mutis, where Caldas wrote, 'Vegetation, which makes my first object, does not keep the laws of level that I have constantly observed in all the elevated hills that I know and have climbed.'[68] Indeed, before this specific epistolary report of August 8, Caldas had already proposed to Mutis on April 21, 1802, the princeps idea of his design:

> Wouldn't it be new, and at the same time beautiful, to divide into 12 zones of an inch in the barometer of width each, all the part of the earth that is able to vegetate? Wouldn't it be new to assign to each floor its limits, and in a laconic and exact way say: inhabits the first zone, inhabits from the third to the fifth zone, and so on?[69]

This princeps idea had experimental antecedents, such as the one that Caldas referred to in an unpublished notebook of 1802, where he recorded the following paragraphs, written on July 28 while in Tabacundo:

> In all the places through which I have traveled on this short trip as in the one I made in [1]801 from Popayán to Quito, I have noticed with the greatest care the quality of the flours and the elevation of the soil that produces them, and I think I can point out the superior term in which this precious plant vegetates, with utility for the farmer. … This matter has seemed important to me and deserves to be looked at in detail. With temperatures as different as they are in the regions [between] Santafé and Quito, I have noticed that wheat is not cultivated below 22 inches of the barometer, thus above 1100 toises … it does not vegetate with happiness in the regions situated under this level. … We know that humidity and heat are the most powerful agents for vegetation, and that these increase in inverse ratio with the elevation of ground over the sea. Will wheat be among those plants that do not happily prosper [in] the tropics, at an atmospheric pressure other than 22 inches of the barometer? In Europe it is cultivated at very low elevations, and in America one does not begin to find it until one reaches a height of 1100 toises above sea level. Will the limit at which it prospers rise to the vicinity of the line?[70]

Caldas' complementary idea of the role of altitude in the geography of plants is clearly defined in this unpublished 1802 manuscript. Humboldt would only properly define and illustrate this role fifteen years later, in his *De distributione geographica plantarum* (1817).[71] In the same unpublished manuscript of 1802, Caldas presented 'a levelling of the peoples that are on the road from Quito and Santa Fe, from 0° 14' south latitude to 4° 36' 0" boreal latitude,' as a complement to his reflections. This 'levelling' may be the 'trifle' mentioned by Caldas in his note to Mutis in 1803. Based on the evidence presented, it may be concluded that the phytogeographic profile of Imbabura was drawn by Caldas at the end of 1803 or at the beginning of 1804, and that the 'matter' was already taking shape in his mind several months before he met Humboldt in Ibarra on December 31, 1801. It is also notable that, when comparing the profiles of Imbabura and Chimborazo, Caldas persisted, despite having seen Humboldt's 1803 sketch, with his original idea of dividing it into twelve parts. Caldas did not replicate any additional graphic element from the Prussian's watercolour drawing. The biogeographical syntheses of Caldas and Humboldt are, therefore, coincident but different. In any case, both appear to have been born in the Andes.

Notes

1 I would like to acknowledge the very kind cooperation of Esther García Guillén and José Luis Fernández Alonso at the Royal Botanical Garden of Madrid, and the attentive reading, comments, and contributions of Ulrich Paessler at the Berlin-Brandenburgische Akademie der Wissenschaften. Preliminary elaborations on the documentary foundations

of biogeography in the writings and drawings of Francisco José de Caldas, were presented in seminars in Colombia, Quito, Paris, and Berlin, where multiple attendees have influenced the content of this text with their interventions, among whom I would like to mention the fruitful exchanges with Florike Egmond and Peter Mason, as well as those with the editors of this volume, Mark Thurner and Jorge Cañizares-Esguerra, who shared particularly useful comments in both formal and informal communications.

2 S. T. Jackson, 'Introduction: Humboldt, Ecology and the Cosmos,' in S. T. Jackson, ed., *Alexander von Humboldt and Aimé Bonpland. Essai on the geography of plants* (Chicago, IL: The Chicago University Press, 2009), 13–14.

3 R. K. Merton, 'Resistance to the Systematic Study of Multiple Discoveries in Science,' *European Journal of Sociology* 4:2 (1963), 237–282.

4 Brown University, 'Biogeography,' http://biomed.brown.edu/Courses/BIO48/29. Biogeography.HTML. Accessed June 10, 2021.

5 For a revision of these and other pioneering works on biogeography, see M. C. Ebach, *Origins of Biogeography: The Role of Biological Classification in Early Plant and Animal Geography* (Dordrecht: Springer, 2015); D. M. Williams and M. C. Ebach, *Foundations of Systematics and Biogeography* (New York: Springer, 2008); M. V. Lomolino, B. R. Riddle, and J. H. Brown, *Biogeography* (Sunderland: Sinauer, 2006); M. V. Lomolino, D. F. Sax, and J. H. Brown, *Foundations of Biogeography. Classic Papers with Comments* (Chicago, IL: University of Chicago Press, 2004).

6 Alexander von Humboldt and Aime Bonpland, *Essai sur la géographie des plantes; accompagné d'un tableau physique des régions équinoxiales, fondé sur des mesures exécutées depuis le dixième degré de latitude boréale, jusqu'au dixiéme degré de latitude australe, pendant les années 1799, 1800, 1801, 1802 et 1803* (Paris: Levrault, Schoell et Compagnie, [1805] 1807), vi.

7 Humboldt and Bonpland, *Essai sur la géographie des plantes*, ix.

8 Humboldt and Bonpland, *Essai sur la géographie des plantes*, viii.

9 Humboldt and Bonpland, *Essai sur la géographie des plantes*, 71; E. Guhl, ed., *Alexander von Humboldt. Ideas para una geografía de las plantas* (Bogotá: Jardín Botánico José Celestino Mutis, 1985), 58.

10 J. L. Giraud-Soulavie, *Histoire naturelle de la France méridionale. Seconde partie. Les végétaux. Tome prémier. Contenant les principes de la Géographie physique du règne végétal, l'exposition des climats des Plantes, avec des Cartes pour en exprimer les limites* (Paris: Quillau, 1782), t. 1, plate II, 264–265. For a revisión of botanical and barometric works of Giraud-Soulavie, see M. N. Bourguet, 'Landscape with Numbers: Natural History, Travel and Instruments in the Late Eighteenth and Early Nineteenth Centuries,' in M. N. Bourguet, C. Licoppe, and H. O. Sibum, eds., *Instruments, Travel and Science: Itineraries of Precision from the Seventeenth to the Twentieth Centuries* (London: Routledge, 2002), 97–126.

11 Humboldt and Bonpland, *Essai sur la géographie des plantes*, 79.

12 See J. R. Forster, 'Excerpts from Observations Made during a Voyage Round the World, on Physical Geography, Natural History, and Ethic Philosophy,' in Lomolino, Sax, and Brown, eds., *Foundations of Biogeography*, 19–24. For further elaboration of this idea, see Forster, *Observations Made during a Voyage Round the World* (London: G. Robinson, 1778), 160–161 and 176–177, URL: https://archive.org/details/NHM6732.

13 Humboldt and Bonpland, *Essai sur la géographie des plantes*, ix. Humboldt might have been referring to Willdenow's *Phytographia* (1794).

14 See A. Buttimer, 'Alexander von Humboldt and Planet Earth's Green Mantle,' *Cybergeo: European Journal of Geography, Epistémologie, Histoire de la Géographie, Didactique*, document 616 (2012), 26, http://journals.openedition.org/cybergeo/25478.

15 J. W. Goethe, Letter to Alexander von Humboldt, Weimar, April 3, 1807, in Buttimer, *Alexander von Humboldt*, 28.

16 See Buttimer, *Alexander von Humboldt*, 25–28. Italics added.

17 Humboldt and Bonpland, *Essai sur la géographie des plantes*, 138–139.

18 For a map of Humboldt's itinerary from 1799 to 1804, see 'Karten zu Humboldts Reisestationen durch die amerikanischen Tropen (1799–1804),' https://www.avhumboldt. de/?p=18024.

19 Spanish translation cited in Alberto Gómez Gutiérrez, ed., *Humboldtiana neogranadina* (Bogotá: Colegio de Estudios Superiores de Administración - Pontificia Universidad Javeriana - Universidad de los Andes - Universidad del Rosario - Universidad Eafit - Universidad Externado de Colombia, 2018), v. 2, 137, https://bibliotecanacional.gov. co/es-co/colecciones/biblioteca-digital/humboldtiana/Documents/index.html#book/

20 On Mutis and Cinchonae, see https://raccefyn.co/index.php/raccefyn/issue/view/95.

21 See E. Posada, ed., *Obras de Caldas* (Bogotá: Imprenta Nacional, 1912), 75.

22 For a recent approach to Caldas, see https://raccefyn.co/index.php/raccefyn/issue/ view/30.

23 Humboldt and Bonpland, *Essai sur la géographie des plantes*, 44–45. Italics added.

24 This manuscript is preserved as part of the Ms456 folder in the Historical Archives of the Central Library of the Musée National d'Histoire Naturelle in Paris.

25 Alexander von Humboldt, Bibliothèque Centrale, Musée National d'Histoire Naturelle, Ms456, ff. 8–9 [1805–1806]. Translated from the French by the author. The sentence included in italics was crossed out in the original manuscript.

26 F. J. Caldas, Letter to Santiago Arroyo, Quito, January 21, 1802, in A. Bateman and J. Arias de Greiff, eds., *Cartas de Caldas* (Bogotá: Academia Colombiana de Ciencias Exactas, Físicas y Naturales, 1978), 130–131.

27 For a revision of the dedications in the Spanish, French and German editions, see Alberto Gómez Gutiérrez, 'Alexander von Humboldt y la cooperación transcontinental en la Geografía de las plantas: una nueva apreciación de la obra fitogeográfica de Francisco José de Caldas,' *Internationale Zeitschrift für Humboldt-Studien* 17:33 (2016), 22–49.

28 P. Vila, 'Caldas y los orígenes eurocriollos de la geobotánica,' *Revista de la Academia Colombiana de Ciencias Exactas, Físicas y Naturales* 11:42 (1960), xvii.

29 Vila, 'Caldas y los orígenes,' xviii.

30 F. J. Caldas, 'Memoria sobre el plan de un viaje proyectado de Quito a la América septentrional, presentada al célebre director de la Real Expedición Botánica de la Nueva Granada, don Jose Celestino Mutis,' in J. Acosta, ed., *Francisco José de Caldas. Semanario de la Nueva Granada* (Paris: Lasserre, [1802] 1849), 559.

31 Caldas, 'Memoria sobre el plan de un viaje,' in Posada, ed., *Obras de Caldas*, 68–72. Italics added.

32 Alexander von Humboldt, 'Barometrische Nivellirung der Gegend zwischen Cartagena und Santa Fé,' *Allgemeine geographische Ephemeriden* X:3 (1802), 210–212. Also, two manuscript copies of this 'Barometrical levelling,' drawn by Humboldt in 1801, were sent to Caldas by his friends Antonio Arboleda (1770–1825) and Santiago Pérez de Arroyo (1775–1845). See 'Nivellement barométrique du terrain depuis Carthagene á S[an]ta Fe' (1801) in the Archivo del Ejército de Tierra de Madrid, Ar.J-T.7-C.1_10, and in the Archivo del Ejército de Tierra de Madrid, Ar.J-T.7-C.1_10Bis.

33 Real Jardín Botánico de Madrid, AJB, Division III, M00529. To localise the depicted regions, please visit the satellite map of Humboldt's journey near the equator, at http:// geoatico.net/v2/geo-atico-es/index-map.html#.

34 R. Andress and M. Nieto Olarte, *Diario de viajes de Francisco José de Caldas* (Seville: CSIC, 2013); J. W. Appel, 'Francisco José de Caldas: A Scientist at Work in Nueva Granada,' *Transactions of the American Philosophical Society* 84:5 (1994), 1–153; L. C. Arboleda, S. Díaz-Piedrahita, and R. Molinos, eds., *Francisco José de Caldas* (Bogotá: Molinos Velázquez Editores, 1994); J. Arias de Greiff, A. Bateman, A. Fernández Pérez, and A. Soriano Lleras, eds., *Obras completas de Francisco José de Caldas* (Bogotá: Universidad Nacional de Colombia and Imprenta Nacional, 1996); J. Arias de Greiff, 'Algunos documentos, desconocidos unos, y poco conocidos otros, pertinentes a don Francisco José de Caldas y Tenorio,' *Boletín de Historia y Antigüedades* 61:704 (1974), 187–200; J. Arias de Greiff, 'Francisco Josef de Caldas y Thenorio,' in Arboleda, Díaz-Piedrahita, and Molinos, eds., *Francisco José de Caldas*, 11–21; J. Arias de Greiff, 'Caldas: inquietudes, proyectos y tragedias,' in Arboleda, Díaz-Piedrahita, and Molinos, eds., *Francisco José de Caldas*, 37–53; J. Arias de Greiff, 'El método de Caldas para medir la elevación de las montañas,' in D. Valencia Restrepo, ed., *Ensayo de una memoria sobre un nuevo método de*

medir las montañas por medio del termómetro (Medellín: Universidad de Antioquia, 2016), 99–113; J. Cañizares-Esguerra, 'How Derivative Was Humboldt? Microcosmic Nature Narratives in Early Modern Spanish America and the (Other) Origins of Humboldt's Ecological Sensibilities,' in L. Schiebinger and C. Swan, eds., *Colonial Botany: Science, Commerce and Politics in the Early Modern World* (Philadelphia: University of Pennsylvania Press, 2005), 148–165; A. Castrillón, 'Fitogeografía, paisaje y territorialidad al comienzo del siglo XIX,' *Boletín Cultural y Bibliográfico* 34:46 (1977), 60–84; S. Díaz-Piedrahita, *Nueva aproximación a Francisco José de Caldas. Episodios de su vida y de su actividad científica* (Bogotá: Academia Colombiana de Historia, 1997); C. E. González-Orozco, M. C. Ebach, and R. Varona, 'Francisco José de Caldas and the Early Development of Plant Geography,' *Journal of Biogeography* 42:11 (2015), 2023–2030; M. Nieto Olarte, *La obra cartográfica original de Francisco José de Caldas* (Bogotá: Universidad de los Andes, 2006); M. Nieto Olarte, *Americanismo y eurocentrismo. Alexander von Humboldt y su paso por el Nuevo Reino de Granada* (Bogotá: Ediciones Uniandes, 2010); M. Nieto Olarte, ed., *Memoria histórica sobre la vida, carácter, trabajos científicos i literarios, i servicios patrióticos de Francisco José de Caldas* (Bogotá: Instituto Caro y Cuervo, 2016); Posada, *Obras de Caldas*; M. A. Puerta Olaya, and J. M. Escobar Ortiz, 'Botánica y topografía: el problema de la nivelación de las plantas en la historiografía científica sobre Francisco José de Caldas,' *Historia y Sociedad* 33 (2017), 77–109; H. Schumacher, *Caldas. Un forjador de la cultura* (Bogotá: Empresa Colombiana de Petróleos, [1884] 1986); D. Valencia Restrepo, 'Contribución de Caldas a la fundación de la geografía de las plantas,' *Revista Aleph* LII:185 (2018), 5–25.

35 Joaquín Acosta included this footnote: 'Remember that when Caldas was writing this, he still did not know the picture of the geography of the plants that [a year] later he sent from Guayaquil to Bogotá to Mr. Mutis the Baron of Humboldt.'

36 George Augustus William Shuckburgh (1751–1804), British mathematician and explorer, who published the *Observations Made in Savoy to Ascertain the Height of Mountains by the Barometer* (1777).

37 Johann Georg Tralles (1763–1822), German physicist, astronomer and mathematician who participated in the meetings that established the Decimal Metric System in Paris.

38 F. J. Caldas, Letter to José Celestino Mutis, Ibarra, August 8, 1802, in Bateman and Arias de Greiff, eds., *Cartas de Caldas*, 193–194. Italics added.

39 F. J. Caldas, Letter to José Celestino Mutis, Ibarra, September 23, 1802, in Bateman and Arias de Greiff, eds., *Cartas de Caldas*, 195–196.

40 See facsimile edition published in Seville in Andress and Nieto Olarte, eds., *Diario de viajes de Francisco José de Caldas.*

41 F. J. Caldas, in Andress and Nieto Olarte, eds., *Diario de viajes de Francisco José de Caldas*, 70–82. Italics added.

42 F. J. Caldas, Letter to José Celestino Mutis, Otavalo, November 7, 1802, in Bateman and Arias de Greiff, eds., *Cartas de Caldas*, 200–202. Italics added.

43 F. J. Caldas, Letter to Alexander von Humboldt, Otavalo, November 22, 1802, in Bateman and Arias de Greiff, eds., *Cartas de Caldas*, 208–209. Italics added.

44 F. J. Caldas, Letter to Alexander von Humboldt, Otavalo, November 17, 1802, in Bateman and Arias de Greiff, eds., *Cartas de Caldas*, 210–211. Italics added.

45 Caldas, Letter to Alexander von Humboldt, Otavalo, November 17, 1802, in Bateman and Arias de Greiff, eds., *Cartas de Caldas*, 211.

46 F. J. Caldas, Letter to Alexander von Humboldt, Otavalo, November 17, 1802, in U. Moheit, ed., *Alexander von Humboldt. Briefe aus Amerika, 1799-1804* (Berlin: Akademie Verlag, 1993), 199.

47 F. J. Caldas, Letter to Alexander von Humboldt, Otavalo, November 17, 1802, at Staatsbibliothek zu Berlin: http://digital.staatsbibliothek-berlin.de/werkansicht?PPN= PPN779884310&PHYSID=PHYS_0927&view=overview-toc&DMDID= DMDLOG_0001.

48 F. J. Caldas, Letter to José Celestino Mutis, Quito, April 21, 1803, in Bateman and Arias de Greiff, eds., *Cartas de Caldas*, 219. Italics added.

49 F. J. Caldas, Letter to José Celestino Mutis, Quito, April 21, 1803, in Bateman and Arias de Greiff, eds., *Cartas de Caldas*, 219.

50 This *Memoria* and the corresponding 'Profile' dated April 21 in 1803 where thus 'formed in 1802' as the result of Caldas' trip from Popayán to Quito in 1801.

51 F. J. Caldas, Letter to José Ramón Leyva, Santafé, September 30, 1802, in Bateman and Arias de Greiff, eds., *Cartas de Caldas*, 274–280. Italics added.

52 F. J. Caldas, 'Notas a la *Geografía de las plantas* de Alexander von Humboldt,' in F. J. Caldas, ed., *Semanario del Nuevo Reyno de Granada* (Santafé de Bogotá: Bruno Espinosa de los Monteros, 1809) n. 16, 124–126. Italics added.

53 Caldas, 'Notas a la *Geografía*,' 125. Italics added.

54 The first publication of this drawing, once engraved, was the 1826 edition of Alexander von Humboldt and Aimé Bonpland's *Relation historique*. See Jackson, ed., *Alexander von Humboldt and Aimé Bonpland*, plate 3.

55 Jackson, ed., *Alexander von Humboldt and Aimé Bonpland*, plate 3, and Alexander von Humboldt, 'Voyage vers la cime du Chimborazo, tenté le 24 juin 1802. Geographie des plantes dans les Andes de Quito,' in Alexander von Humboldt and Aime Bonpland, eds., *Voyage de MM. Alexandre de Humboldt et Aime Bonpland. Atlas Geographique et Physique, pour Accompagner la Relation Historique* (Paris: J. Smith, [1824] 1831), plate IX. See also D. Rumsey, Historical Map Collection, at https://www.davidrumsey.com/luna/servlet/detail/RUMSEY~8~1~292857~90064432

56 F. J. Caldas, in Posada, ed., *Obras de Caldas*, 72.

57 For example, in '*Nivelación* de algunas plantas que cultivamos en las cercanías del Equador' (1802), '*Nivelación* de 30 especies en el Imbabura' (1803), and '*Nivelación* de las quinas' (1804).

58 F. J. Caldas, Letter to Santiago Arroyo, Quito, October 6, 1801, in Bateman and Arias de Greiff, eds., *Cartas de Caldas*, 107.

59 F. J. Caldas, Letter to Antonio Arboleda, Quito, October 6, 1801, in Bateman and Arias de Greiff, eds., *Cartas de Caldas*, 111.

60 F. J. Caldas, Archivo del Ejército de Tierra de Madrid, Ar.J-T.7-C.1_10Bis.

61 F. J. Caldas, Letter to Santiago Arroyo, Quito, October 21, 1801, in Bateman and Arias de Greiff, eds. *Cartas de Caldas*, 115.

62 On Caldas' circe, see Renan Silva, *Los Illustrados de Nueva Granada, 1760-1808. Genealogía de una comunidad de interpretación* (Medellín: Eafit, 2002).

63 Anonymous, 'Discurso sobre el calendario rural del Nuevo Reyno,' in J. T. Lozano and J. L. Azuola, eds., *Correo curioso de Santafé de Bogotá*, 35 (1801), 137.

64 M. Nieto Olarte, *Orden natural y orden social: ciencia y política en el Semanario del Nuevo Reyno de Granada* (Bogotá: Universidad de los Andes, 2008), 1.

65 Alexander von Humboldt and C. Kunth, *Géographie des plantes* (Prospect) (1826), at Staatsbibliothek zu Berlin Preußischer Kulturbesitz, Nachl. Alexander von Humboldt, gr. Kasten 13, Nr. 26, Bl. 1–2. [Communicated by Ulrich Paessler].

66 Alexander von Humboldt, Letter to Jean-Baptiste Joseph Delambre, Lima, November 25, 1802, in Moheit, ed., *Alexander von Humboldt. Briefe*, 205. Italics added.

67 F. J. Caldas, Letter to José Ramón Leyva, Santafé, September 30, 1809, in Bateman and Arias de Greiff, eds, *Cartas de Caldas*, 276.

68 F. J. Caldas, Letter to José Celestino Mutis, Ibarra, August 8, 1802, in Bateman and Arias de Greiff, eds., *Cartas de Caldas*, 193.

69 F. J. Caldas, in Posada, ed., *Obras de Caldas*, 71–72.

70 F. J. Caldas, *Relación de un viaje à Ybarra y demás pueblos circinvesinos al Nordeste de Quito hecho en 1802* (manuscript), ff. 1–27.

71 Alexander von Humboldt, *De distributione geographica plantarum secundum coeli temperiem et altitudinem montium: Prolegomena* (Paris: Libraria Græco-Latina–Germanica, 1817).

5

AN ARCHAEOLOGY OF MUTIS'S DISAPPEARING GIFT TO HUMBOLDT

José Antonio Amaya[1]

TRANSLATED BY MARK THURNER

[epi] I humbly confess that I count myself among those who enjoy investigating what our great minds consider mere miseries. My eyesight is short, if not my mind myopic, and so minuscule things present themselves more readily to my eyes. I am interested in the multitude of microscopic events that produce pity in those historians and critics who claim, and are acclaimed to be, enchanted with themselves because they have the wings and eyes of the eagle.
– Auguste Jal[2]

During their stay in Santafé de Bogotá, Humboldt and Bonpland were presented by José Celestino Mutis (1732–1808) with what tradition has called a 'collection,' naturally of botanical materials. In this case, the indeterminacy of the word *collection* is perhaps explained by the fact that the material in question has not been located for more than a century.[3] In any case, no complete material description has been drawn up. Still, this collection is not as silent as it may seem, and traces of it persist in the American botanical works of Humboldt and Bonpland, as may be seen in *Plantes équinoxiales* (1808; 1809) and *Monographie des melastomacées* (1816; 1823).

The existence of these traces has lent impetus to our research into the materials used to prepare those publications. In our search, we have not neglected other relevant documents initially not intended for publication, such as the travel diaries and correspondence, which it turns out also include bits and pieces that may be assigned to Mutis. In a similar fashion, our task has required that we comb the herbaria, field notes and even preliminary drawings, most kept by the *Muséum National d'Histoire Naturelle de Paris* (MNHNP). Thus defined, the field of our archaeology extends temporally from 1801 to 1846. As will become evident, within this field, we have noted a marked contrast between the spontaneity of the early, involuntary 'traces' that emerge in Humboldt and Bonpland's American notes, and the subsequent,

DOI: 10.4324/9781003231479-6

conscientious preparation of works published after their return to Europe. Finally, it should be noted that our exploratory archaeology of the substrata of the works of Humboldt and Bonpland cannot be carried out in its entirety from Colombia. In part for this reason, and despite our having compiled multiple forms of scattered evidence, this chapter only opens a door, merely outlines an itinerary.

It is somewhat surprising that the materials received in Santafé de Bogotá by Humboldt and Bonpland from Mutis have left such a negligible mark on the works they published in Europe, and apparently none in the gift-giver's files. This fact is noteworthy, particularly given the existence of two detailed catalogues that reveal Mutis's customary practices when gifting a collection: the *Inventario de las descripciones y observaciones para la 'Flora de Bogotá' de José Celestino Mutis que se conservan en el Real Jardín Botánico* (Madrid) and the *Catálogo de los envíos de Historia Natural remitidos por José Celestino Mutis a Suecia*.[4] However, certain passages of a letter from Humboldt to Bonpland, dated June 10, 1805, in Rome, have called to our attention a disenchanting possibility. There, Humboldt confesses that 'I make promises to others without keeping my word' and then adds, 'the dead make mistakes, and so you can publish the *Hoitzias* yourself.' It was an allusion to the late director of the Royal Botanical Garden of Madrid, Antonio José Cavanilles (1745–1804) and his botanical work.[5] This passage and others have prompted me to pursue the issue of the disappearance of Mutis's collection of botanical plates and the related question of the sources of Humboldt and Bonpland's American botanical knowledge.

Any attempt to reconstruct the content of the disappearing gift must begin by noting that they proceed from the massive *Flora de Bogotá* project, which means that they were not materials prepared at random. It can be said that Mutis inspired, advanced, and coordinated in his *Flora* one of the most ambitious projects of its kind anywhere in the world during the second half of the eighteenth century. Begun in 1760 on the initiative and with resources provided by Mutis himself, the project became official only on November 1, 1783, when King Charles III (1716–1788) created the Royal Botanical Expedition at the request of Viceroy Antonio Caballero y Góngora (1723–1796).

As the nineteenth century dawned, however, the international Republic of Letters was still eagerly awaiting the results. It is clear that Mutis's Botanical Expedition privileged collecting and graphic presentation over the processing and publication of botanical data. This fact suggests that, at the time of the gift to Humboldt and Bonpland, Mutis still had a lot to *say* about the *Flora*, literally speaking. Given the physical absence of the collection of plates, for which we have searched unsuccessfully for more than thirty-five years,[6] all that we have to work with are the preserved words about the plates. Fortunately, these words do contain truths which may eventually lead us to the originals. It is precisely with those preserved words that we have attempted to reconstruct and infer the content of the collection and the relations among its parts, as well as their significance for the history of science in the New Kingdom of Granada at the start of the nineteenth century. What we seek in this effort is a new clue or key that one day will help establish the weight that corresponds to Mutis and his *Flora* in the botanical opus of Humboldt and Bonpland.

The Unlikely Encounter

Multiple contingencies surrounded Humboldt's journey to Santafé de Bogotá. Originally, neither Mutis nor that part of the territory of the New Kingdom of Granada today known as Colombia and Ecuador was on his itinerary. The postponement of Captain Nicolas Baudin's circumnavigation of the globe (1754–1803), followed by the impossibility of joining a team of scientists sent by Napoleon Bonaparte (1769–1821) to Egypt (1798–1801), led Humboldt to attempt, as it turns out based on erroneous reports, to join Baudin's voyage at Lima's port of Callao. In both cases, Humboldt and Bonpland, his naturalist secretary of French nationality and surgeon of the French Navy, were profiled as members of a team under French military command.[7] Such details help us to understand the terms of the passport granted by King Charles IV (1748–1819), authorising them to visit 'the Americas and other overseas possessions of their dominion,' where they could 'freely collect plants and seeds,' among other objects. Mutis was among the officials obliged to give them 'all the favours, assistance and protection they need,' without placing 'any obstacles' whatsoever upon them, all for the sake of 'enriching the Royal Cabinet of Natural History, and the Royal Botanical Gardens' at Madrid.[8] The Hispanic monarch had supposed that the itinerary conceived by Humboldt at his own risk and expense would be fleeting, since his final objective was to join Baudin's voyage. Traditionally, scholars have emphasised the king's liberality for having granted such a *carte blanche* to Humboldt. The instructions did not oblige him to travel in the company of Spanish or Creole scientists, as was customary at the time, and they did not provide any details of the foreseeable contacts with peninsular or overseas scientific centres that one would presume indispensable. In short, the traditional interpretation tends to ignore the importance at the time of the French military connection and the intended rendezvous with Baudin.

In any case, it can be said that Humboldt's trip to Spanish America (June 5, 1799, to August 1, 1804) was little more than a substitution. Moreover, in 1801 Humboldt was not interested in Spanish America in itself, nor was botany his priority. He was interested in his theory of the cosmos. In addition, at the time his image of Mutis the naturalist was far from positive;[9] indeed, at one point, he was seized by the fear that the works of the man from Cádiz would be published before his own. At this point, Humboldt learned in Cuba that he had missed the trade winds that would have allowed him to reach Guayaquil via Panama. Given the impasse, he determined to reach Lima by land. It was then that it occurred to him to visit Mutis. Under these circumstances, and after staying in Cartagena de Indias from March 30 to April 19, 1801, Humboldt and Bonpland chose to ascend along the Magdalena River up the foothills of the Andes toward Bogotá.[10] The 'motives' are given in a passage written by Humboldt himself, wherein the sender is already involved in the reconstruction of his trip, evidently striving to appear in control of his destiny:

> The keen desire to meet the great botanist, Don José Celestino Mutis, who was a friend of Linneaus, and lives today in Santa Fe de Bogotá; and to compare our herbaria with his; and the curiosity to ascend the immense Andes Mountain Range that extends from Lima (on the north side) to the mouth of the Atrato River, in the Darien Gulf, in order to be able to make, according to my personal observations, a map of all of South America, from the Amazon River to the north, prompted me to prefer the overland route from Quito and beyond Santa Fe and Popayán, rather than the sea route beyond Porto Bello, Panama and Guayaquil.[11]

The announcement of the visit of the young scientists energised Mutis, who at the time was nearing seventy years of age. Mutis suffered from the ailments of his age. He had been without adjuncts to his Royal Expedition since 1794, and he continued to be reticent to request substitutes.[12] His links with the Royal Botanical Garden had been interrupted since 1784, that is, practically since the official creation of his Expedition in 1783. His correspondence with the illustrious Valencian botanist Cavanilles, begun in 1786, had, with the publication of *Caryocar amygdaliferum* in 1797, barely satisfied his repeated requests.[13] However, despite or perhaps because of this publication, Mutis seemed reluctant to publish with Cavanilles. It is true that the manuscript he sent had hardly been edited, but the corresponding plate (54 × 38 cm) in full colour was reproduced in black-and-white and reduced by more than 80% (22 × 17.5 cm). Furthermore, by then Mutis's international ties had ebbed, despite the demands of the Frenchman Charles Louis L'Héritier de Brutelle (1746–1800) and the Englishman James Edward Smith (1759–1828),[14] among other naturalists.[15]

His local supporters in New Granada included the viceroy and the archbishop of Santafé de Bogotá, the rector of the Colegio del Rosario, the notable Jorge Tadeo Lozano (1771–1816), and José Ignacio de Pombo (1761–1815), a merchant from Popayan established in Cartagena. Despite this local support, his Botanical Expedition was the object of opposition and disapproval. Precisely in 1801, a couple of projects related to the so-called Botanical House (Casa Botánica) were being prepared. On one front, Francisco Antonio Zea (1766–1822), deputy director of the Expedition in commission in Paris, together with other Creoles and undoubtedly with the support of the metropolitan authorities, was preparing a reform aimed at reorienting the Santa Fe centre towards chemistry and agriculture. On another front, a group of New Granada soldiers were advocating the degradation of botany from the leadership position that it had held in New Granada since 1783, to the benefit of mining.

To reaffirm its threatened authority, nothing was more appropriate for Mutis' Casa Botánica than the endorsement of the travellers. Mutis knew that if Humboldt heeded the opposition's insinuations,[16] the consequences would be projected *ad vitam æternam.*[17] Aware of this threat, he mobilised all his influence and persuasive power to welcome the visitors into his social network. In Santafé, Mutis had been living at the headquarters of the Expedition since 1791;[18] this circumstance allowed him to graciously accommodate his visitors at his private residence, 'with patio, garden and kitchen, and damask sofas, the *non plus ultra* of American luxury.'[19]

Despite what we have said so far, an ambivalence haunts Mutis's hospitality. To be sure, his relationship with his guests was impregnated with ulterior motives, and this since before they arrived at Santafé. The director of the Royal Expedition could not agree to the exploration of his adopted land by foreigners without the company of national savants. He surely must have disapproved of a royal passport that did not provide protection for his own work. Under these circumstances, he joined the protest of a group of Popayan lawyers led by Santiago Pérez de Arroyo y Valencia (1773–1845), chaplain of the Colegio del Rosario, and made up of Miguel Pombo (1773–1816) and Francisco José de Caldas (1768–1816). This group challenged the conditions under which the metropolitan government had authorised Humboldt's journey, based on the jurisprudence that had emerged after the Hispano-French (La Condamine) geodesic expedition (1735–1746) and the botanical expedition to Peru and Chile (1777–1788). They pointed out that Jorge Juan and Antonio de Ulloa had been added to the first expedition, just as Hipólito Ruiz (1754–1816) and José Antonio Pavón (1754–1844) had been to the second. Arguments of a patriotic bent against exclusively foreign expeditions had been used by Caballero and Góngora before José de Gálvez (1720–1787), Minister of the Indies, to gain King Charles III's support of the Botanical Expedition.[21] The group now proposed to repair the apparent laxity of the Crown in this regard, specifically by requesting the integration of Caldas into Humboldt's team. In this group's view, the material and intellectual wealth of the country properly belonged to them by right. The time would come, they believed, when the New Kingdom would contribute to global knowledge with her own representation of her territory and its human and natural wealth. Notably, the group refused to ask Viceroy Pedro de Mendinueta (1736–1825) to support their effort, since they did not consider him a 'man of lights.' Instead, and knowing that officials were obliged to abide by the terms of the royal passport, they chose to go to Mutis. The wise man accepted the proposal even though he still did not know Caldas personally, being familiar with his works and projects. The enlightened sense of territorial sovereignty that operated in the Creoles and the Creolised Mutis nearly a decade before the first cry of independence (1810) is perhaps surprising. However, the recent historiography of civil geography in the colony is now revealing that the independence movement had cultural and scientific origins.

On July 6, 1801, Humboldt and Bonpland reached Santafé de Bogotá, where they had planned to spend just a few days. However, 'poor Bonpland had again … a fever on the route from Honda to Santafé, which forced [them] to stay in the latter city for two whole months.'[22] This unlikely encounter is at the origin of a friendship that will stimulate both protagonists, without of course forgetting Caldas. It will lead to the dissemination of the first engraved portrait of Mutis (see Figure 5.1), and to the publication of the first biography in French of the savant from Cádiz.[23]

FIGURE 5.1 Portrait of José Celestino Mutis. Dedicatoria, Alexander von Humboldt and Aimé Bonpland, *Plantæ æquinoctiales* (Paris, 1808).[20]

Mutis was a reserved man, almost secretive. During the time he led the expedition, apart from Francisco Martínez[24] and Humboldt and Bonpland, it appears that no one had access to his cabinet, not even his closest collaborators.

> His mysterious and distrustful character always kept him silent and retired. He never began the promised confession, he never lifted the veil, nor did he introduce me to his sanctuary [his cabinet]. He always kept me in the dark about the state of things, and I have only come to know them superficially after his death.[25]

If, even in Humboldt's testimonies, he was considered selfish, sullen and misan-thropic, how did the visitor manage to turn the 'reserved and very grumpy' Mutis into someone communicative and effusive?[26]

Everything indicates that Humboldt was the first to break through. He began by presenting his scientific samples and observations (botany, zoology, anthropology, mineralogy, astronomy, physical geography, plant geography, etc.) collected during his journey and destined to expand the limits of human knowledge and thereby win for himself the glory that would result from the feat of his travels. Humboldt then revealed his career plans, which he predicted would allow him to join the prestigious Institut National de France (IF) as a foreign correspondent[27] and which would allow Bonpland to collaborate with the famous MNHNP.[28] Such ambitions evoked in Mutis the motives that had led him to New Granada in the first place. In turn, the latter recounted the adventures that he had had to overcome in America to lead the Botanical Expedition. Despite Mutis's imperial convictions and Humboldt's quality of *citoyen du monde*, the two savants recognised one another. They belonged to the same lineage of scientific travellers; their projects reflected in the same mirror.

The Exchanges

The permanence of Humboldt and Bonpland in Santafé generated documents that ratify the relative autonomy of Mutis vis-à-vis metropolitan Spain. These same documents give us a good sense of the scope and limitations of the *Flora de Bogotá* project. During the extended visit of Humboldt and Bonpland, Mutis had the opportunity to personally engage with European scholars. Gaining recognition for his work was a priority. He began by talking about his discovery of the Santafé Cinchona, the corresponding therapeutic investigations, the cultivation, harvesting and transportation of the precious bark, the affronts suffered to make it recognised, and the writing of his *Arcano de la quina*. The Prussian, as well as the Frenchman, were taken by surprise by this scholarship, unknown in Europe.

Bonpland, who 'must be seen as the botanist of the Humboldt expedition,'[29] was eager to record this knowledge in his 'Observations communiquées par le Doct[eur] Mutis.' As a convalescent during a large part of his stay in Santafé, the ailing Bonpland took advantage of the company of Doctor Mutis; meanwhile, Humboldt 'measured the surrounding mountains …, visited the Guatavita lagoon, the Tequendama waterfall …, the gem salt mines of Zipaquirá, etc.'[30]

Eventually persuaded of their capacity to grasp the scope and importance of his work, Mutis decided to open his cabinet and library to the European savants. He authorised them to compare the plants they had collected with those of the Expedition herbarium, take as many notes as they wished, and to examine and even tabulate the plates of the *Flora de Bogotá*. In short, he let them retrieve information that no other publisher or contributor had, or ever would have, access to. The comforts of the accommodation and especially the excellent working conditions in Santafé surprised the guests: the wealth of the herbarium, the organisation of the

Office of Painting, the profusion of the library, all gathered in a singular research centre. Humboldt eloquently recorded his esteem in a letter to Cavanilles, writing that Mutis 'treated us in Santa Fe with that frankness which seemed the peculiar character of Banks;[31] he unreservedly showed us all his riches in Botany, Zoology and Physics.'

In addition to perceiving the ageing fragility of Mutis's health, Humboldt did not miss the scientific and technical weaknesses of the botany practised at Santafé. 'The immense Flora of the New Kingdom ... will never be published in its entirety,' he concluded. He also noted the preponderance of accumulation over data management. In this regard, he pointed out that Mutis 'continued to accumulate materials for his work ... without concentrating on a fixed project of publication.' In addition, he listened with benevolent scepticism to his plans to publish the *Flora of Bogotá* in Santafé.

> Accustomed to overcoming obstacles that seemed insurmountable, [Mutis] gave himself up with pleasure to the idea of one day setting up a printing press in his home and teaching the indigenous people to make engravings of what they had learned to paint with such success.[32]

Research has shown that in 1790, Mutis had abandoned writing his daily notes and descriptions of the plants kept in his herbarium.[33] From that point onward, he recorded his erudition on herbarium sheets and colour plates, which he had to estimate as much or more than the descriptions. Humboldt penetrated this aspect and rightly concluded that a large part of the text of the *Flora of Bogotá* was solely supported by Mutis's memory. Indeed, and as the failure of his succession reveals, only he could relate the full range of dispersed information that his expedition had been accumulating. Humboldt also noted that Mutis would not be returning to Spain, because he was 'too attached to the establishments he had created, and because he loved the country that had become his second homeland too much.'[34] Finally, by 1821 Humboldt did not think it plausible that the *Flora of Bogotá* would ever be published in Spain.

Mutis recognised in this fortunate young man, brimming with health and energy, the man that he himself had not been able to be. Mutis had lacked sufficient resources to finance his projects in the 1760s and 1770s, or indeed to travel across Hispanic America with the aim of publishing a natural history of the region. Listening to these confidences, Humboldt assuaged his pain and expressed admiration. Reassured by everything he was discovering and having recovered his self-confidence, Humboldt hinted to Mutis his desire to publish, upon his return to Europe and with the collaboration of the IF, the novelties of the *Flora* without altering the iconography, which he considered fully sufficient for the identification of the plants.[35] Mutis, 'First Botanist of His Catholic Majesty on the Expedition to America,' was prevented from publishing outside national frontiers. Humboldt offered him an unexpected opportunity. Mutis understood that Paris was then the capital of science and the arts. Consulting his own library had made it possible for him to appreciate the relative quality of his own iconography. His botanical plates surpassed anything that had been published until then, as his correspondents had been corroborating since the mid-1770s.[36] What is certain is that both in the formation of the library and in the consolidation of its style of botanical iconography, Mutis's Casa Botánica was the product of its intimate contact with America. In no case was it mediated by an imagined official metropolitan influence. In this case, the traditional, centre–periphery scheme normally deployed to understand such enterprises is reversed.

Mutis immediately understood that upon his return to Paris the doors of the MNHNP, dependent on the IF, the first scientific centre in the world at that time, would open to the Prussian. The man who liked to support young talent agreed to Humboldt's request to send with him the collections to be donated to the IF, thereby helping Humboldt to position himself better in this prestigious centre. In addition, the gesture of personally submitting the collections via Humboldt would give visibility to Mutis himself before the IF, where his name was shuffled as a possible corresponding member.[37] The old Mutis began to cherish the idea that the young Humboldt, already linked to the most vigorous scientific networks in Germany and France, could become the witness of his work, the guarantor of its continuity, the one who would contribute to perpetuating his name. In this way, a fortunate coincidence invited Mutis to come out of the silence that his official commitments had imposed upon him; his independent demeanour did the rest.

In this field he was not without experience. Naturalists in the Lutheran (Linnaeus father and son, Bergius and Thunberg) and the Anglican (Sir James Edward Smith) botanical traditions, without forgetting the Abbe Antonio José Cavanilles, had taken care, between 1767 and 1797, to publish about seventy-nine of his plants (see Table 5.1). Many of these represented new genera (*Mutisia, Escallonia, Bejaria*, among others). These plants had made his name known, even in Spain itself. Some had appeared with engravings, monochromes without exception, some made on sheets awash in India ink, and others retouched or 'illuminated,' and always prepared under the direction of Mutis. Some of these European prints had, more or less faithfully, reproduced the American originals, while others had reduced them by rearranging their content. Linnaeus had received thirty-two unsigned gouaches from the Bogotá artist García del Campo, who worked for Mutis from the mid-1770s to 1784. Linnaeus did not publish any of the plates that Mutis sent to him in a private capacity, however.

For his part, Smith, who acquired Linnaeus's cabinet in 1784, published twenty-three of Mutis's plants in *Plantarum icones hactenus ineditae*, a series of three fascicles that appeared in London between 1789 and 1791, this after its author's failed attempt to succeed Linnaeus as Mutis's primary European correspondent. Smith introduced some variations by associating the plates with observations corresponding to herbarium specimens sent by Mutis to Linnaeus. *Plantarum icones* was illustrated by the celebrated English artist James Sowerby; the beauty and accuracy of this work made it an obligatory reference. Despite this treatment, the book did not account for the novelties introduced by Mutis after 1783, when he was appointed

TABLE 5.1 Botanical collections sent to Europe by J.C. Mutis, 1760–1802[39]

Place and date	Destination	Recipients	Specimens	Descriptions	Plates	Seeds	Plants published	Plates published / publicadas
Cádiz, Santafé de Bogotá, Suratá, Sapo, Mariquita, 1760–1790	Sweden	Linneus father and son; H.J. Gahn; P.J. Bergius; C.P. Thunberg	≥270	>51	38	>49	78	15
Santafé de Bogotá, 31-1-1777	Spain	Real Gabinete de Historia Natural (Madrid)	86	–	43	–	–	–
[Santafé de Bogotá, 1784]		Casimiro Gómez Ortega	–	–	2	–	–	–
Santafé de Bogotá c. [1797–1798]		Antonio José Cavanilles	–	1	1	–	1	1
[¿Santafé de Bogotá?], entre 1789 y 1796		José de Ezpeleta, Viceroy of New Granada	–	–	4	–	–	–
Junio de 1802	France	Institut national de France (París), by recipient A. von Humboldt	'Una gran cantidad'; 'muestras de herbario en flores'; 'cortezas secas de quinas'	–	107	'Muestras de herbario en semillas'	?	?

head of the Expedition. In that year, and as the person in charge of a work under royal patronage, Mutis adapted with increasing success the watercolour and tempera to illuminate the Grand-Aigle paper. He further determined to expand the format of the sheets from folio (42 × 28 cm) to large folio size (54 × 38 cm). For his part, Bergius, who had also sought to emerge as Linnaeus's successor, published two of his plants, and Thunberg two more. With this rich experience and talent in selecting his correspondents, a rare privilege for a naturalist based in America,[38] Mutis judged that Humboldt and Bonpland guaranteed scientific seriousness and promised the faithful reproduction of his iconographic work, both in terms of size and colour.

Although these foreign publications did not fully account for the quality achieved in Santafé, each editor without exception had assigned the corresponding credits, in accordance with the guidelines then in force. Mutis considered sacred respect for the priority of a discovery. For this reason, and before presenting them with a 'large quantity of dry specimens,' with a 'work on the genus *Cinchona*,' and with about one hundred magnificent drawings in large folio, representing new genera and species from his manuscript *Flora de Bogotá*, the host and his guests had to agree on how the plants from the collections of the Royal Botanical Expedition of the New Kingdom of Granada would be cited. The matter is important, considering the Expedition's many taxonomic novelties, as Humboldt and Bonpland explicitly admit again and again in their correspondence (at least while they remained in America). Notably, because after 1790, Mutis ceased to describe his plants, opting instead to concentrate on representing them in iconography, the value of the testimony of Humboldt and Bonpland is heightened, since it establishes the existence of novelties.

Humboldt and Bonpland were bound to respect the precedence or priority of Mutis and his collaborators as collectors, descriptors, iconographers, and, more generally, botanists. Given what we know about his practice in this area, only the explicit acceptance of his precedence and terms can explain Mutis's magnanimity. It no doubt also reveals his realisation that publishing in Santafé in the medium term was out of the question. We may conclude then that before leaving Bogotá, Mutis entrusted to Humboldt and Bonpland the largest collection of botanical illustrations that he would ever transmit to any publisher.

Nothing indicates that Mutis had requested permission or even informed Viceroy Mendinueta or indeed Cavanilles of his gift. Most likely, Spanish officials would never have authorised the publication in Paris of such a mass of work produced under royal patronage. Neither Caldas nor Sinforoso Mutis (1773–1822) appears to have been informed of the matter. Mutis probably recommended to Humboldt reservation. It is striking that Humboldt limited himself to communicating to Cavanilles,

> [W]e have sent to the National Institute of France a curious collection of the New Granada cinchonas, which consisted of well-chosen barks, beautiful exemplars in flower and fruit, and magnificent illuminated drawings, in a large folio that the generous Mutis has given us,

omitting in this instance the collection of (one hundred?) colour plates and the herbarium. The letter in question was published in Madrid, Paris, and London.

On September 8, 1801, Humboldt and Bonpland resumed their journey through the Andes south to Popayán, Quito, and Lima. They continued their relationship with Mutis by correspondence.[40] For his part, Mutis continued to wait for the opportunity to place Caldas at the side of the travellers. Amid all the adhesions and praise, Mutis remained faithful to the alliance with Pérez de Arroyo. It is also easy to imagine that the presence of Caldas in Humboldt's entourage would be seen by Mutis as a guarantee that his scientific achievements would be respected. At first, Mutis noted Humboldt's repeated and public praise for Caldas's personality and work, issued even before the Prussian met the Creole in Ibarra on December 31, 1801. Then, given the evident enthusiasm with which they had been working, Mutis formulated the request to Humboldt, by then in Quito, issuing him a blank bank draft to cover his foreseeable expenses. Humboldt appears to have been irritated by the petition.[41]

On January 6, 1802, Humboldt and Bonpland arrived in Quito, where they learned to their dismay that Captain Baudin 'had taken the route ... through the Cape of Good Hope.' They found consolation in 'looking at [his] herbariums, ... [his] drawings ... and did not [regret] at all having travelled to countries that, for the most part, have never been visited by naturalists.' Resigned to missing the voyage around the world, Humboldt now determined that 'in September or October 1803' he would be 'in Paris!' From this point on, Humboldt became obsessed with the idea of 'conserving the manuscripts he possesses and getting them published.' In his correspondence he now appears impatient to know the fate of the collections received from Mutis. Thus, on November 25, 1802, after announcing to the astronomer Jean-Baptiste Joseph Delambre (1749–1822), at the time secretary of the IF in Paris, of the dispatch of the two well-known boxes, Humboldt wrote: 'It is very sad to remain in such uncertainty about the fate of these objects, as well as the collections of rare grains that we sent three years ago to the *Jardin des Plantes* in Paris!' It appears that Humboldt remained without news of the Mutis collections until the end of his journey, recording growing anxiety in his letters. In this vein, at least five other direct and indirect communications are known.[42]

Sharing this anxiety to publish with his IF correspondents, Humboldt transmitted his desire to settle in Paris. There he would prepare the publication of his travel narrative, with the collaboration of Bonpland. Adorned with formulas of courtesy and excessive consideration, his letters during this period seek to cultivate his network of relationships with the most outstanding scientists of Paris. With a similar purpose, his brother Wilhelm compliments Cuvier, professor of comparative anatomy at the MNHNP, in this fashion: Alexander

> hopes to embark in Havana for Spain; from there he will go to Paris, where the impatience to meet you again, sir, and your wise colleagues, who before his departure showered him with kindness and gestures of friendship, naturally moves him more than anything.[43]

For his part, King Frederick-William III (1770–1840) extended to him 'without hesitation the permission to remain until next summer in France and Italy.'[44]

Assured of being able to work on the results of his journey, Humboldt arrived in Paris on August 25, 1804. In a letter dated December 18, 1804, he offered to the MNHNP leadership the collections he had brought with him, which undoubtedly included the gift from Mutis. The heading of the letter is illuminating:

> The generous benevolence with which you have deigned to *receive* ... *the Cinchona* and *the coloured drawings of the plants of Santafé* (which Mr. Bonpland and I have dared offer to you), grants me hope that you will excuse the freedom that I take when I write these lines.
>
> (italics by the author)

This letter confirms that by December 18, 1804, all the collections received from Mutis, including the full-colour plates, which numbered about one hundred, with the usual attached documents, had reached Paris, more precisely the MNHNP, which integrated them into their collections. Although it is true that several forms of specific evidence demonstrate that the materials on the cinchona, the corresponding plates and dried specimens had reached Paris, the letter of December 18 is the only known evidence that remains of the reception at the MNHNP of the hundred or so plates that Mutis entrusted to Humboldt. Together with the Decree of March 13, 1805, of Napoléon Bonaparte (1769–1821) it can be considered that by this last date, all the materials given to Humboldt by Mutis were in the MNHNP.

It will have been noticed that in this allusion to the collection gifted in 1801, the whereabouts of most of which are currently unknown, the names of 'José Celestino Mutis' and 'Flora de Bogotá' are conspicuous for their absence. The MNHNP report recommending acceptance of the Humboldt collections is signed by the recipients of the 'Mutis collections,' that is, by Jussieu, Lamarck, and Desfontaines. This report, which is scrupulously silent on the two boxes sent by Mutis, now apparently incorporated into the set of collections offered by Humboldt to the MNHNP, does not question its receipt nor its conservation.[45] In 1821, Humboldt confirmed this information in part, stating that 'the well-executed illuminated drawings [of the *Cinchona* genre] that Mr. Mutis gave me in Santafé ... have been deposited ... in the *Jardin des Plantes* in Paris.'[46] However, at the time the boxes were received and transmitted to the MNHNP, the IF appears to have been free of the obligation to keep a correspondence record; the MNHNP likewise failed to register the corresponding arrival. For his part, the Colombian botanist José Jerónimo Triana (1828–1890) assures that 'it seems proven that these drawings were not donated to the museum's herbarium; only the bark and the samples were sent by Humboldt.'[47] All of the above has been feeding an unfounded controversy over the receipt of the Mutis materials.

Nevertheless, the issue of its loss or, more precisely, its location, remains to be resolved. Humboldt, who held the Mutis gift-collections in the highest esteem, particularly the plates, as we have seen, never mourns their loss. If it had been lost, he would have surely announced it, either in the preface to *Plantes équinoxiales*, a book dedicated to Mutis, or in *Monographie des melastomacées*, or in the editions of the biography consecrated to Mutis.

Although the investigations that we have carried out to locate the mentioned plates, in the *Bibliothèque Centrale* and in the Laboratory of Phanerogamy at the MNHNP have been unsuccessful, we believe that the search should continue. Around 1806, an American traveller and professor at Harvard University, passing through Paris, referred in his *Diary* to having found Bonpland 'copying' a plate of Mutis's in the *Jardin des Plantes*.[48] The inflection 'copying' requires clarification. It is certainly not a tracing, which otherwise would have betrayed the unmistakable style of the collection, obviating any claim to originality; Bonpland would have provided perspective to the illustration while preserving the pertinent scientific observations, thus modifying it notably.

The consummation of an encyclopaedic publishing project such as Humboldt's, at the time intended to revive a spiritually and politically depressed Europe sclerosed by the Revolution, could conceivably only occur in Paris.[49] Only there did the material resources tailored to Humboldt's collections exist; only there did an intellectual ambition of European magnitude reign. Along with Arago, Humboldt presided for twenty years over what was then known as the world's first scientific body, as he himself referred to it.[50] French logistics at the service of the production and dissemination of knowledge was embodied in the Institut de France. Paris offered a centralist research policy; scientific newspapers and magazines; institutions for the training of specialised scientists, engravers, and naturalist painters; publishing houses with their finances, translators, proofs readers, and distribution services; printing houses with specialised labour; factories of paper, ink, mobile characters, and so on. Together, these elements made Paris an empire capable, in a certain way, of 'colonising' the Republic of Letters.[51]

To carry forward his project, Humboldt negotiated collaborations among the savants and scientific centres of the French capital: from Laplace to Cuvier, passing through Delambre and Gay Lussac, from the Longitudes Office to the Polytechnic School, passing through the Jardin des Plants and the Louvre. The production of his American works would constitute the largest publishing enterprise of his time: more than thirty large volumes *in-quarto* and *folio*, which exceeded even the imperial *Description of l'Egypt*. The work will feed a small army of young French physicists, mathematicians, draftsmen, engravers, and cartographers, and drain the resources of three publishing houses. When in 1809 Napoleon proposed to expel Humboldt, who was accused of being a spy for Prussia, Jean-Antoine Chaptal (1756–1832), chemist and former minister of the interior of the directory, dissuaded him arguing that his departure would paralyze science in the capital.[52] In short, there were few obstacles that could curb Humboldt's publication plans in Paris.

Mutis and His *Flora de Bogotá* in the Works of Humboldt and Bonpland

Searching for Mutis and *Flora de Bogotá* in Humboldt's work exceeds the limits of this chapter. However, to understand the expectations that Mutis had placed in Humboldt when he gave him those extraordinary gifts, it is essential to take a step back. We must look at the credits granted without exception to Mutis by his European editors. The Linnaeuses, father and son; Smith; and Cavanilles had all been rigorous and detailed when citing the naturalist from Cádiz, specifying his authorship in the minting of new generic names, in the assembly of herbarium sheets, in the lifting of descriptions and in the preparation of iconographies, as well as in various comments included in his letters and manuscripts. In publishing the *Gomozia*, for example, Linnaeus specified that it was a 'genus of Mutis.' In addition, he was always careful to cite the Mutis plants backed with illustrations prepared in Santafé, as with *Passiflora adulterina*: 'American plates of Mutis, (hereinafter icon. Mutis. Amer.) v. 1 t. 16.' In this case, the credits also include research prompts for Mutis and naturalists in general, as in *Begonia urticae*: '*Begonia* species are still very unknown and must be described by an American botanist.' *Psychotria emetica* illustrates the case of a chain of credits established by Mutis himself:

> my great friend Catotz [¿?], Governor of the province of Girón, sent me a flower sample, diligently collected in Cañaveral, next to the mouth of the Magdalena River, in an attempt to determine with certainty if it is the true ipecacuana, as was believed by the testimony of an empirical observer.[53]

When Smith published 23 plants of Mutis's in the herbarium of Linnaeus, he identified authorship and specified his sources, as in *Calceolaria perfoliata*:

> my representation was drawn from the specimen of the herbarium of Linnaeus. The parts related to the flower are taken from the Mutis drawing. I would have liked to represent them from the dry plant, but it lacks sufficient details to do so.[54]

In this sense, the case of *Castilleja fissifolia* is also eloquent:

> Mutis made the description with insufficient care and Linnaeus did not understand it well, mainly in relation to the lower lip of the corolla, which in our image is painted in a much more careful way than in the representation of the plant sent by Mutis (it should be known that the flower was submerged in boiling water, and with this it managed to expand).[55]

Finally, by publishing the *Caryocar amygdaliferum*, Cavanilles implicitly acknowledged his role as mere editor of the 'illustrious José Celestino Mutis, who observed, described, and took care to draw it with vivid colours.'[56] These examples illustrate contemporary practices of recognition and citation, on which the very possibility of access to new materials depended, at a time when cabinet botanists were the subsidiaries of the collaboration of correspondents and collectors located in different parts of the world.

From his arrival in Paris, Humboldt began publishing his journey although to be sure many elements had preceded him. Long before his return, a good number of his letters with information on the content of the 'Mutis collections' had already been published in various European capitals. A copy of his letter to Delambre of November 25, 1802, was published in 1803; the letter addressed to Cavanilles on April 22, 1803, appeared in Madrid (1803) and in Paris (1804) shortly after the return of the travellers and in London (1805). The letter to the IF dated June 21, 1803, was published in Paris in 1804. Thus, relevant information was known to the European scientific community well before 1805, when Humboldt began to publish his *Voyage aux régions Equinoxiales du Nouveau Continent, fait in 1799, 1800, 1801, 1802, 1803 and 1804* (1805–1834), in thirty-five volumes. To assume this community was composed solely of naturalists would be to ignore the editors and institutional actors, which included astronomers, for example, at a moment that did not know the specialisations or disciplines of our age. It is precisely in this context of 1805 that an enigmatic passage from a letter from Humboldt to Bonpland appears:

> As for the Satire that they say is taking shape, it may be that it is a good invention of M. Zea. It should be avoided; although, in joking, if that were the case.... I would be most pleased if M. Zea translated it for me, although I would settle for a look at the proofs. It would be unwise to send you the pages, since no one will be able to notify you, given the lentitude of the Spanish.[57]

Had the Second Professor of the Royal Botanical Garden of Madrid caricatured the originality of the first run of Humboldt's botanical work? The truth is that when the two volumes of *Plantes equinoxiales* appeared in 1808–1809, the coronation of José Napoleón I (1768–1844) as the king of Spain and the Indies and the presence of Zea, unconditional supporter of the emperor, at the head of the Royal Botanical Garden, extinguished any fickleness on the part of Madrid to enforce its rights over the scientific results of the voyages to its American territories.[58] With the establishment of the French Empire, it is unsurprising that no criticism came to disturb the publication of those botanical works paid, by imperial decree, with the 'pension funds,' and advanced on a collection registered in the MNHNP, also by imperial will. Fortunately, today nothing calls for silence.

The period during which Humboldt's American works were published extends from 1804 to 1834. As noted earlier, searching for Mutis and *Flora de Bogotá* in Humboldt goes beyond the limits of this work. However, *Plantes équinoxiales*, excepting the plates of the quinas, barely alludes to the Mutis gift-collections or more generally to what their authors lived and learned in Santafé. The corresponding information consigned in the travel diaries and correspondence of each one and subscribed jointly appears non-existent. However, during the preparatory work, more precisely in 1805, Humboldt ordered Bonpland in these terms: 'Make a list of the people to be praised perpetually, and praise at the same time Née, Zea, Mutis, Cavanilles, Sessé, Pavón and Ruiz, Tafalla and Olmedo,' and he added, '[I]f you approve of the figure of old Mutis, I will place him somewhere in my work, since the issue is dedicated to him.'[59] Indeed, the dedication of *Equinoxial Plants* is accompanied by an engraving of Mutis with the following legend: 'Don José Celestino Mutis Director of the Botanical Expedition of the New Kingdom of Granada, Astronomer of the King in Santafé de Bogotá. A faint sign of gratitude and admiration. A. de Humboldt [and] Aimé Bonpland.'

Gratitude for what? Discounting the *Cinchona condaminea*, the reader is left disoriented.[60] Indeed, the inscription of the engraving has no relation to the content of the work except perhaps the preface, which appears intended to redeem Humboldt and Bonpland from possibly abusive self-attributions of the discovery of new genera and species, when they write:

> There is no doubt that many plants are in our possession that are preserved in the herbaria of our friends, Messrs. Mutis, Ruiz, Pavón, Cervantes, Mociño and Sessé. Having botanized in countries that enjoy a similar climate, it is natural that we have found the same plants. For us it will be a pleasant duty to recognize what we owe to these famous botanists; however, it will not be our responsibility if sometimes, ignoring their works, we give new names to genres for which they may have destined others, long before us.[61]

'Ignoring their works!' What happened to the notes taken in Mutis's cabinet? The full colour prints received as a gift, where did they end up? What happened to the information orally transmitted by Mutis? What about the consultations in the herbarium and the library in Bogotá? The publication of letters sent to Europe, with entire passages devoted to the 'Mutis collections,' was also relegated. Perhaps in this case the term 'ignorant' should be understood not to mean he who 'has no news of something' but he who 'has chosen to ignore.'

It becomes clear here that the topos of 'Mutis' is one thing in the American letters of Humboldt and Bonpland, and quite another in their published works in Europe. An example of the former is illustrated by a letter from Bonpland to Mutis dated in Riobamba on June 27, 1802. Here the agreements concluded in Bogotá are evoked:

> Since our separation, sir, how many times have we talked about you, both among ourselves and with persons you know! How many times have we taken up the immense works that you are preparing for posterity; this especially when we remember Quindío ... and in the soil of Quito, where at every step we have found genera and species of the immortal *Flora de Bogotá*, which has reminded us of the discoveries of a friend as generous as he is kind.[62]

An example of the latter is illustrated by the preface to *Plantes equinoxiales*, where Humboldt and Bonpland place Mutis as one naturalist among others, without revealing that they had examined the material and intangible heritage of the Botanical Expedition. Now, they cite his name in a negative sense in two of the three cases in which it appears in the work; thus, 'Mutis … could not penetrate the Andes of Quindío,' and 'this palm is found on the peaks … of Quindío, places where Mutis did not carry out his wise investigations.'[63] This, in short, is how a collective and lasting amnesia is constructed.

Contrary to Humboldt's perception, not everyone in the Spanish overseas colonies lived 'on the moon.'[64] At the beginning of 1810, Jose Caldas, then in charge of the astronomical work of the Expedition, and familiar with at least the Preface to *Plantes equinoxiales*, wrote in the February 10 edition of the newsletter *Continuación del Semanario del Nuevo Reino de Granada* the following:

> The delays have been disastrous for the *Flora de Bogotá* such that [Nikolaus Joseph von] Jacquin [*Selectarum Stirpium Americanarum* (1763)], the Flora of Peru [Ruiz y Pavón], that of Mexico [Sessé and Mociño], Nee, Haenk, Humboldt, have snatched a part of her riches …; it is about foreigners completing their conquests of the *Flora de Bogotá* … It may be that Humboldt and Bonpland publish some genera that are common to the *Flora de Bogotá* and their *Plante[s] Equinoct[x]iales*, but the fact that Mutis made his discoveries long before the arrival of these travellers; his complete descriptions; his superb plates; his long and careful observations; all this gives him a decided superiority over what Humboldt and Bonpland can publish.[65]

Caldas is direct in his criticism but notably he also reveals his ignorance here, suggesting that Humboldt *voluntarily* 'snatched' a part of the *Flora de Bogotá* while publishing 'some genera that are common to the *Flora de Bogotá*.' Caldas, who since Mutis's death on September 11, 1808, was co-director of the Casa Botánica, appears to believe Humboldt and in so doing reveals, as was the case with Sinforoso Mutis as well, his ignorance of the gift and the mutually agreed upon commitments it had entailed. The editor of *Continuación* was well aware that the newsletter circulated in France,[66] where his criticism was destined to interpolate Humboldt and Bonpland.[67] Be that as it may, the piece did not elicit a reaction, and those who subsequently took up the matter, including Mariano Lagasca Segura (1776–1839) and Antonio Lorenzo Uribe Uribe (1900–1980), appear to ignore Caldas's protest.

Caldas does not limit himself to the issue of the precedence of the *Flora of Bogotá* over European publications. He revives and radicalises the complaint aired during Humboldt's visit nine years before: the king had been wrong not to assign a national savant to the Prussian's expedition. In the New Kingdom of Granada, this error had to be corrected. Humboldt's recent publications revived the slight Caldas had experienced when Humboldt refused to allow him to join his entourage. The slight looks all the more violent when we consider that Mutis had intervened in the matter, believing that his gift to Humboldt would tip the balance in Caldas's favour. Caldas now calls for the appropriation of the country's knowledge by and for the people of New Granada. His claim has a political character that affirms the ownership claim of the native-born on their territory and its wealth: 'it is about foreigners completing their conquest of the *Flora of Bogotá*.' By raising the primacy of a native-born New Granada identity, Caldas also rejects the transatlantic Spanish 'national' identity, that is, the nation that embraces Spain and the American kingdoms. Now, Caldas challenges the passports granted to foreign travelling naturalists, including in the category of foreign non-American or peninsular Spaniards. Furthermore, its considerations are intended to be retroactive, as we have noted, applying equally to 'Jacquin, the Flora of Peru, that of Mexico, Nee, Haenk and Humboldt.'[68] On the eve of the first independence, there is little doubt that Caldas declares the work of Mutis to be the property and instrument of glory of New Granada; he promotes Mutis in order to affirm the sovereignty of the people over its territory.

In 1827, Lagasca, a disciple of Cavanilles, enters the scene. His appearance may be explained by his knowledge of the iconographic work of Mutis. The political background also plays a role here. Humboldt and Lagasca had known each other since 1799. In 1809 the Prussian supported him, albeit in vain, before Joseph Bonaparte for the post of Director of the Royal Botanical Garden. Lagasca only accepted the appointment upon the restoration of Fernando VII in 1813.[69] From a young age, Lagasca had been interested in the collections of Santafé, which he undoubtedly came to know in the publications of the Linnaeuses, father and son, and those of Smith. More direct information came from the mouth of Zea, whose testimony Cavanilles collected in 'Materials for the History of Botany,' published in the *Anales de Historia Natural de Madrid* in 1800.[70] 'Materials' indicates that Mutis 'has finally managed, after forty years of work, to complete the Flora of New Granada, which today consists of four thousand drawings, and as many descriptions.'[71] At the head of the Royal Botanical Garden, Lagasca fought the diplomatic battle that led to the transfer of the cabinet of the Bogotá Botanical Expedition to Madrid on October 3, 1817. Since that moment, he was responsible for its installation, arrangement, conservation, inventory, study, and publication attempt, before leaving Spain for political reasons in 1823.

From his exile in London Lagasca went to Humboldt, presenting himself as the author of the 'rescue' of the materials of the *Flora de Bogotá* from the hands of Creoles and foreigners who had tried to appropriate them.[72] In his capacity as the first Spanish curator of the materials of the *Flora de Bogotá*, his words are those of the embattled expert. He narrates the travails he suffered thinking about the loss of Mutis's works after his death, the steps taken to 'deposit them' in the Royal Botanical Garden, and the sleepless nights that its inventory had required. He points out that 'of the first six thousand [images], half are in black and white, and the other half are magnificently coloured; … It will be seen that the six thousand represent some three thousand different species of plants.' He claimed that Mutis prepared numerous volumes and, relying on the figures communicated to him by Zea, confirmed the disappearance of numerous descriptions.[73]

In Humboldt's eyes, these statements called Lagasca's expertise into question. Indeed, during his visit to Santafé Humboldt had perceived, as we have seen, the non-existence of a true publishing project for Mutis combined with an intimate attachment to his adopted home in the New Kingdom. In addition, apart from the information relating to the cinchona recorded by Bonpland in his *Diary*, Humboldt made the following critical note: 'M. Mutis possède près de 2000 dessins folio +, chaque plante une fois colorée et 2 fois en noir donc in triplio. Il compte qu'un des exemplaires noir doit rester à S[anta] Fé.'[74] Without having yet read the biography of Mutis by Humboldt, and ignoring the collection of plates that Mutis had presented to him, in his subsequent letter Lagasca shows himself more direct, and sure of the sharpness of his eye. He snaps at Humboldt:

> I close this letter, assuring you that I am firmly persuaded that several of the plant drawings you published in *Plantae aequinoctiales* and *Monographia Melastomae et Rhexiae* are copies of those of the *Flora de Bogotá*, although generally more or less cropped to accommodate the dimensions of the work.[75]

As in the case of Caldas, although this time on the side of the Peninsula, 'national' pride prevailed. Lagasca emphasises that Mutis is Spanish, like himself, and that Humboldt's glory owes much to the generosity of Spain. For this reason and to do justice to the memory of Mutis, he demands that Humboldt publicly acknowledge his debt to the savant from Cádiz,[76] a request that will remain a dead letter. However, despite the imprecision of his arguments, today everything indicates that his suspicion was correct.

The critiques of Humboldt's publications by Caldas and Lagasca are undoubtedly nourished by a vehement patriotism that, to some extent, coloured their respective scientific analyses. The Peninsular and the Creole, since independence now belonging to different sovereign nations, at turns display signs of isolationist if not derogatory sentiments. However, their criticisms also had scientific pretensions and are expressed through more circumspect observations. These observations can and should be examined by a research project consisting of a team of experts.

Such measured observations are found in volumes 30 and 31 of the *Flora of the Royal Botanical Expedition of the New Kingdom of Granada*, for example, published by Uribe Uribe in 1976 and 1983, and dedicated to melastomatáceas. These have been consulted bearing in mind the testimony of Lagasca and the unconscious paraphrasing of certain historians. Notably, nothing in these publications indicates that Uribe Uribe was aware of the collections offered to Humboldt and Bonpland, despite the fact the donation has been a topic repeated ad nauseam in the historiography of Mutis and his expedition. The Colombian botanist repeatedly expresses his perplexity at what he describes as 'strange similarities' between plates of Mutis and those described and represented by Humboldt and Bonpland in *Monographie des Melastomacées* that appeared in 1816 (v. 1) and 1823 (v. 2 Rhexies).[77] The disappearing gift continues to haunt science.

Because Words Always Contain Truths

Our archaeology of the collections of Mutis gifted to Humboldt and Bonpland grants primacy to the traces of Mutis's conversations with his guests and the guided access he granted them to his cabinet. In addition, it considers attributes of the gifted objects themselves: mutual linkage, quality, quantity, beauty, originality, size, scientific novelty, and precision. It should not be forgotten that Humboldt dedicated the first edition of *Equinoctial Plants* to Mutis and paid him justice when he shared with him the authorship of the report 'Account of the Cinchona Forest of South America.' Here, Humboldt stipulates that 'the diagnoses that I have added have not been taken from published works, but are the product of my own observations, on the one hand, and of the instructive conversations I had with Mr. Mutis, on the other.'[78] Second, as we have already mentioned, the director of the expedition allowed Humboldt and Bonpland privileged access to his cabinet. In this regard, a statement of singular importance is preserved: Mutis 'has compared his plants with ours and, finally, has allowed us to take all the notes that we wanted to obtain on the new genera of the flora of Santafé de Bogotá.'[79] The conjugation of the verb in 'we wanted' is significant, because it suggests the growing interest of the visitors in the collections of the Expedition, based on a retrospective appreciation of their discussions with Mutis. The adverbial 'finally' suggests that the desired access was made possible by the triumph of a reciprocal seduction. Such a reading provides insight into Mutis's expectations when he made the gift to Humboldt and Bonpland.

In the collections and the manuscripts, the travellers found empirical support that allowed them to better understand the statements, proposals, suggestions, intuitions, and criticisms that Mutis had been sharing with them aloud. Reducing the cabinet of the Botanical Expedition to a microcosm of the nature of the Viceroyalty would ignore the importance of the Mutis library, which the visitors profitably consulted and compared, no less, with that of Joseph Banks (1743–1820) in London. 'We have,' write Humboldt, 'compared our herbaria with those of M[onsieur] Mutis, and we have consulted many books in the immense library of that great man.' The primary purpose of the comparison was to establish the respective priorities. On this point there can be no mistake. The terms with which Humboldt reassured Delambre are eloquent, assuring him that 'we are persuaded that we have new genera and new species.'[80] This he was able to establish only after comparing his collections with those of Mutis, assuming due consideration to the achievements and discoveries of the naturalist from Cádiz. For the rest, the passage reveals that Humboldt's fears had already subsided. As was customary at the time and place, Humboldt and Bonpland show respect for the collection and its graphic representation, no matter how voiceless they may seem, for the handwritten descriptions and, naturally, for the more or less successful classification.

Humboldt consulted in detail the iconography of the Expedition, which was in its culminating moment: volume, format, quality of the paper, style, scientific content, and thematic diversity. These plates were of superior quality to anything produced in Europe on the subject in the early 1800s. If we consider that the Prussian lacked experience in publishing botanical manuscripts, it is likely that he acted like a sponge. He consigned his analysis in the semblance of Mutis that he published in *Biographie universelle* that began to circulate in Paris on the early date of 1821, being reprinted there in 1846. The corresponding substrate is none other than the notes that he took during his stay in Santafé. The text establishes the pre-eminence of the Mutis plates, which Humboldt recognises as 'the most luxurious and largest-scale collection of plants ever executed.' Appreciation was equivalent to confessing that once having seen Mutis's iconography, one was condemned never to forget it and to try to imitate (or copy) it. Humboldt's interest in this iconography is proportional to Paris's financial and technical limitations in botanical illustration.[81]

In the dissections of floral details reproduced in the Mutis plates, the visitors gained a thorough botanical knowledge of the seasonal cycles of hundreds of plants. As is known, Mutis formed his herbaria and built his descriptions and iconographies with information progressively gathered from specimens observed in different stages of growth, foliage, flowering, or fruiting. Sometimes his mastery involved a study of the evolution of the flower over the necessary timespan. In different plates he represented stages of development of a flower until it became a fruit, which means that, more than an expedition in the literal sense of the term, Mutis directed a botanical institution analogous to the Parisian *Jardin des Plantes*.[82] Indeed, this genre of comparative illustration-based study was unimaginable on expeditions of short or very short stays in different locales. Thus, the reserve of the old Mutis toward travellers who could only observe specific moments in the development of the plants.

Mutis is known to have gifted Humboldt and Bonpland with 'a large quantity of dried specimens,' which Bonpland confirmed by referring to 'the plants that [Mutis] gave us in the last days of our delay in Santafé.' This collection contained scientific information of the greatest importance, hence the determination of the travellers to take it with them on their itinerary to Popayán, Quito, and Lima, and to refrain from sending it to Europe with the remaining gifts from Mutis which, due to the quality and beauty of the full-colour plates, were intended to wow the MNHNP naturalists. In this way, the travellers enjoyed the privilege of having a magnificent collection of equinoctial plants during the rest of their pilgrimage through the Andes. 'Some' of these plants were 'determined by the names' written in Mutis's hand, and all went to enrich Bonpland's 'little herbarium.'[83] It should be noted that here Mutis behaves exactly as he had done with Linnaeus from 1760 until the latter's death in 1778, communicating materials many times without classification or graphic representation. In the first place, Bonpland's letter to Mutis, dated from Popayán on November 26, 1801, reveals the existence of an epistolary exchange between Mutis and the travellers after their departure from Santafé.[84] This letter systematically alludes to a 'we' that denotes the sender's secretary. Under these circumstances, the content responds to the concerns of the latter, related to his future publications. The author does not cease to shy away from the teacher ('my bad descriptions') and to promise him samples ('of the plants that seem rarest to me, chosen from our collections to send to you'), barely veiling his insatiable thirst for information on the *Flora de Bogotá*.

> I hope that before our departure for the Philippines I can clear up the many doubts I have about most of the plants that we are collecting every day, and about a countless number of plants that form part of the immortal *Flora de Bogotá*.

It is evident that he reaffirms the commitment signed in Santafé ('our intention is not to steal'), observing immediately after: 'there is such a great analogy between the vegetation of Purasé and that of Santa Fé and Quindiu, that undoubtedly on this trip we will have described many of your plants,' which is nothing more than the argument that Humboldt uses in the preface to *Equinoctial Plants* to wave aside any suspicion of abuse in the attribution of the title of discoverer of certain novelties.[85] By integrating the plants Mutis identified with his 'little herbarium,' Bonpland undoubtedly identified them with a number from the series from his own collection. In these circumstances, he probably destroyed the order of the collection of samples offered by Mutis, perhaps compromising the reestablishment of provenance. Thus, for example, he writes here:

> We have found the plant n. 2017 on our journey to the Desert of Purasé at an altitude of 2000 t[oesas]. We had taken it as a new genus and then, upon opening the plants of Santa Fé, we found it there: that plant cannot be anything other than a new genus of the *Flora of Bogotá*, or at least a new species, which in any case belongs to you. … We request of you the name that you have given to this plant … and a description of the leaves.

For us, it is impossible to know to what number Bonpland refers: the number attributed by Mutis to the gifted collection, the number in the herbarium of the *Flora of Bogotá*, the number in Bonpland's 'little herbarium,' or the number in his *Register of Botanical Notes* taken during his itinerary. However, the reference numbers Humboldt and Bonpland assigned to the collections made in the present territories of Colombia and Ecuador between May 1801 and March 1802, ran from 1592 to 2257. Prudence is advised until an investigation of the Bonpland documents, kept in the Manuscripts and Archives Section of the *Bibliothèque Centrale* of the MNHNP, establishes the series of numbers applied to the plants collected during their trip to the interior of the New Kingdom. For now, it can only be said that the numbers that Bonpland cites in his communication to Mutis (1833, 1859, 1860, 1906, 1911 and 2017) belong to this series. However, Bonpland also cites the number 1982, which does not appear in his *Registre des notes botaniques*. Indeed, in Manuscript 2534, the number 1919 is followed by the number 2000, with no trace of theft of folios. Given that the diagnoses lack data and that the series of manuscripts conserved in the MNHNP is a continuous series, it seems difficult to determine if the numbers 1920 to 1999 existed, or if they were reserved for another notebook not yet located. Secondly, in their communication to the IF dated June 21, 1803, Humboldt and Bonpland announce the dispatch of 'two boxes accompanied by letters,' here called the first box and the second box. The first contained a 'work on the genus cinchona,' which Humboldt also calls 'a curious collection of the quinas of New Granada.' The information on these 'seven species' of *Cinchona* established by Mutis was represented by 'dissected skeletons,' 'beautiful specimens of flowers and fruits,' 'herbarium samples in flowers and grains,' 'bark samples,' 'well-chosen barks,' 'coloured drawings … with the anatomy of the flower so different from the stamina,' 'illuminated drawings of seven species of Chinchonas,' and 'magnificent illuminated drawings of folio size.'[86]

As can be seen, this 'work' does not claim to include any manuscript by Mutis on the cinchona. However, related information of the first importance is given by Bonpland in his *Registre des notes botaniques*, the result of the exchanges that he and Humboldt had with Mutis in relation to his knowledge of the cinchona (ff. 91–95). The information in question is an introduction that gathers the history of the discovery by Mutis of the cinchona tree, together with his work on the *Cinchona* species in the New Kingdom of Granada. Unnumbered, the manuscript is preserved in the plant diagnosis series of his *Registre*. There it can be read:

> Mr. Mutis was the first to establish the generic character of the *Cinchona*; he examined a large number of species from Quito, Santafé, and Santa [Marta]; for this reason, it is he more than any other botanist who is in a position to distinguish the following characteristics …

This is followed by a series of descriptions of the specific characteristics of different species of *Cinchona*. At this point, only a comparative reading by an expert could establish whether these notes transcribed by Bonpland were taken up in whole or in part in Humboldt's publication, *Account of the Cinchona Forest of South America*.

The content of the second box has attracted much attention. It is described as follows: 'More than sixty drawings of plants in beautiful colours, made by its best painters, Matís, Rizo, Cortés f [the First, without a doubt].'[87] 'About one hundred magnificent drawings in large folio, representing new genera and new species of [the] manuscript *Flora de Bogotá*.' 'One hundred folio-size drawings, representing new genera and new species of the flora of Bogotá.' 'More than 100 illuminated drawings.' 'He [Alexander von Humboldt] tells me that he sent to Messrs [Antoine-Laurent de] Jussieu and [André] Thouin a collection of … 120 *in-folio* plant drawings, which Mutis gave to him …'[88]

Although the quoted passages are generally equivalent and almost never contradictory, some clarifications are in order. First, botanical descriptions are conspicuous by their absence in this reconstruction. In this sense, the gift confirms the certainty that encouraged Mutis, namely, that it was perfectly possible to describe and classify plants from their plates, especially when they were delivered under the conditions that have been indicated. For his part, 'Humboldt, touched by this unexpected degree of perfection' in the Mutis plates, shared this criterion, to the point of assuring Caldas 'that the brush has rendered the descriptions unnecessary, and if the manuscripts were to be lost, Jussieu, or another skilled professor, could describe the plant as perfectly as if he saw it alive.'[89]

One should avoid feeling perplexed at the differences in numbers in the citations on the plates in the second box: 'over sixty'; 'about a hundred'; 'One hundred;' 'more than 100;' '120.' This simply reflects the informal nature of the entry in Humboldt's *Diary* dated shortly after September 8, 1801. It is unlike his direct communications to the IF, although these are not always rigorously equivalent in this sense either. It is clear to Humboldt and Bonpland that the principal attributes of the collection of plates are its novelty and beauty. The first fruits of a generic and specific range of the 'Bogotá manuscript flora' are confirmed over and over again in their communications to the IF ('about one hundred magnificent large-folio drawings, representing new genera and new species of their manuscript, *Flora de Bogotá*, 'one hundred folio-size drawings, representing new genera and new species of the flora of Bogotá').

In accordance with the terms of these letters, Mutis presented the visitors with a select sample of his iconography so that Europe could see in the publication of Humboldt and Bonpland the scientific and aesthetic quality of his work. The delivery of signed plates is striking, which reflects Mutis's 'desire to promote American talent and reinforces the idea that the concerted publication should respect the painters' credits. The amount given also matters. This would represent between 3.3% and 5.3% of the total of 'drawings already finished' around 1801 by the Expedition, according to the calculations published by Humboldt himself: '30 painters have been working for Mutis for 15 years; he has 2,000 to 3,000 in-folio size drawings, which are miniatures.'[90] These are large folio originals of which Mutis undoubtedly kept at least the two corresponding monochrome copies. It should also be noted that these are finished originals, to the extent that apparently nowhere is there any reference to the auxiliary strips in which the cut-outs of the flower or fruit had been represented before integrating them into the lower part of the original. Finally, a subsequent archaeology in Mutisian iconography will have to specify whether between September 8, 1801, and June 1802, when the boxes were dispatched by Mutis, the Office of Painters produced illuminated copies of the plates selected for the IF.[91]

Final Considerations and New Challenges

How to conclude a study whose objective is none other than to urge that it be continued? It must be said, at the risk of insistence, that this chapter has never wanted to insinuate that the American botanical works of Humboldt and Bonpland are invaded by the work of Mutis. The question raised here simply asks whether the space reserved for Mutis by Humboldt and Bonpland is proportional to his contribution to the knowledge that the Prussian and the Frenchman acquired in Bogotá. In establishing the practices of those who previously edited plant materials and information provided by Mutis, it is evident that at the time, publication did not in any way or means establish priority. To conclude as much would be to anticipate anachronistically what is the coin of the realm among no small number of scholars today: 'publish or perish.' After having reviewed the genesis of the Humboldt–Mutis–Bonpland meeting and having reconstructed its development, the interactions it engendered are now clearer. Nevertheless, essential aspects remain to be investigated, such as the transformation of Humboldt's projects, or the reforms introduced in the Botanical Expedition in Bogotá after his visit.

Despite having documented the receipt of Mutis's gifted collections by its final recipients at the MNHNP (Jussieu, Lamarck, and Desfontaines) the fact is that it has not been possible to trace the whereabouts of most of the collection. The silence that the scientific community of the time reserved on this matter is striking, since in previous decades it had repeatedly expressed impatience to learn about the work of Mutis. Certainly, it will be necessary to continue reading more closely the works of Humboldt and Bonpland, in order to establish the place occupied by Mutis, even if it is only the visible part of the iceberg. If this chapter has offered anything new to the debate, it will have been to suggest that an archaeology of scientific objects submerged in the publication projects of Humboldt and Bonpland remains a project worth pursuing.

Notes

1 The initial phase of the research for this chapter corresponds to a project presented with Beatriz González Aranda, prepared with funds from Project No. 805176 of the Bogotá-DIB Research Division of the National University of Colombia. During this phase, the students of the Methods course in the Department of History at the National University of Colombia, 2010 edition, participated. Later, the ecologist Ananay Arango Matiz made the botanical checks that appear in this study. Finally, the preliminary results of this study were presented in the series *Viajeros por Colombia: Pasado y presente*, organised by the Gilberto Alzate Avendaño Foundation in September 2011. A preliminary version of this chapter appeared under the title 'Como débil muestra de admiración y gratitud. José Celestino Mutis en la Obra de Humboldt y Bonpland. Estudio preliminar,' in Olga Restrepo, ed. *Proyecto Ensamblado en Colombia*, v. 1 of *Ensamblando estados*, (Bogotá: Universidad Nacional de Colombia, year), 77–100. Finally, this work was presented at the international symposium of the LAGLOBAL network, *The Invention of Humboldt*, FLACSO, Quito, August 12–13, 2019. The Royal Botanical Garden of Madrid kindly authorised reproduction of plates 2542 and 2553 of the digitised drawings of the Royal Botanical Expedition of the New Kingdom of Granada (1783–1816), directed by José Celestino Mutis. The Manuscripts and Scientific Archives Service of the Bibliothèque Centrale du Muséum National d'Histoire Naturelle (Paris) allowed us to reproduce 6 pages of MS 2534, thanks to the management of Monsieur Michel Lille, who acted at the request of the author. The content of this article was discussed regularly with Monsieur Lille, who also purveyed bibliographic material from Paris. I also wish to thank Dr. Javier Fuertes Aguilar of the Real Jardín Botánico-CSIC, Madrid for bibliographical suggestions.

2 *Dictionnaire critique de biographie et d'histoire*, 1864.

3 José Jerónimo Triana, *Nouvelles études sur les quinquinas: d'après les matériaux présentés en 1867 à l'Exposition Universelle de Paris; et accompagnés de facsimile des dessins de la quinologie de Mutis suivies de remarques sur la culture des quinquinas* (Paris: F. Savy, 1871), 2.

4 José Antonio Amaya, *Mutis, apóstol de Linneo. Historia de la botánica en el virreinato de Nueva Granada, 1760-1783* (Bogotá: Instituto Colombiano de Antropología e Historia, 2005).

5 Charles Minguet, ed., *Cartas americanas*, trad. Marta Traba (Caracas: Italgráfica, 1993 [1980]), 150–151.
6 Doña Ivonne Eiseley de Ojeda (1905–2011), founder of the Association of Friends of José Celestino Mutis (Madrid) was not successful either nor was Don Félix Muñoz Garmendía.
7 'When I embraced you for the last time on the Rue Helvétius in Paris at the moment when I was thinking of leaving for Africa and the Great Indies, I had only a faint hope of seeing you again and sailing under your orders.' Humboldt to Baudin, Cartagena de Indias, April 12, 1801, in Minguet, ed., *Cartas americanas*, 80.
8 Minguet, ed., *Cartas americanas*, 248–249.
9 In this regard, Humboldt notes in his *Diaries*:

> I announced [to Mutis, on the way to Santafé, in the middle of 1801] some new genres by Schrebersche and Swartzische, those that we had seen in the Rio del Magdalena and of which we could assume that he would hardly know them by name.

Colombian Academy of Exact, Physical and Natural Sciences and Academy of Sciences of the German Democratic Republic, *Alexander von Humboldt in Colombia: excerpts from his diaries / Alexander von Humboldt in Kolumbien; Auswahl aus seiner Tagebüchern* (Bogotá: Publicismo y Ediciones, 1982), 109. Johann Christian Daniel von Schreber (1739–1810), a disciple of Linnaeus, had dealt with (1789–1791) the eighth edition of his teacher's *Genera plantarum*. Olof Swartz (1760–1817) had published *Florae Indiae Occidentalis* in 3 t. (1797–1800).
10 Minguet, ed., *Cartas americanas*, 130.
11 Alexander von Humboldt to Wilhelm von Humboldt, Contreras de Ibagué, Nuevo Reino de Granada (4°, 5′ de latitud norte), 2 September 1801, in Minguet, ed., *Cartas americanas*, 83–84.
12 José Antonio Amaya, 'Una Flora para el Nuevo Reyno. Las relaciones de Mutis y sus colaboradores con la botánica madrileña (1790-1808),' in Diana Obregón Torres, ed. *Culturas científicas y saberes locales: Asimilación, hibridación, resistencia* (Bogotá: Universidad Nacional, 2000), 108.
13 Cavanillas' *Icones* was the most outstanding botanical work of the Spanish Enlightenment in content and extension, and for the illustrations of plants that appear there.
14 Amaya, *Mutis, apóstol*, v. 2, 607; María Pilar San Pio Aladrén y Paloma Collar del Castillo, *Catálogo del fondo documental José Celestino Mutis del Real Jardín Botánico* (Madrid: Real Jardín Botánico-CSIC, 1995), 174, entrada 1538.
15 'I will have received twenty visits from the most famous botanists in Europe, no more, because they know that I have seen and know your grace; Mr. Smith, in particular, an Englishman, from the Royal Society of London, has told me that he has written to you for a long time and awaits your reply with the greatest desire … he wishes, it seems, to verify with you the part of the herbarium of [Linnaeus] in which your merciful figure appears as the star among the stars.' Jean Baptiste Le Blond, n/p, n/d, in Guillermo Hernández de Alba, ed., *Archivo epistolar del sabio naturalista don José Celestino Mutis* 2a ed., 4 v. (Bogotá: Instituto Colombiano de Cultura Hispánica, 1968–1975), v. 4, 19. The chronic and topical date of this letter could be deduced from the one that Le Blond himself, *médecin naturaliste du roi*, sent to Juan Jiménez, a merchant in Santafé: 'the one included for Dr. Mutis, I trust he will send it to him forthwith,' Paris, October 16, 1786. Archivo General de la Nación, Colecciones, EOR, caja 122, carpeta 7, f. 3–4, electronic communication from Professor Daniel Gutiérrez Ardila, June 16, 2011.
16 José Antonio Amaya, 'Cuestionamientos internos e impugnaciones desde el flanco militar a la Expedición Botánica,' *Anuario Colombiano de Historia Social y de la Cultura* 31 (2004): 75–118.

17 The seldom cited Humboldt *Diaries* relative to the New Kingdom abound in allusions to Mutis's station in the cultural panorama of Santafé, a subject scarcely dealt with in studies on the New Granada Enlightenment. The following fragment shows how Humboldt took advantage of Mutis's contested authority in the viceregal capital for his own purposes:

> In reality, I had to flatter the old man not just a little, making the public of Santa Fé see that a man from the far north had come solely to visit a scholar whom a part of that public treated with affected indifference. Hence the unlimited friendship, the sacrifices, and expenses that Mutis lavished on us. It was in his own interest; it was in the interest of his party... Mutis sent his secretary [José María Carbonell (1778–1816)] to Facatativá in order to show me once more that he was coming out to receive me through his friends and that he would not let me be seduced by others or accept any house other than his own.

Academia Colombiana, *Alexander von Humboldt en Colombia*, 109–110. A similar interpretation of these fragments is found in Olga Restrepo Forero, 'Naturalistas, saber y sociedad en Colombia,' in O. Restrepo, L. C. Arboleda y J. A. Bejarano, eds., *Historia social de la ciencia en Colombia*, v. 3, *Historia Natural y Ciencias Agropecuarias* (Bogotá: Instituto Colombiano para el Desarrollo de la Ciencia y la Tecnología Francisco José de Caldas, Colciencias, 1993), 72–73.

18 As for the headquarters of the Botanical Expedition, it occupied one of the 'best and largest [houses] in this capital.' JC Mutis to Ignacia Consuegra, Santafé, October 14, 1791. Hernández de Alba, ed., *Archivo epistolar*, v. 2, 64.

19 Academia Colombiana, *Alexander von Humboldt en Colombia*, 111.

20 'The man who knew how to display such amazing activity during forty-eight years of work in the New World was endowed by nature with the most fortunate physical constitution. He was of tall stature, noble features, grave bearing, easy and polite manner. Their conversation was as varied as the object of their studies. He often spoke with passion, although he also practiced the art of listening, which Fontenelle held in such high esteem, and was already considered a rarity in his day.' Alexander von Humboldt, 'Biografía de José Celestino Mutis,' trans. José Antonio Amaya, in Frank Holl, ed., *El regreso de Humboldt* (Bogotá: Ministerio de Cultura de Colombia and Embajada de la República Federal de Alemania, 2001 [1821]), 59–65.

21 Hernández de Alba, ed., *Archivo epistolar*, v. 1, 116–120.

22 Alexander von Humboldt to Wilhelm von Humboldt, Contreras de Ibagué, in Minguet, ed., *Cartas americanas*, 85.

23 Humboldt, 'Biografía de José Celestino Mutis,' 59–65.

24 Francisco Martínez, dean of the Cathedral of Santafé, described the 'character' of Mutis as 'very reserved' and 'extremely touchy and almost inordinate in his fixation on scientific perfection.' Informe secreto al ministro Pedro Acuña, Santafé, 19 de mayo de 1793, in Enrique Pérez-Arbeláez, Enrique Álvarez López, Lorenzo Uribe Uribe y Eduardo Balguerías de Quesada, *La Real Expedición Botánica del Nuevo Reino de Granada*, v. 1., *Flora de la Real Expedición Botánica del Nuevo Reino Granada* (Madrid: Ediciones Cultura Hispánica, 1954), 131.

25 Francisco José de Caldas to José Ramón Leyva, secretario del virreinato de Santafé y juez comisionado para los Asuntos de la Expedición Botánica, Santafé, September 30, 1808. Academia Colombiana de Ciencias Exactas, Físicas y Naturales, *Cartas de Caldas* (Bogotá: Colciencias, 1978), 281.

26 Academia Colombiana, *Alexander von Humboldt en Colombia*, 109.

27 Humboldt will be welcomed by the IF as a Foreign Associate on May 14, 1810. Minguet, ed., *Cartas americanas*, 340.

28 Although he studied the collections donated to the MNHNP, Bonpland never joined this centre. On December 15, 1817, being established in Buenos Aires, he will be elected a corresponding member of the Academy of Sciences, Section of Botany. See *Rapport sur la proposition faite par MM. Humboldt et Bonpland de déposer dans les collections du Muséum des échantillons de toutes les plantes recueillies par eux dans l'Amérique méridionale*, January 1, 1805, signed by Jussieu, Lamarck, Desfontaines, in Ernest-Théodore Hamy, ed., *Lettres américaines d'Alexandre de Humboldt (1798–1807) précédées d'une notice de J.-C. Delamétherie et suivies d'un choix de documents en partie inédits* (Paris: E. Guilmoto Hamy, 1905), 232–233, and Imperial Decree of March 13,1805, in Minguet, ed., *Cartas americanas*, 257.

29 Walter Lack, 'The botanical field notes prepared by Humboldt and Bonpland in Tropical America,' Taxon 53 (2004), n. 2, 501–510. See also in this regard Humboldt's letter to the MNHNP Professors-Administrators, Sans-Souci, June 12, 1851: 'I must say that the true merit of the botanical work carried out during the course of the expedition [1799 to 1804] They do not belong to me, but to the courageous zeal of M. Bonpland.' Henri Cordier, 'Papiers inédits du naturaliste Aimé Bonpland conservés à Buenos Aires,' *Comptes rendus des séances de l'Académie des Inscriptions et Belles-Lettres* (1910), n. 6, 461.

30 A. von Humboldt to Wilhelm von Humboldt, Contreras de Ibagué, Minguet, ed., *Cartas americanas*, 85.

31 In *Magasin Encyclopédique: ou journal des sciences, des lettres et des arts* (1802), n. 13, v. 6, 238, we read: 'M. de Humboldt ne peut comparer la collection botanique de D. J. Celestino Mutis, qu'à celle de sir Joseph Banks de Londres.'

32 Humboldt is mistaken. The painters of the Botanical Expedition that he met were without exception creole, mestizo or mulatto; none are known to have been 'indigenous.'

33 Amaya, *Mutis, apóstol*, v. 1, 50–75.

34 Humboldt, 'Biografía de José Celestino Mutis,' 61–62.

35 In his letter to Wildenow, dated February 21, 1801, in Havana, Humboldt writes: 'I have already told you, my dear …, that I intend to publish my plants myself, after I return. However, if you found, in the two boxes that Frazer could send you, new species that catch your eye, you will of course have [them] available, to incorporate into your new edition of the species, just not too many or all of them. On the contrary, Bonpland and I will be very honored to be cited by you in your work. I say intentionally, 'not too many and not all,' because it is impossible to describe on dissected specimens what we have drawn out of nature. Minguet, ed., *Cartas americanas*, 74–75. It refers to the sixth edition of *Species plantarum* by Linnaeus, advanced under the direction of Wildenow (Berlin, 1831–1833).

36 'Magnificent are the paintings,' Linnaeus said to Mutis in his letter of thanks, dated May 20, 1774, for the 'First Collection' in which he had sent 19 plates of plants to the gouache of Chinese ink by Pablo Antonio García del Campo (1744–1814), among other materials. Hernández de Alba, ed., *Archivo epistolar*, v. 4, 27. 'When I saw the pictures, I was greatly admired that he could have very excellent draftsmen in America, superior to Europeans,' Pehr Jonas Bergius (1730–1790) said to Mutis, thanking him for sending 3 drawings of plants to Stockholm at the end of 1784. Amaya, *Mutis, apóstol*, v. 2, 506.

37 This is how it appears in the *Actes de l'Assemblée de l'Institut de France* in the early 1800s.

38 The author relationship in the so-called centre, correspondent in the so-called periphery is generally highlighted as an asymmetric relationship; however, here it is perceived that the correspondent, in this case Mutis, chose with whom to establish an epistolary relationship (Olga Restrepo, personal communication, 2011).

39 Amaya, *Mutis, apóstol*, v. 2; Miguel Ángel Puig-Samper, Luis J. Maldonado y Xosé Fraga, 'Dos cartas inéditas de Lagasca a Humboldt en torno al legado de Mutis,' *Asclepio* 56 (2004), n. 2, 75; Academia Colombiana, *Alexander von Humboldt en Colombia*, 76; Ulrike Moheit, ed., *Alexander von Humboldt: Briefe aus Amerika 1799–1804* (Berlin: Akademie Verlag, 1993), 235 and 245; J. C. Mutis to C. Gómez Ortega, Santafé, March 31, 1784, in Hernández de Alba, ed., *Archivo epistolar*, t. 1, 184.

40 Humboldt and Bonpland's correspondence with Mutis, after their departure from Santafé, consists of no less than fourteen letters, seven from Humboldt, five from Mutis, and two from Bonpland. No communication is known dated in Europe.

41 Mutis's letter to Humboldt requesting that Caldas join his expedition has not been found. However, a letter is preserved that allows its existence to be inferred:

> What is this my beloved Baron? What! Will a proposal made with the greatest sincerity and frankness be able to alter our constant friendship? Was it my fault that Caldas had become so enthusiastically fond of the illustrious Baron to think of following him through the two Americas? Could I have proceeded with greater frankness than that indicated in the expressions in my letter, and sending it open so that you may read the response and the bank draft for Caldas? And wouldn't it be my true intention to add a student that I thought would be to your liking? Break your silence my grace and continue to bestow your correspondence on your beloved friend as if such a thing had not happened.

Mutis to Humboldt, Santafé, May 21, 1802, in Moheit, ed., *Alexander von Humboldt*, 172.

42 The five are: Humboldt to Cavanilles, Mexico City, April 22, 1803; Humboldt to Carl Ludwig Willdenow, Mexico City, April 29, 1803; Humboldt to Georges Cuvier, Rome, May 28, 1803; Humboldt to IF, Mexico City, June 21, 1803; Humboldt to Delambre, Mexico City, July 29, 1803.

43 W. von Humboldt to Georges Cuvier, Rome, May 28, 1803. Hamy, ed., *Lettres américaines*, 224.

44 Federico Guillermo III to Humboldt, Potsdam, September 25, 1804. Minguet, *Cartas americanas*, 258.

45 *Rapport sur la proposition faite par MM. Humboldt et Bonpland*. Hamy, ed., *Lettres américaines*, 230–233.

46 Humboldt, 'Biografía de José Celestino Mutis,' 36.

47 Triana, *Nouvelles études sur les quinquinas*, 2.

48 José Antonio Amaya and Miguel Ángel Puig-Samper, *Mutis al natural; ciencia y arte en el Nuevo Reino de Granada* (Bogotá: Museo Nacional de Colombia, 2008), 62. The *Diary* is held in the Bibliothèque Centrale of the MNHNP.

49 Michael Dettelbach, 'La science omnivore d'Alexander von Humboldt: Peut-on défendre à l'homme d'avoir le désir d'embrasser tout ce qui l'environne?' *La Recherche* (1997), n. 302, 90. The author notes: 'As suggested by this passing threat, the climate of the Napoleonic Paris was not as favourable as one might think to the encyclopaedic Humboldt.' Paradoxically, while the influence of Paris spread throughout Europe, spirits there remained damp. François Guizot, at that time a young journalist, recalls in his *Memoirs* (1858–1867, v. 1, 7) the collapse of public life during the Empire: 'The drunkenness of 1789 had completely disappeared. … The aridity, the coldness, the withdrawal of the feelings and personal interests were the daily bread; France, tired of errors and strange excesses, eager for order and common sense, fell into this quagmire again.' See Guy Chaussinand-Nogaret, *Comment peut-on être intellectuel au siècle des Lumières?* (Bruxelles: André Versaille, 2011).

50 E. du Bois-Reynaud, 'Alexandre de Humboldt,' *La Revue scientifique de la France et de l'étranger* 32 (1883), n. 19, 183.

51 The idea of a 'scientific colonisation' is developed in Oliver Lubrich, 'The Secret of Composition: Humboldt's Kosmos as Postcolonial Space,' *Journal of German Philology*, 25 (2017), 25–44.

52 Dettelbach, 'La science omnivore,' 90. Duviols and Minguet date Napoleon's order of expulsion and the intervention of Chaptal to 1810. *Humboldt savant-citoyen*, 64.

53 Carl von Linné, *Supplementum plantarum systematis vegetabilium editionis decimae tertiae* [1774]: *Generum plantarum editionis sextae* [1764], et *Specierum plantarum editionis secundae* [1762–1763], *Editum a Carolo a Linné* [el Joven] (Brunsvigae: Impensis Orphanotrophei, 1782), 129, 145, 408, 420.

54 James Edward Smith, *Plantarum icones hactenus ineditæ, plerumque ad Plantas in Herbario Linnæano conservatas delineatæ*, 1 tomo en 3 fasc. (Londini: J. Davis and Parisiis: Lud. Nic. Prevost, 1789–1791), fasc. 1, icon 4.

55 Smith, *Plantarum*, fasc. 2, icon 40.

56 Ant. Iosephi Cavanilles, *Icones et descriptiones plantarum: quæ aut sponte in Hispania crescunt, aut in hortis hospitantur* (Madrid: ex regia Typographia, 1797–1798), 38. Translations from the Latin of the quotes were prepared by José Noel Olaya and Jorge Arango.

57 Minguet, ed., *Cartas americanas*, 149.

58 Zea represented the intruding government as mayor of Malaga. When the Napoleonic troops left Spain, he marched to Paris with them. From Paris he went to the Antilles and from there to the Orinoco to join Bolívar in 1816.

59 *Gazette Nationale* ou *le Moniteur Universel* 2, n. 326 (14 de agosto de 1803), 1445–1477; *Annales du Muséum National d'Histoire Naturelle* 11, v. 2 (1803), 170–180; *Magasin Encyclopédique: ou journal des sciences, des lettres et des arts* 8 (1803), t. 6, 537–549, y en Berlín: *Gilbert's Annalen der Physik* 16 (1804), 475–489. Minguet, ed., *Cartas americanas*, 149–151; Hamy, ed., *Lettres américaines*, 190–195.

60 Humboldt and Bonpland, 1808: 29–34 and 28–29, drawing.

61 Alexander von Humboldt et Aimé Bonpland, *Plantes équinoxiales recueillies au Mexique: dans l'île de Cuba, dans les provinces de Caracas, de Cumana et de Barcelone, aux Andes de la Nouvelle-Grenade, de Quito et du Pérou, et sur les bords du Rio-Negro de l'Orénoque et de la rivière des Amazones*, 2 vols. (Paris: F. Schoell, 1808), v.

62 Hernández de Alba, ed., *Archivo epistolar*, t. 3, 32.

63 Humboldt y Bonpland, *Plantes équinoxiales*, ii, 3.

64 A. von Humboldt to Willdenow, Mexico City, April 29, 1803. Minguet, ed., *Cartas americanas*, 113.

65 Francisco José de Caldas, *Continuación del Semanario del Nuevo Reino de Granada*, Memoria 2ª (Bogotá: February 25, 1810), 18.

66 Today the National Library of France retains a copy of *Continuación del Semanario del Nuevo Reino de Granada*, Santafé de Bogotá, 1810–1811, FRBNF33325785.

67 Extracts from Humboldt's Diaries present 19 'notes in the margin' that make explicit references to the *Seminario* and its *Continuacion*. Only one of these notes specifies the date: '1842,' which makes it possible to assure that Humboldt read this criticism before the second publication of his biography of Mutis in 1846.

68 Caldas, *Continuación*, 17–20.

69 Puig-Samper *et al.*, 'Dos cartas,' 71–72.

70 José Antonio Amaya and Julián Leonardo Rendón Acosta, 'Veintiuna líneas que cambiaron la Historia de la Ciencia en Nueva Granada y su relación con la Metrópoli. Análisis de la descripción de la Flora de Bogotá de Francisco Antonio Zea a Antonio José Cavanilles,' *Historia Crítica* (January 2017) 63, 33–52.

71 Antonio José Cavanilles, 'Materiales para la historia de la botánica,' en Joaquín Fernández Pérez, ed., *Anales de Historia Natural, 1799-1804* (Madrid: Comisión Interministerial de Ciencia y Tecnología, Ediciones Doce Calles, 1993[1800]), 28.

72 Puig-Samper *et al.*, 'Dos cartas,' 77.

73 Lagasca to Humboldt, London, April 30, 1827. Puig-Samper *et al.*, 'Dos cartas,' 80–83.

74 Mr. Mutis has about 2000 drawings on folio; each plant once coloured [is copied] 2 times in black, therefore, in triplo. He considers that one of the black copies should remain at Santa Fé (*Registre des notes botaniques prises par Aimé Goujaud, dit Bonpland (1773-1858) et Alexander von Humboldt (1769-1853) pendant leur voyage en Amérique du Sud, 1799 à 1804*, Manuscripts and Scientific Archives Service of the *Bibliothèque Centrale* de MNHNP, MS 2534, f. [93], left margin).

75 Puig-Samper *et al.*, 'Dos cartas,' 80–86.

76 Lagasca to Humboldt, London, May 3, 1827, with a *postscript* of July 25, 1831. Puig-Samper *et al.*, 'Dos cartas,' 86.

77 In particular, two species deserve our attention due to the similarity of the graphic characteristics of the images and the acuity with which an omission by Humboldt and Bonpland of the plates of Mutis and his *Flora* is revealed. The case of *Monochaetum Bonplandii* tempts one to attribute to Mutis the clairvoyance of giving to Humboldt and Bonpland a deliberately erroneous plate. However, it is more likely that this error was unconscious, since it appears on a plate preserved in the Mutis archive itself.

78 Alexander von Humboldt, 'On the Cinchona Forest of South America,' in Aylmer Bourke Lambert, ed. *An Illustration of the Genus Cinchona: Comprising Descriptions of All the Officinal Peruvian Barks, Including Several New Species* (London: John Searle, Longman, Hurst, Hees, Orme and Brown, 1821), 36.

79 *Anales de Ciencias Naturales*, 6, n. 18 (1803), 281–287; its translation into French in: *Annales du Muséum National d'Histoire Naturelle* (Paris) 4 (1804), 475–478, and into English in: *Annals of Botany* (London) 1 (1805), 573–576.

80 *Gazette Nationale ou le Moniteur Universel* 2, n. 326 (August 14, 1803), 1445–1477; *Annales du Muséum National d'Histoire Naturelle* 11, t. 2 (1803), 170–180; *Magasin Encyclopédique: ou journal des sciences, des lettres et des arts* 8 (1803), t. 6, 537–549, and in Berlin, *Gilbert's Annalen der Physik* 16 (1804), 475–489.

81 Wilfrid Blunt and William Thomas Stearn, *The Art of Botanical Illustration* (Woodbridge, Suffolk: Antique Collectors' Club in association with the Royal Botanical Gardens, Kew, 1994), *passim* 194–209.

82 Javier Fuertes Aguilar, 'La colección de láminas de Mutis del archivo de la Expedición Malaspina,' in María Pilar de San Pío Aladrén and María Dolores Higueras Rodríguez, eds., *La armonía natural. La naturaleza en la expedición marítima de Malaspina y Bustamante 1789-1794* (Madrid: Jardín Botánico, Lunwerg and Caja Madrid, 2001), 85.

83 It is surprising that Caldas never refers to these specimens. It can be assumed that Bonpland and Humboldt kept this group of plants out of Caldas's reach, which did not happen with Carlos Montúfar, who learned that Humboldt was looking for the Loja quinas to compare them with those of Mutis. In his diary, Montufar notes that 'Loxa … is the place where the best Quinas known to us are extracted for the Royal Cabinet; the desire to see this plant so useful to humanity, and to compare it with the samples that the Baron has brought from Santa Fe and about which he has conversed with the botanist of the Expedition, Don Jose Celestino Mutis, was one of the objects of our journey to Loxa.' Reinhard Andress and Silvia Navia, 'Das Tagebuch von Carlos Montúfar: Faksimile und neue Transkription,' *HiN* XIII, 24 (2012), 21–74, the quote is from p. 22 of the manuscript, according to the transcript.

84 Minguet, ed., *Cartas americanas*, 255–256.

85 Humboldt and Bonpland, *Plantes équinoxiales*, v.

86 *Anales de Ciencias Naturales* 6 (18) (1803), 281–287; its translation into French in: *Annales du Muséum National d'Histoire Naturelle* (Paris) 4 (1804), 475–478, and into English in: *Annals of Botany* (London) 1 (1805), 573–576; *Annales du Muséum National d'Histoire Naturelle* 3 (1804), 396–404.

87 Academia Colombiana, *Alexander von Humboldt en Colombia*, 76.

88 W. von Humboldt to Cuvier, Rome, May 28, 1803. Hamy, ed., *Lettres américaines*, 225.

89 Caldas, 'Discurso en elogio de J. C. Mutis,' in Eduardo Posada, ed., *Obras de Caldas* (Bogotá: Imprenta Nacional, 1912), 103.

90 Humboldt, 'Biografía de José Celestino Mutis,' 501.

91 'This collection [of drawings] and the Chinchonas left for Cartagena de las Indias around June of this year. M[onsieur] Mutis took it upon himself to see to it that they continued on to Paris.' Humboldt to Delambre, Lima, November 25, 1802. Minguet, ed., *American Letters*, 101.

6

INCAS, PYRAMIDS, AND AMAZONS

Notes on Humboldt's Equatorial Encounters

Neil Safier

The Andean ruins of *Ingapirca*, a well-preserved architectural vestige of at least two ancient South American civilisations, lie just north of the city of Cuenca in the Province of Cañar, in what is today southern Ecuador.[1] Like most Incan ruins that were regularly visited in the aftermath of the Spanish invasion of South America, these sites left successive waves of travellers with a sense of amazement bordering on stupefaction: that a people without the assistance of iron tools could possibly have managed to form such architecturally elegant and physically functional fortresses in stone. 'In order to provide this regular and uniform convexity to all of these stones, and polish their interior faces so perfectly where they come together,' one eighteenth-century visitor wrote,

> what work, what industry must have been required in addition to our instruments, and this in a people who had not a single iron tool, and could not cut stone that was harder than marble except with the axes of stone, nor flatten them other than rubbing them together?[2]

This observer, the French Academician Charles-Marie de la Condamine, had spent a significant portion of his many-year sojourn in the Andes working to construct his own physical monuments out of the raw materials of the region. In his case, it was not a royal palace delimiting political territory, nor a military fortress, but rather a more metaphorical construct of intellectual space: a set of pyramidal markers that would adorn two endpoints of a baseline marking the site where one of the greatest eighteenth-century scientific problems had been resolved, or nearly resolved, and which had attracted the attention of Alexander von Humboldt when he passed through this region more than a half-century later.[3]

DOI: 10.4324/9781003231479-7

Attuned, quite possibly, to the dangers that could accrue when constructing such monuments on South American soil, Charles-Marie de la Condamine chose to begin the narrative of his visit to the ruins of Cañar (*Ingapirca*) with a cautionary tale, one that he had collected from Garcilaso de la Vega's own *Comentarios reales*, or history of the Incas, published in the first decade of the seventeenth century. La Condamine wrote that despite the extraordinary skills the Incas had exhibited in transporting massive pieces of stones from one location to another (which, on this occasion, might have been related to the construction of the Cuzco fortress of Sacsahuayman), one of the rocks fell into the valley and killed between three and four thousand bystanders.

La Condamine was likely not seeking to warn passers-by about the potential danger of being in an Incan construction zone. The Incas had been gone for centuries, and what remained of their architectural jewels largely lay in ruins. But he did nonetheless wish to provide a sort of caveat to his own descriptions of their treasures:

> I feel I should warn the Reader, that the description that I am going to make of the Ruins near Cannar may give a certain impression of the materials, the form, and perhaps the solidity of the Palaces and Temples constructed by the Incas, but not their magnificence nor their extent.[4]

Indeed, La Condamine went on to describe the extraordinary transformations undertaken by the early Peruvians under the leadership of the first Inca, Manco Capac, who had 'brought the Peruvians out from the depths of Forests, where they lived scattered according to their own traditions, and assembled them into a Society, giving them Laws and a smattering (*teinture*) of various arts.' According to La Condamine, once again basing his analysis on Garcilaso de la Vega's earlier text, Manco Capac instructed his followers in what clothing they should wear, how they should cultivate the earth, and how small channels could be used to irrigate their agricultural fields, a technology for which the Incas were duly famed. These were the extraordinary achievements of the first Inca ruler, and La Condamine praised these and other features of Inca life as he commented in the *Mémoires de l'Académie des Sciences de Berlin* on the vestiges of these 'ancient monuments of the Incas.'[5]

It was not with La Condamine's words echoing in his mind, however, that Alexander von Humboldt visited the province of Cañar and the site of 'Inca Pirca' in July 1802, even though he later wished that he would have read La Condamine's 1746 'Mémoire' prior to examining the site himself. We learn from Humboldt that he only read La Condamine's short text printed by the Royal Berlin Academy of Sciences and Belles-Lettres after he had already returned to Europe, despite mentioning in his personal journal 'the famous palace and fortress of the Inca that La Condamine described under the name of the fortress of Cañar.'[6] Some years later, when he wrote his official description of Cañar published in the *Vues des cordillères et Monumens des peoples indigènes de l'Amérique* (1813), he made explicit mention of La Condamine's 'mémoire,' to which, he lamented, he had not had access before his own visit to the Incan site. Nevertheless, in retrospect he used his own on-the-spot observations to challenge some of the conclusions La Condamine had reached regarding European and indigenous architectural practices in the region.

It deserves mention that Humboldt barely mentioned another set of eighteenth-century observers who had visited Cañar: the Spanish naval officers Antonio de Ulloa and Jorge Juan. Although he did refer to their *Dissertación histórica y geográfica sobre el meridiano de demarcación entre los dominios de Portugal y de España* (1749), which dealt primarily with the geopolitics of the border disputes between Spain and Portugal in the run-up to the 1750 Treaty of Madrid, it is striking that Humboldt largely avoided references to the much larger and more substantive account of their journey through South America: the *Relación história de un viage a la América meridional* (1748). In addition to recounting their many years of scientific experiments as part of the Franco-Hispanic expedition to Quito, which sought to determine the shape of the earth at the equator, Ulloa and Juan also provided information on Ingapirca and the controversy surrounding the pyramids of La Condamine (see Figure 6.1).

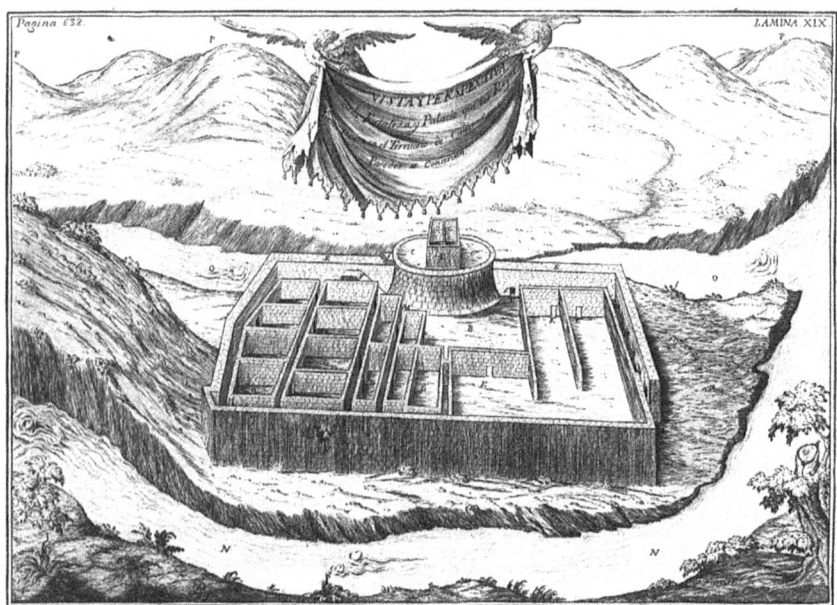

FIGURE 6.1 'Vista y perspectiva de la Fortaleza y Palacio que los Reyes Yngas tenían en el Territorio de Cañar cuias Paredes se Conservan,' in Antonio de Ulloa and Jorge Juan, *Relación histórica del viage a la América meridional*, Primera parte, Tomo segundo (Madrid: Antonio Marin, 1748), fold-out plate 18, following p. 632. Courtesy of the John Carter Brown Library.

Humboldt did cite Ulloa in his *Vues des cordillères*, but he did so in particular instances and in particular ways, always emphasizing small (and often erroneous) interpretive interventions rather than privileging large-scale accounts by these thinkers. Whether Humboldt neglected to affirm the broader contributions by Ulloa and Juan purposely or simply because he did not always have access to their five-volume work, is not clear. But it does bear questioning why such important accounts were largely left out of many of Humboldt's more widely circulating accounts, comparatively speaking.[7]

Visiting the ruins of Ingapirca in 1802 was for Humboldt part of a larger diversion on the way to a far more important goal: seeing the cinchona plants of nearby Loja, renowned for their febrifugal qualities and made even more famous by La Condamine's 1737 'mémoire' written for publication in the *Mémoires de l'Académie Royale des Sciences de Paris* and published there in 1738. During the same month that he passed through Cañar, Humboldt would visit Loja over the course of five days, where he discussed the various forms of cinchona with the botanist Vicente Olmedo, a man whom Humboldt, in fact, seemed to hold in quite low esteem. But before arriving in Loja, along the route that headed south from Riobamba to Cuenca, Humboldt made some astonishing discoveries, including in one instance 'the magnificent remains of a road built by the Incas of Peru.' 'This path [*chaussée*],' he wrote, 'adorned with large fitted stones, can be compared to the most beautiful Roman roads I have seen in Italy, France, and Spain.'[8] His fascination with the technical abilities of the Incas was palpable, and like La Condamine, Humboldt took great interest in the tools the Incas had used to forge these large structures from stone. Querying how a people without access to iron had created instruments or utensils with which to work the stone, Humboldt later put forward the hypothesis that the Incas had used copper, and in particular, hardened copper, for their stone-cutting projects, likening their work to the Greeks who, he observed, had also used copper or bronze for their weapons and tools.

How much, we might ask then, was Humboldt's eye trained by those travellers he read before or during his journey? How much 'eyewitness observation' are we to impute to Humboldt? If, as Michael Dettelbach has written, we need to open the 'black box of Humboldtian science,' what role did earlier travellers play in forging his ideas? As he travelled, Humboldt frequently returned to issues that previous travellers had raised in their published accounts, often after his own consultation of these printed works once back in Europe. But he also incorporated local legend into his accounts as well. Near Cañar, for instance, he discussed the historic mountain site of Inga-Chungana, where there existed a kind of fissure in the rock that may have been used as a game or sportive diversion for the Inca ruler. Humboldt points out, however, that according to 'avaricious' European legend, the Incas had hidden all of their treasure in this mountain upon hearing of the arrival of the conquistadores. The earliest Europeans to arrive in this place, again according to legend, met a young boy wearing the imperial insignia of the Inca, and who calling them 'impious foreigners,' asked why they insisted on disturbing what the Inca's 'artful ways' had hidden in the mountain. At which point, not surprisingly, the miners fled in terror and the mountain fissure closed up again.

The editorial practices of Humboldt in portraying his observations in print have also become relevant in recent years, thanks to a renewed emphasis on the cultures of notetaking and annotation across a broad swath of (especially) early modern scholarship.[9] In discussing the 'invention' of Humboldt, it is highly relevant to understand the role of Humboldt's erudite apparatus, that is, his own practices of annotation and note-taking, in building his reputation as a trustworthy interpreter and recorder of others' scholarship and observations. As Marie-Noëlle Bourguet has recently written, it is important to recognise that Humboldt himself was conscious of the desire, on one hand, to write down everything worthy of observation, and at the same time the impossibility to record everything that should and could be remembered. Humboldt provided in many ways a unique portrait of the individual notetaker, one who was

> mû par le souci compulsive d'inscrire sur le papier tout ce qui pouvait server sa réflexion.... Écrire, prendre note semblait chez lui une posture si familière et constitutive de son identité savant que plusieurs portraits, tracés par des artistes différents et en des circonstances ou à des époques diverses de sa vie, le représentent dans cette attitude.[10]

All the while, Humboldt acknowledged that

> il est difficile de tout noter.... On note à la hate tous les faits particuliers, les mesures, les descriptions de la nature – et toutes les observations plus générales, et donc plus intéressantes, précisément sur la culture des hommes, sur leur vie sociale se trouvent écartées.[11]

Humboldt's reliance on the selective observations and sometimes lopsided conclusions of others needs to be contextualised in order to assess adequately the extent to which his citation practices were, consciously or conscientiously, manipulated in order to buttress a reputation of lone scientific genius or well-read scholar-explorer.

Four months before following La Condamine's footsteps through the province of Cañar, Humboldt had visited the pyramids of Yaruquí, those monuments to measurements that La Condamine had so carefully erected sixty years beforehand. These structures on the plain of Yaruquí outside Quito had been built to function as permanent markers of an astronomically determined baseline that had been conceived to determine a global phenomenon: the shape, or curvature, of the Earth's surface. But for Humboldt, these pyramids stood out ironically as symbols of decay. 'We might imagine ourselves among the Turks,' he wrote, 'when we find the most important monument to the progress of the human spirit ransacked before us.'[12] At the same time, he felt particularly proud of La Condamine and Pierre Bouguer's contribution to this region.

The works of Bouguer and La Condamine have had an extraordinary influence on the Americans from Quito to Popayan. ... The Audiencia of Quito may have been able to destroy the Pyramids, but they were incapable of drowning the spark of genius that rises up from time to time in this land, and that shines brightly along the trail that Bouguer and La Condamine have blazed.

(See Figure 6.2)

For La Condamine, these constructed monuments were conceived to serve as material markers of transcendental truths that had been arrived at through empirical observations on South American soil. Each tetrahedral pyramid was to sit upon a square base of between five and six feet in height, constructed from rocks taken from nearby ravines and covered in a brick exterior of local manufacture. These physical monuments functioned as markers of intellectual, if not territorial, imperialism and symbols of an increasingly global scientific hegemony on the part of institutions such as the one that had sponsored La Condamine's travels: Paris's Académie Royale des Sciences. These pyramidal monuments were not only a literal imposition of European architectural and scientific codes onto the South American landscape. They also effectively erased the contributions of those who assisted most directly in their construction, not to mention the local participants that had contributed to the astronomical measurements themselves.[13] The attempt to commemorate the results of the Quito expedition *in situ* brought into sharp relief the ways in which Europeans accommodated themselves and their projects to local conditions as they sought to make a plain near Quito conform to the universal aspirations of Enlightenment science all the while diminishing the participation of those actors 'from below' who made these universalizing activities possible (see Figure 6.3).

This mid-century moment also signalled a transition between a period in which great achievements were commemorated on the spot through construction and inscription, a carryover from earlier traditions that saw the construction of monuments as an homage to individual monarchs or deities, and the more conventional procedure of setting down the results of scientific experiment through published works. Humboldt, for instance, believed that mountains would make more lasting monuments than pyramids and that the results of instrumental observation should not be embodied physically but in print. The volcano of Cayambé that straddled the equator near Quito was, in Humboldt's words, 'one of those eternal monuments by which nature has marked the great divisions of the terrestrial globe.'[14] For Humboldt, empirical science should follow suit in using these sites as markers of scientific truth. But La Condamine knew that monuments to individual achievement had symbolic value as well, especially since they could be forged out of local materials and captured iconographically in print. He thereby employed the locals of Quito in helping him inscribe the scientific results of his journey *onto* the American landscape, using the pyramid, a historical symbol of permanence and stability, as the ideal surface on which to commemorate their activities on a highland Andean plain.

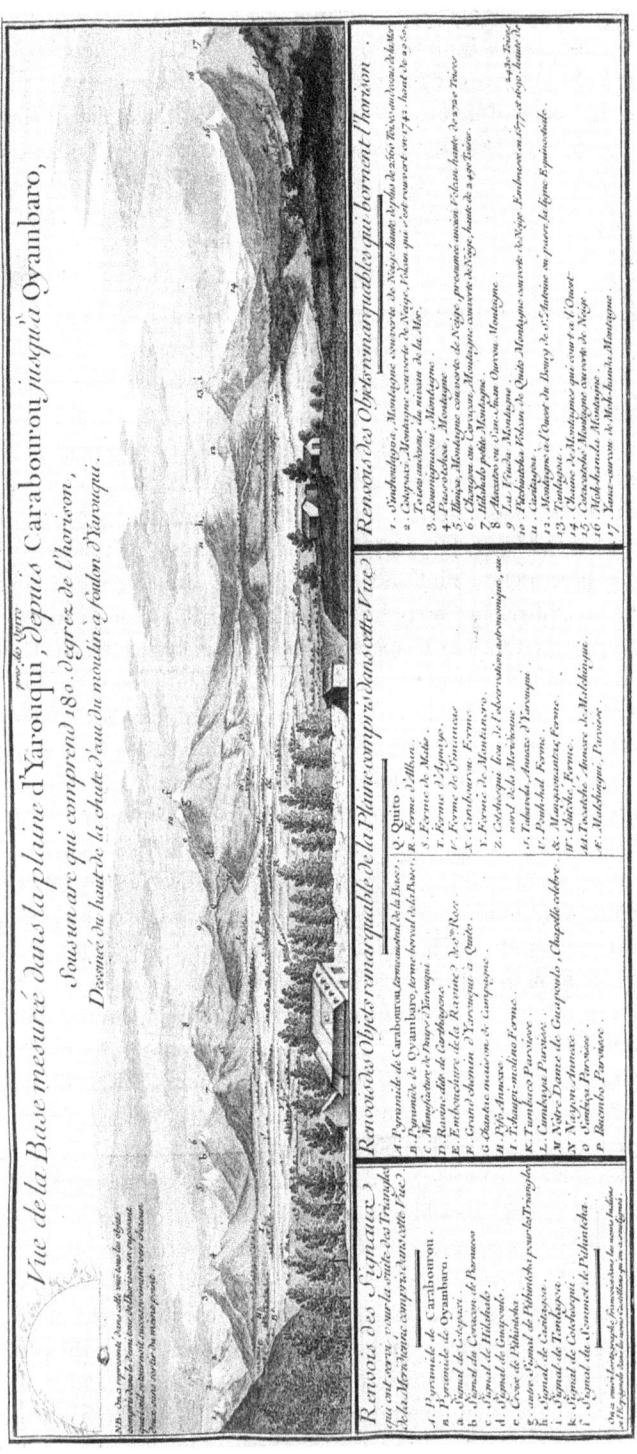

FIGURE 6.2 'Vue de la Base mesurée dans la plaine d'Yarouqui pres de Quito ... sous un arc qui comprend 180. degrez de l'horison,' in Charles-Marie de La Condamine, *Journal du voyage fait par ordre du Roi, a l'equateur servant d'introduction historique a la Mesure des trois premiers degrés du Méridien* (Paris: l'Imprimerie royale, 1751), fold-out plate; following p. 20. Courtesy of the John Carter Brown Library.

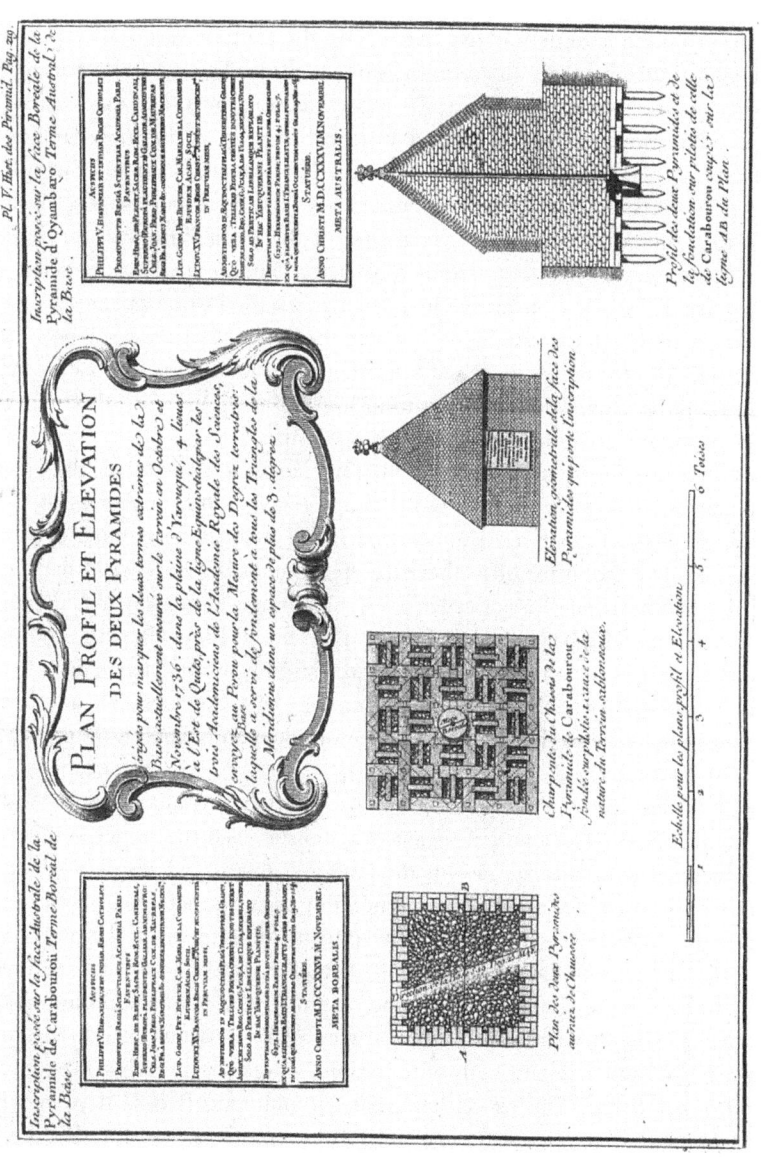

FIGURE 6.3   'Plan Profil et Elevation des deux Pyramides', in La Condamine, *Journal du voyage fait par ordre du Roi*, fold-out plate 5, following p. 218. Courtesy of the John Carter Brown Library.

Eventually, these pyramids became embroiled in a legal controversy between the Spanish Crown, represented by the Audiencia of Quito, and those who wished to erect them: the French Academicians sent to Quito. The two Spanish naval officers Ulloa and Juan also reacted critically to the pyramids and their proposed inscriptions. They argued that La Condamine had failed to secure the 'required permission' of the Audiencia to place commemorative markers on the plain of Yaruquí, and they rejected the status they were accorded on the stone tablets. They were also enraged that the participation of the Spanish Crown was symbolised by a fleur-de-lis, age-old symbol of the Bourbon monarchy and an icon that only in the most oblique sense could be associated with the new Bourbon King installed on the Spanish throne. The contours of the struggle that ensued between the French and Spanish plaintiffs before the Audiencia reveal the nature of the extra-European juridical space in which experimental observations took place. The French Academicians, on the other hand, and La Condamine in particular, claimed to be operating in a jurisdictional vacuum, a juridical sphere to which the Audiencia's temporal powers did not extend. All of these issues raised important questions about the nature of conducting scientific inquiry in a place far from Europe, and whose jurisdiction, in fact, should reign when scientific questions were at issue.

As I have written elsewhere, Ulloa and Juan by and large passed over the episode of the pyramid controversy in their *Relación histórica*. Although this ideological battle with the French academicians occupied them during much of the time they spent in Quito, it may have been Andrés Marcos Burriel's wish that the Spaniards' respond to what Burriel called La Condamine's 'disturbance' (*bullicio*), but he hesitated as to whether or not such a response should appear in print. In the final printed version of the *Relación histórica*, there was only a single paragraph that referred to the conflict over the pyramids, which may have been a feeble attempt to smooth over in print what had, in fact, been a turbulent and contentious process before the Audiencia.[15]

Humboldt visited these pyramids on a passage through Yaruquí in 1802. And when he stood near the Hacienda of Oyambaro and surveyed the ruined outlines of La Condamine's pyramids, he saw in the scattered shards and illegible tablets confirmation of his own idea that by commemorating scientific activities physically, rather than in print, the Frenchman had allowed feeble monuments to be built on metaphorically shaky ground. Even if these markers were intended to allow future generations to determine the accuracy of the original measurements, Humboldt suspected that the region's native populations would take anything of value from these pyramids, and the incessant rains of the tropics would soon finish off the job. 'In the land of the Indians,' he wrote, 'far greater monuments [would have been] required for [the pyramids] to remain intact.'[16] But beyond the suggestion that special kinds of commemorative markers needed to be constructed to dissuade local populations from taking what was valuable within them, Humboldt felt even more powerfully that the broken marble tablets, the shards of a broken fleur-de-lis, and the illegible inscriptions in Latin he saw when visiting the site were evidence for the demise of the human spirit.

Eventually, these pyramids would be destroyed, and efforts to rebuild them would be perpetuated over the course of the next two centuries.[17] Following a polemical debate with the Spanish officers, Antonio de Ulloa and Jorge Juan, who had accompanied the French academicians, the Audiencia (possibly with the assistance of the local populations) saw to it that the pyramids were reduced to rubble. But during the time of Humboldt's passage, the president of the Audiencia had wished to rebuild the pyramids as well, and he considered their destruction an act of 'vandalism.' Humboldt concurred with Selva Alegre's assessment but feared that rebuilding the structures might not lead to any improvement: in the event of their reconstruction, he wrote, 'one must fear that the same Indians had been able to learn about the sheets of silver, [and] that they had excavated the site like a *huaca*.' 'On the other hand,' he continued, 'in a country where everything that is considered effort is avoided, [the mayor] and the Indians would have been too lazy to dig down into the interior of foundations that were so well cemented.'[18] Humboldt concluded the recounting of his visit to Oyambaro with a note that he finally received a copy from the Audiencia of the act authorizing their destruction, where there was an effort made not to touch the base of the pyramids that would allow them to be rebuilt without affecting the scientific content embodied by their exact location. But everything was different from how it had been portrayed: the pyramid's interior was distinct from how La Condamine described its construction, the inscription inside the pyramid was not the same as that which had been written outside it, and the silver plate had been stolen as well. The mayor, Piedrahita, said that soon enough the rain would finish off what the Indians had started.

The diatribe against local populations with which Humboldt ended his reflection on the commemorative markers of Yaruquí, that in 'the land of the Indians … very large monuments would be needed' in order to remain intact, serves as a typically Humboldtian punctuation mark placed strategically to exoticise and destabilise the socio-cultural universe in which non-European science played itself out in South America. That is, by denigrating the ambitions of the local populations and elevating at the same time the erudite reflections of La Condamine and his crew, Humboldt seems to create a natural hierarchy by which those who have legitimate material needs are relegated to a social sphere that places them necessarily in a different plane from those who may be conducting more symbolic activities with regard to the commemoration of scientific processes. Despite the very real desire of the French academicians to place a physical monument that recognised their accomplishments on South American soil, similar material interests on the part of local populations seem to be viewed through Humboldt's optics as undermining and destructive of the American landscape. This in stark contrast to the manner in which Humboldt discussed the roadways that stretched across the 'llano del pullal,' likening the 'magnificent ruins of a trail built by the Inca of Peru' to 'the most beautiful Roman roads that I have seen in Italy, France and Spain,' as discussed earlier.[19] This may have been a direct response to Raynal's assertion that Incan road technologies had been fictively imagined by early chroniclers, especially since Humboldt was largely critical of Raynal's *Histoire des Deux Indes* and its more egregious assertions (some of which, we now know, were penned by Diderot). This oscillation between the praising of pre-Columbian polities and their exquisite material achievements in certain places and the less august manner by which contemporary, eighteenth- or nineteenth-century local populations were frequently portrayed by Humboldt in print, points to an instability in the approach Humboldt took when assessing the contributions of native communities vis-à-vis their European scientific collaborators that he may have inherited from earlier Spanish chroniclers, including Ulloa. But whatever the motivation, Humboldt's manner of lauding Incan practices when he deemed fit and harshly criticizing their activities in the contemporary moment may be relevant when considering how the achievements of Creole and Afro-descendent populations, in real or fictional terms, were portrayed in other Humboldtian accounts as well.

Humboldt often seems to denigrate the Spanish academicians, Ulloa and Juan, as well. In several places within the *Voyage aux regions équinoxiales du nouveau continent*, the Prussian seems to imply that the instrumental observations undertaken by Ulloa and Juan were regularly unsystematic, if not necessarily unreliable. 'Quelques observations barométriques rapportées, comme au hazard, par Ulloa, m'ont appris cependant que de l'embouchure du Rio Chagre à l'embarcadère de Cruces il y a une différence de niveau.'[20] Along with Richer, La Condamine, and Bouguer, Ulloa and Juan employed '[barometric] instruments which were quite imperfectly purged of air,' leading elsewhere to erroneous observations.[21] When describing the Inca's House at Callo (Quito), Humboldt emphasises that 'the drawing that Ulloa provided of the Inca's house shows so little of the original building plan that one would almost be tempted to believe that it is completely fictitious.'[22] It is highly relevant, however, that Humboldt reveals in the following paragraph that he had access to 'the plate contained within Ulloa's travels,' which he showed to a group of several 'elderly friars' from the Augustinian order as a comparison to his own (seemingly more accurate) drawings. While Humboldt did refer to Ulloa as a 'celebrated astronomer' when they were received amongst a group of Haitian émigrés by a nephew of Ulloa's in Trinidad, and whose authority 'carries great weight,' it is clear from the relative paucity of citations, and usually negative ones at that, that Humboldt's penchant was to cite French and German authors when he was able and that he denigrated Iberian contributions to natural knowledge despite having clear access to their materials.

Turning away from the archaeological ruins at Cañar, Yaruquí, and Callo, we move toward the more ephemeral and legendary spheres of Humboldt's journeys. There is no better place to explore Humboldt's interest in debunking, or supporting, mythological vestiges of America's past than the Amazon River region and the associated histories of gynecocratic societies that predated the arrival of European explorers to the Americas. As is well known, the initial exploration of the Amazon basin was linked to two myths that circulated since the arrival of Europeans in the New World.[23] Early in the sixteenth century, rumors had spread through Central and South America of a 'país de la canela,' a site of mythical and geographical speculation where gold, spices, and other natural treasures were thought to abound.

Europeans were particularly interested in a specimen resembling Indian cinnamon (*Cinnamomum zeylanicum*) that was reputed to exist in the eastern foothills of the recently conquered Andes. Gonzalo Pizarro and Francisco de Orellana set out to explore this region in 1541. On the same expedition, Pizarro was determined to find the domain of an indigenous warrior-chieftain named 'El Dorado,' who was reputed to dip himself each day in a lake of gold to impress his subjects and frighten his enemies. In his attempt to find this city of gold, Pizarro undertook the first European exploration of the Amazon River. During his journey, Orellana also claimed to see a group of bellicose women 'who set about fighting in front of [their Indian subjects] as captains, so valiantly that the Indians did not dare to turn their backs.'[24] A century later, the Portuguese captain Pedro de Texeira chose the reverse itinerary, leaving from the Portuguese city of Pará (present-day Belém) to explore the river westward. In Quito, he met up with the Jesuit Cristobal de Acuña, who accompanied him on his return to Pará. Acuña's text was the first full narrative account of the river's descent, published in 1641 as the *Nuevo descubrimiento del gran Río de las Amazonas*. La Condamine made frequent reference to the 1682 French translation by Gomberville, to which he had regular access as he descended the river.[25]

Far earlier, however, other European travel narratives of the Americas, and the cartographic representations that arose in their wake, conjured a host of provocative images to fire Europe's mythical imagination, and these also influenced the representation of the Amazon River. In his description of Matinino Island, for instance, Columbus referred to a place 'occupied entirely by women, without a single man,' and this tale was soon echoed in the accounts of Amerigo Vespucci, Hernan Cortés, and other early explorers to the Caribbean. Later in the sixteenth century, rumours of American Amazons continued to spread into texts such as Pietro Martire d'Anghiera's *De orbe novo* (1530), Giovanni Battista Ramusio's *Navigationi et viaggi* (1550–1559), and Francisco López de Gómara's *Historia de la conquista de México* (1552). But it was the voyage of Count Francisco de Orellana from Quito to the 'Land of Cinnamon' in 1541–1542 that first propagated the myth of a community of unmarried women warriors living in isolation along the river that would eventually adopt their name. Late in the sixteenth century, the 'River of the Amazons' shared center stage with its first explorer, Orellana, in the Spanish court where the count's discovery was considered an 'important and marvelous' event. A branch of the river was even named in his honour on several maps from this period. But territorial sovereignty in the Amazon basin was still in dispute: Jacques de Vau de Claye's 1579 depiction of South America, for example, emerged in the context of a never-realised French plot to invade the region and rise up militarily against the Spanish and Portuguese. In Vau de Claye's map, a group of female 'Almazones' were shown as potential French allies. They were represented as 'industrious' arms-makers who lived in Arabian tents, constructing bows and arrows and conforming to a cartographer's Orientalist fantasy mixing legends from Asia Minor with the most recent discoveries in New World geography. Nevertheless, their cartographic presence reinforced the possibility of their physical existence and added to the conceptual vocabulary of a mythic civilisation living hidden from view by the banks of the newly discovered river.[26]

Alongside the Amazons, the fabled existence of a 'city of gold' known as Manoa El Dorado was also a well-established myth stemming from the earliest European exploration of South America. Orellana's expedition in 1541 had failed to come up with any evidence of a city of gold beside a golden lake, but his reports of American Amazons did push Europe's imagination eastward from the Andes into the continent's lowland region. It was here in the Amazonian lowlands, in fact, that both Pedro Texeira and Father Acuña testified to having seen native tribes engaged in trade using gold as the medium of commercial exchange. And it was here, two centuries later, that La Condamine attempted to confirm the veracity of El Dorado and inscribe its location cartographically on his map (see Figure 6.4).

Humboldt's description of the Amazon region was in many ways indebted to La Condamine's earlier cartographic discoveries, but he was also quite critical of certain aspects of his map. 'The journey of M. de La Condamine,' Humboldt wrote,

> which has spread so much light on different parts of the Americas, has made the region involving the Caqueta, the Orinoco, and the Rio Negro even murkier. This illustrious savant knew … that the Caqueta was the river that, once it arrived at the Amazon, carried the name Japura. But he not only adopted Sanson's hypothesis [that the Caqueta bifurcated into the Orinoco and the Rio Negro], he tripled the number of its bifurcations. … This imaginary system was represented in the first edition of the attractive map of the Americas, produced by d'Anville.[27]

Humboldt went through the cartographic history of the Amazon region, this in Book 8, chapter 23 of the *Relation historique*, and showed quite clearly how earlier accounts of the region had been flawed.

But the case of the American Amazons is especially worth noting, since it shows that Humboldt was not always so explicit about violating previously held theories, even if they involved myth and fables stretching back into the pre-Columbian past. In his *Relation historique*, after describing La Condamine's cartographic representation of the region, Humboldt explained that since his return to Europe, people in Paris had often asked him if he

> shared [La Condamine's] opinion or if [he] believed, like many of his contemporaries, that [La Condamine] undertook the defense of the Cougnantainsecouima [women warriors of the Amazon] … merely to take advantage of the generous reception of a public session of the academy, and their eagerness for [hearing] new things.[28]

Humboldt did not come down on one side or the other, either in defence of La Condamine or in concert with his Parisian critics. Nevertheless, when describing a report by Father Bartolomé Gili on a conversation he had had with an indigenous

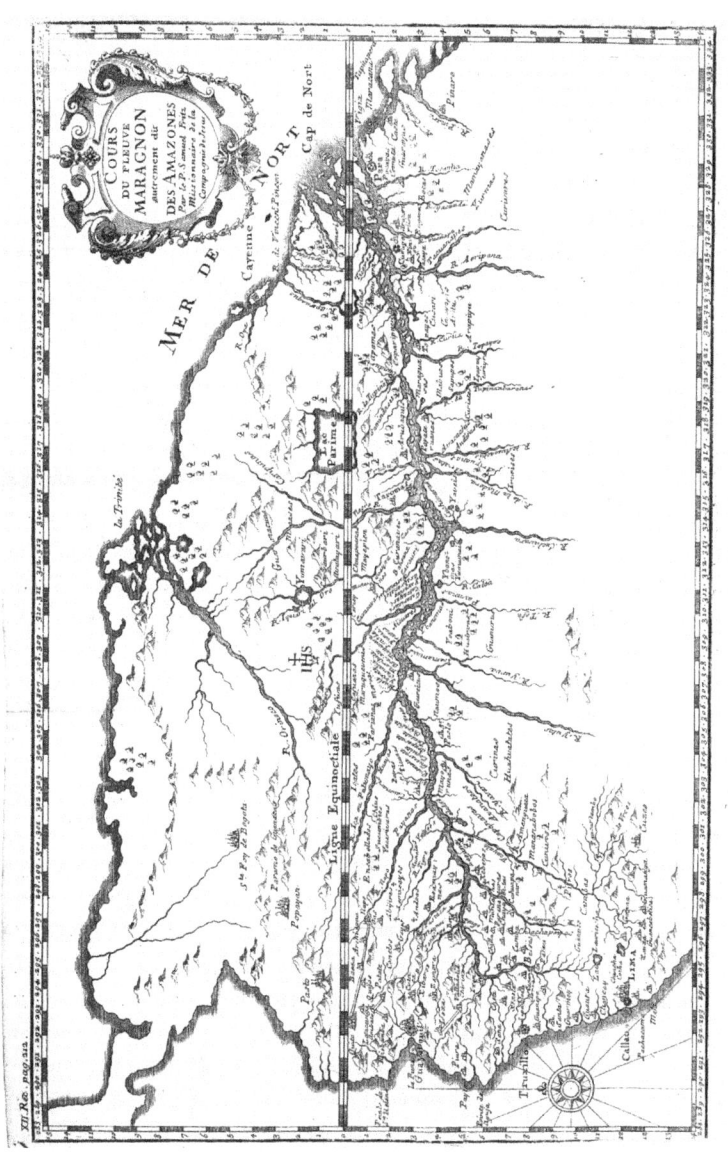

FIGURE 6.4 'Cours du fleuve Maragnon autrement dit des Amazones Par le P. Samuel Fritz Missionnaire,' in *Lettres édifiantes et curieuses, écrites des missions étrangères par quelques missionnaires de la Compagnie de Jésus XII* (Paris: Chez Nicolas le Clerc, 1717), fold-out engraved map. v. 12, following p. 21. Courtesy of the John Carter Brown Library.

resident of the Encaramada region north of the Amazon River, an individual who had proclaimed the existence of a group of 'women living on their own' who fabricated for themselves instruments of war, Humboldt made it clear that he found the missionary's testimony quite credible, in that it conformed with everything else that had been known about the Amazon warriors since the time of Orellana. 'What must we conclude from [Gili's] testimony?' Humboldt asked. And here was his response:

> Not that there are Amazons on the banks of the Cuchivero [river], but that in different parts of the Americas, women, left in a state of slavery into which they had been abandoned by their husbands, had gathered together, much like fugitive black [slaves] in a *quilombo*; that the desire to retain their independence had turned them into warriors; [and] that they had received visitors from some nearby and friendly group, perhaps less methodically than tradition had claimed. It would have been sufficient for this society of women to have acquired some strength in one of the parts of the Guyanas in order for these simple events, which may have repeated themselves in different places, to have been described in a uniform and exaggerated manner. That is the way with traditions. And if the extraordinary uprising of slaves about which I spoke earlier had taken place not near the coast of Venezuela, but rather in the middle of the continent, a credulous people would have seen in each *quilombo* of black maroons the court of king Michael. ... The Caribs of Terre-Ferme [Venezuela] communicated with those of the islands, and it is by this means that the traditions of the Marañon and the Orinoco made their way northward. Prior to Orellana's navigation, Christopher Columbus already believed he had found Amazons in the Antilles.[29]

Here, Humboldt allies the Amazon myth with the cruel realities of African servitude in the New World, using some of the same arguments that La Condamine himself put forward to corroborate for his Parisian readers the stuff of sixteenth-century fable. Indeed, it is striking that Humboldt referred to the similarity between Amazons and fugitive slaves, for it is precisely the argument that La Condamine employed to defend the idea. La Condamine's support for the Amazon myth may have owed a great deal to the Frenchman's own experience sharing a canoe with a fugitive slave from Surinam, who had sought refuge in Portuguese territory across a political border where different laws obtained with respect to the legal treatment of the enslaved.[30] In the case of La Condamine, this cross-border experience was crucial in determining the nature of a myth that had been perpetuated since Orellana and even Columbus. In the case of Humboldt, his research into the nature of the Amazons was carried out in a borderland region as well, between the Guianas, Terra Firma, and Brazil. And he appears to be heavily indebted to La Condamine for what he concluded from these researches in the watery hinterland that would come to be called Amazonia.

What can we conclude, then, from these equatorial encounters with Humboldt, other than stating the obvious fact that the Prussian mineralogist was an avid reader and paid explicit attention to those who travelled before him to places he would see? There are many things to be said about the way he read and the contrast between how he acquired knowledge from the printed page and the manner he gleaned information from other kinds of sources, including oral accounts and Amerindian interlocutors, between contemporary indigenous communities and those more 'classic' polities to which he only had access through the material vestiges of civilisations past. The question of sources is an important one, since having access to texts in situ, as opposed to consulting them after the fact, transforms the nature of the observations that travellers are able to make on the spot. One eighteenth-century Portuguese traveller in Amazonia considered having a 'competent library' as a critical element that allowed travellers to observe objects not only laterally (that is to say, visually) but also to reflect on them vertically (or conceptually) as well. Alexandre Rodrigues Ferreira, the so-called Brazilian Humboldt, wrote that

> putting aside for its appropriate time and place, and for the presence of a competent Library, the complement of History in which one should, according to the Art, describe the Animals, Plants, and Minerals observed and collected, I will be satisfied for right now in looking at [these objects] from the side, as they offer to the Naturalist Traveler a simpler and less complicated aspect.[31]

This was only one phase in a larger process by which naturalists sought to contextualise objects in time and space, historically and geographically, and it is perhaps in this vein that Humboldt succeeded beyond anyone before him. He had the ability to fluently merge distinct registers, eyewitness and textual, but his technique belies the extent to which he made conscious decisions about whom to include and whom to highlight along the way.

It is for this reason that his comment about having wished to know a particular 'mémoire' written by La Condamine some fifty years before allows us to reflect on his own process of reconstructing the past, both materially and bookishly. His approach seems to differ from that of La Condamine, in that he seems more at ease attributing indigenous provenance to the objects he observed during his voyage, whereas La Condamine maintained a much stronger narrative attachment to a European perspective that forced him to speak in often denigrating terms about native technologies. As La Condamine wrote,

> Il semble qu'à l'envi des Torrens, qu'ils voyoient se creuser un lit dans les Rochers, ils ont cherché & trouvé le moyen de se passer de fer pour tailler les pierres les plus dures; mais ces memes qui ont travaillé le Granit, & fore l'Emeraude, n'ont jamais sçû assembler une charpente par des mortaises, des tenons, des cloux, ni des chevilles. Les Volcans & les Mines, dont leur païs est rempli, ont pû leur offrir le spectacle de métaux liquefies, mais quoiqu'ils

aient rëussi à fonder l'or & l'argent, & à les jetter en moule; ils ne sont pas avisés de faire cuire des briques, ni des tuiles, dont ils avoient la matiere. Souvent ils se sont contentés de ce qu'ils ont rencontré d'abord, sans chercher un mieux, dont ils n'avoient pas l'idée.[32]

Humboldt takes pains not to 'cast doubt upon the observations of that famous traveler,' which is to say La Condamine himself, but he does reflect on the longer history of roof-coverings in both American (he makes explicit reference to the 'sloped roofs' of the 'natives in the northwest coast of the Americas') and southern European contexts, attempting as he often does to place a specific set of his own observations from within the Americas in a broader geographical context. Like the stone carvers of the Incan period, Humboldt rarely allowed an earlier observation to pass him by without attempting to shape and smooth that which lay before him.

Humboldt's observations of this region, and his rereading of La Condamine's own observations, were not limited to Ingapirca. Humboldt went on to discuss the 'Boulder of Inti-Guaicu,' which in his reading may have been the impetus for the Inca to construct a residence at Cañar, likening its origins to that of the Vaivasvata of India, referenced in an 1805 publication by William Jones from the *Recherches Asiatiques*. These orientalist comparisons are frequent in many of Humboldt's texts, from the *Vues des cordillères* through *Cosmos*, and place his work in a comparative context that should be considered not only along archaeological and material lines, as do Miruna Achim and Mark Thurner in this volume, but also alongside the comparative linguistic analysis effected by Wilhelm von Humboldt, Alexander's brother.[33] Humboldt ended this vignette with a reference to Benjamin Smith Barton and his study of American languages, and the limited number of roots that could also be found in the languages of the Manchu Tartars, the Mongols, the Celts, the Basques, and the Estonians. He later presented a study of the interior portions of the Inca's house at Cañar, with some slight modifications to what La Condamine had written in his earlier study. Humboldt wrote that the buildings around Cañar

> are not made of the same quartziferous sandstone that covers the clayey schist and the porphyry of Azuay … [n]or are the stones used in the structures at Cañar granite, as Mr. La Condamine believed, but, rather, an extremely hard trappean porphyry within which are embedded both vitreous feldspar and amphibole.[34]

Here, Humboldt's expertise in the categories of geological strata appears to trump the observations of the 'famous traveler' yet again, in that he carefully sourced each of the different elements of this construction, including the quarries that 'furnished the gorgeous stone used in the Inca's house in the plain of Pullal.' But Humboldt also used La Condamine and his colleague Pierre Bouguer's observations about porphyry decorations to argue in favour of the use of tools made from copper mixed with tin, a practice according to Humboldt that 'prevailed over that of iron everywhere in the old continent,' and therefore would not have been unexpected or uncommon in parts of South America as well: 'The Peruvians' *sharp copper* is almost identical to that of the Gallic axes,' Humboldt wrote, 'which chop wood just as well as steel does.'[35]

In this essay, we have been privileged to accompany Humboldt as he visited a wide array of pre-Columbian sites in the Andes: an Incan fortress from pre-Spanish times that lay in a ruined state but was nevertheless reasonably well preserved; a monument to a polemical debate hatched in Europe but commemorated on South American soil, built by native hands and later destroyed (presumably) by the same; and a kind of literary monument to a perennial myth, corroborated through native testimony but whose epistemological legitimacy still remained relatively unsettled in the early years of the nineteenth century. Humboldt used these travels and the accounts that previous European writers provided to the public to reimagine the Americas in a way that was innovative and unifying at the same time, seeking to place what had previously been seen as local traditions and local knowledge into a much wider frame. But it is also true that he arrived at his reimagining by reacting and responding to those who came before him, and it is as much his careful reading of previous sources and his rearticulation of certain itineraries, including indigenous itineraries, about which I have written elsewhere,[36] that allowed him to reach the conclusions he did. As Jorge Cañizares-Esguerra has shown elsewhere, and as Alberto Gómez Gutierrez and José Antonio Maya have demonstrated in their own contributions to this volume, Humboldt worked most often in tandem with Spanish American naturalists like Mutis and Caldas, but he did not always give these and others the credit they deserved.[37] As Amaya has emphasised, Humboldt did not always necessarily keep his word as he engaged with his South American interlocutors. Gómez Gutierrez speaks specifically about Humboldt's penchant for leaving out references to works that may have predated his own cutting-edge theoretical contributions, especially when they were authored by Central or South American authors. The particular case of Antonio de Ulloa and Jorge Juan, to whom Humboldt referred but not nearly to the same degree as he did La Condamine and Bouguer, deserves further consideration as well. Nevertheless, Humboldt did write of Caldas that he was

> a prodigy in Astronomy. Born in the darkness of Popayan, having never traveled farther than Santa Fe, he constructed his own Barometers, Sectors, and quadrants out of wood. … What would this young man have done had he been born in a country with means, where he would not have had to learn everything on his own!'[38]

Humboldt's somewhat condescending description of Caldas, upon whom Humboldt relied for many of his own empirical observations, is in many ways the mirror image of what he was himself: a young man born in a country with means where he was able to learn from the best around him.

An analysis of Humboldt's South American writings, in particular those that relate to the Andean phase of his travels, reveals what seems to be an ever-changing set of criteria for the praising and denigrating of sources and evidence. His appreciation for French travellers is undeniable, but as we have seen he occasionally privileges certain kinds of evidence over others, regardless of nationality or ethnic background, even as indigenous or Iberian observers tend to get short schrift in his estimation. What is clear is that Humboldt's drawings and observations of the Andean world, from Ingapirca to Cuzco and beyond, benefit greatly from the texts and contexts to which he had access, as he eloquently, if sometimes erroneously, captivated his audiences with the vertical landscapes drawn by others before him in poetry and prose.

Notes

1 There is an extensive archaeological and tourist literature associated with Ingapirca. See, for example, Terence N. D'Altroy and Katharina Schreiber, 'Andean Empires,' in Helaine Silverman, ed., *Andean Archaeology* (Oxford: Blackwell, 2004), 270.

2 La Condamine, 'Pour donner cette convexité reguliere, & uniforme, à toutes ces pierres, & meme pour polir si parfaitement les faces interieures par où elles se touchent,' one eighteenth-century visitor wrote, 'quel travail, & quelle industrie ont du suppleer à nos instruments, chez des Peuples qui n'avoient aucun outil de fer, & qui ne pouvoient tailler des Pierre plus dures que le marbe qu'avec des haches de Caillou, ni les applanir qu'en les usant mutuellement par le frotement?' All translations, unless otherwise noted and cited, are mine.

3 For a description of the commemorative pyramids La Condamine helped construct in the mid-eighteenth century, and of the controversy that ensued before, during, and after their construction, see Neil Safier, *Measuring the New World: Enlightenment Science and South America* (Chicago, IL: University of Chicago Press, 2008), chapter 1. See also Ernesto Capello, *City at the Center of the World: Space, History, and Modernity in Quito* (Pittsburgh, PA: University of Pittsburgh Press, 2011) for later historical and political manipulations of these early commemorative markers.

4 Charles-Marie de la Condamine, 'Mémoire sur quelques anciens monumens du Pérou, du tems des Incas,' in *Histoire de l'Académie royale des sciences et belles lettres* (Berlin: Ambroise Haude, 1746 [1748], 441. ('Je crois devoir prévenir le Lecteur, que la description que je vais faire des Ruines voisines de *Cannar*, peut bien donner une idée de la matiere, de la forme, & peut-être de la solidité des Palais, & des Temples bâtis par les Incas, mais non de leur magnificence, ni de leur etendue.')

5 La Condamine, 'Mémoire,' 435–456, *passim.*

6 Alexander von Humboldt, 'Viaje de Riobamba a Cuenca y estadía en Cuenca,' in Segundo E. Moreno Yánez, ed., *Alexander von Humboldt. Diarios de viaje en la Audiencia de Quito*, trans. Christiana Borchart de Moreno (Quito: Occidental Exploration and Production Company, 2005). La Condamine writes briefly about Cañar in his *Journal du voyage fait par ordre du Roi* (Paris: Imprimerie Royale, 1751), 80–81.

7 For more on Ulloa and Juan's image of Ingapirca and its role in Humboldt's own depiction, see Georgia de Havenon, 'Humboldt and the Inca Ruin of Cañar,' in Esther Pasztory, ed., *Visual Culture of the Ancient Americas: Contemporary Perspectives* (Norman: University of Oklahoma Press, 2017).

8 Alexander von Humboldt, 'The Peruvian Monument of Cañar,' in Vera M. Kutzinski and Ottmar Ette, eds., *Views of the Cordilleras and Monuments of the Indigenous Peoples of the Americas* (Chicago, IL: University of Chicago Press, 2012), Plate XVII. All translations of Humboldt's *Vues des cordillères*, unless otherwise noted, are from this edition.

9 Ann Blair, *Too Much to Know*; Richard Yeo, *Notebooks, English Virtuosi, and Early Modern Science* (Chicago, IL: University of Chicago Press, 2014). See also the special issue of *Intellectual History Review*, 'Note-Taking in Early Modern Europe,' 20.3 (2010), edited by

Richard Yeo with contributions from Ann Blair, Paul Nelles, Jacob Soll, Marie-Noëlle Bourguet, Margaret Sankey, and Michael Bennett.

10 Marie-Noëlle Bourguet, *Le monde dans un carnet: Alexandre von Humboldt en Italie (1805)* (Paris: Le Félin, 2017), 24.

11 SBB-PK, Nachlass A. von Humboldt (Tagebücher), VII a/b f. 1924, as cited in Bourguet, *Le monde dans un carnet*, 25.

12 Alexander von Humboldt, 'Besuch der Pyramiden von Yaruquí,' in Margot Faak, ed., *Reise auf dem Río Magdalena, durch die Anden und Mexico* (Berlin: Akademie-Verlag, 1990), 2: 71.

13 The *Conspectus Longitudinum et Latitudinum Geographicarum, per Decursum Annorum 1799 ad 1804* (Paris, 1808) includes Humboldt's astronomical observations alongside those of the engineer Carlos Cabrie and José Antonio Caldas, among others. The scientific theories of these two men, elaborated by Colombian Creoles, are thought to have profoundly influenced Humboldt's theories of biogeography and climate science.

14 Alexander von Humboldt, *Vue des cordillères, et monumens des peuples indigènes de l'Amérique* (Paris: F. Schoell, 1810), 242, Plate XLII.

15 There are two manuscript texts, in Spanish, held by the Biblioteca Nacional that attempt to defend the Spanish position against La Condamine. See the 'Historia de las pirámides de Quito, corregida, añadida, y con reflexiones,' Biblioteca Nacional de España, MS 8428.

16 Humboldt, 'Besuch der Pyramiden von Yaruquí.'

17 Capello, *City at the Center of the World.*

18 Humboldt, 'Besuch der Pyramiden von Yaruquí.'

19 Humboldt, 'The Peruvian Monument of Cañar,' Plate XVII, 128.

20 Humboldt, *Voyages aux régions équinoxiales du nouveau continent*, 2:125 (Liv. IX, cap. XXVI).

21 Humboldt, *Voyages aux régions équinoxiales*, 2:313 ('Notes').

22 Humboldt, *Views of the Cordilleras*, Plate XXIV, 195.

23 This section loosely follows my discussion of Amazon mythology and early exploration in Safier, *Measuring the New World*, chapter 2.

24 Gaspar de Carvajal, *Relación del nuevo descubrimiento del Rio Grande de las Amazonas (1541-42)* (Quito: Gobierno del Ecuador, 1992), 127.

25 Marin le Roy, Sieur de Gomberville, *Relation de la rivière des Amazones, traduite par seu Mr de Gomberville de l'Academie Françoise. Sur l'original espagnol du P. Christophle d'Acuña Jesuite. Avec une dissertation sur la rivière des Amazones pour servir de préface* (Paris: Claude Barbin, 1682).

26 For a complete assessment of some of the earliest myths associated with the Amazon River region, see Enrique de Gandía, *Historia crítica de los mitos de la conquista americana* (Buenos Aires: J. Roldán, 1929).

27 Humboldt, *Voyages aux régions équinoxiales* (Book 8, chapter 23).

28 Humboldt, *Voyages aux régions équinoxiales*, 2:484–485. ('On m'a souvent demandé à Paris … si je partageois l'opinion de [La Condamine], ou si je croyais, comme plusieurs de ses contemporains, qu'il n'avoit entrepris la defense des *Cougnantainsecouima* … que pour captiver, ans une séance publique de l'Académie, la bienveillance d'un auditoire un peu avide de choses nouvelles.')

29 Humboldt, *Voyages aux régions équinoxiales*, 2:487–488. ('Que faut-il conclure de ce récit de l'ancien missionnaire de l'Encaramada? non qu'il y a des Amazones sur les rives du Cuchivero, mais que, dans différentes parties de l'Amérique, des femmes, lasses de l'état d'esclavage dans lequel elles sont tenues par les hommes, se sont réunies, comme les nègres fugitifs, dans un *palenque*; que le désir de conserver leur indépendance les a rendues guerrièrs; qu'elles ont reçu de quelque horde voisine et amie des visites, peut-être un peu moins méthodiquement que ne le dit la tradition. Il suffit que cette société de femmes ait acquis quelque force dans une des parties de la Guyane pour que des événemens très simples, qui ont pu se répéter en différens lieux, aient été dépeints d'une manière uniforme et exagérée. C'est le propre des traditions; et si l'émeute extraordinaire

d'esclaves dont j'ai parlé plus haut avoit eu lieu non près des côtes de Venezuela, mais au milieu du continent, un peuple crédule auroit vu dans chaque *palenque* de nègres marrons la cour du roi Miguel. … Les Caribes de la Terre-Ferme communiquoient avec ceux des îles, et c'est par cette voie sans doute que les traditions du Maragnon et l'Orénoque se sont propagées vers le nord. Avant la navigation d'Orellana, Christophe Colomb croyoit déjà avoir trouvé des Amazones dans les Antilles.')

30 Safier, *Measuring the New World*, 268–272.
31 On the bookish practices of eighteenth-century travelers to the Amazon River region, see Neil Safier, "Every day that I travel … is a page that I turn': Reading and Observing in Eighteenth-Century Amazonia,' *Huntington Library Quarterly* 70.1 (2007):103–128.
32 La Condamine, 'Mémoire', *Histoire de l'Académie royale des sciences et belles lettres*. 446.
33 See Thurner and Achim essays in this volume; on Wilhelm von Humboldt's comparative linguistics, see among many other titles, Paul Robinson Sweet *Wilhelm von Humboldt: A Biography, 1767–1808; 1808–1835* (Columbus: Ohio State University Press, 1978).
34 *Vues des cordillères*, Plate XX, 139.
35 *Vues des cordillères*, Plate XX, 142.
36 Neil Safier, 'The Confines of the Colony: Boundaries, Ethnographic Landscapes, and Imperial Cartography in Iberoamerica,' in James D. Akerman, ed., *The Imperial Map: Cartography and the Mastery of Empire* (Chicago, IL: University of Chicago Press, 2009).
37 Jorge Cañizares-Esguerra, 'How Derivative was Humboldt? Microcosmic Narratives in Early Modern Spanish America and the (Other) Origins of Humboldt's Ecological Sensibilities,' in *Nature, Empire, and Nation: Explorations of the History of Science in the Iberian World* (Stanford, CA: Stanford University Press, 2006), 112–128; and the contributions to this present volume by José Antonio Amaya and Alberto Gómez Gutiérrez.
38 As cited in Luis Carlos Arboleda, 'Caldas, matematización de la naturaleza y sentimiento telúrico,' *Revista de la Academia Colombiana de Ciencias Exactas Físicas y Naturales* 40 (154):6–7.

7

HUMBOLDT'S MAGIC MOUNTAIN

Juan Pimentel

TRANSLATED BY PETER MASON

> That which seems unattainable has a mysterious attractive power; we wish that all should be explored, or at least attempted, if not obtained. Chimborazo has been the wearisome object of all inquiries addressed to me since my first return to Europe.
>
> Alexander von Humboldt, 'On Two Attempts to Ascend Chimborazo', 1837

Humboldt's ascent of Chimborazo on June 23, 1802, has always attracted a great deal of attention. Accompanied by Bonpland, Montúfar, and various indispensable guides and assistants, Humboldt had hoped to reach the summit of what was considered at the time the highest peak in the world. The attempt won him fame in his day, and it has continued to do so ever since. Indeed, Humboldt's attempted scaling and representation of his 'magic mountain' remain hot tickets in the Humboldt industry.

Andrea Wulf's best seller, *The Invention of Nature* (2015), opens with the famous ascent, calculated to whet the reader's curiosity and thirst for adventure.[1] The iconic image of Chimborazo included in the *Essay on the Geography of Plants* is featured on the cover of the literary success of Daniel Hehlmann, who devoted the central chapters of his novel *Measuring the World* (2005), later filmed under the same title by Detley Buck (2012), to the Andean volcano.[2] Two prestigious German authorities on Humboldt, Oliver Lubrich and Ottmar Ette, have recently edited *Ueber einen Versuch den Gipfel des Chimborazo zu ersteigen* (2006),[3] while Lubrich has published several versions of an article in German, English, and Spanish on the failure of the ascent and the fascination with the vacuum that Humboldt felt in the face of the chasm that, he tells us, blocked his path.[4] The English version was published in a volume edited by Caroline Schaumann, who also has given pride of place to Chimborazo in her recent study of the emergence of mountaineering *Peak Pursuits* (2020), in which Lubrich's postmodern Humboldt becomes a pioneer

DOI: 10.4324/9781003231479-8

of the Anthropocene.[5] The recent exhibition on Humboldt at the Smithsonian American Art Museum (2020–2021) to celebrate Humboldt's influence on the arts and sciences in the United States also highlighted Chimborazo. The majestic image of its massive snow-covered cone above the Tapia Plateau or that of the distribution of vegetation on its flanks in relation to altitude and atmospheric conditions are key elements that are present in nearly every exhibition or visual presentation of Humboldt. I myself devoted a chapter to the subject in 2003.[6] The present contribution is intended as an update of those pages or, as it were, and in deference to Humboldt, my second attempt to ascend Chimborazo.

Clearly, as in Neruda's poem, we are not now what we then were ('nosotros, los de entonces, ya no somos los mismos'). There is no stepping into the same river twice, as Heraclitus put it. The ways we consider and deal with historical objects of knowledge surely change over time. And yet 'the classics,' Humboldt certainly is one, seemingly endure and adapt to everything and address us all. Every epoch feels called and authorised to interpret and accommodate them to its own sensibilities. In his metabiography, Nicolaas Rupke has uncovered the multiple lives of 'Humboldt' in German intellectual and institutional history. The subversive democrat of the Prussian court was transformed first in the chauvinist *Kultur* of the Weimar Republic, then as the exponent of the Aryan suprematism of National Socialism, later as the anti-slavery socialist, and finally and inevitably, as the pioneer of globalisation and the ecology movement – green Humboldt, pink Humboldt, rainbow Humboldt.[7] Is there a neutral, transparent, colourless anything? If there ever was, we would not be able to recognise it. Like Fabrizio del Dongo in *La chartreuse de Parme*, we would wander through the field of battle unable to distinguish Waterloo.

Still, and despite or because of their historical versatility, some things remain, some narratives retain their power, some images do not lose their symbolic charge. The ascent of Chimborazo is one of these. The image of the Andean volcano serves to present a modern, postmodern, colonial, or postcolonial Humboldt, always immense, in his quest for immortality or even in his disdain for worldly glory. There can be no doubt that as individuals and members of a generation, we tend to invent or discover Humboldt as though we were the first to glimpse or reach that level (mountaineers and historians alike often pursue the myth of solitude, the primacy of discovery and the authority it lends). But science, whether in the lab or in the field, is a collective enterprise, a social activity. So perhaps we should start by abolishing the notion of individual discovery, those giddy moments associated with the elevated, isolated genius situated, as the exhibition in the Smithsonian phrases it, 'beyond this world.'

The Fame of the Mountains

Mountains are often associated in our memory with those who painted, described or, literally or metaphorically, took possession of them. Vesuvius has many authors, but perhaps Pliny the Elder, whose death was hastened by his desire to watch its eruption at close quarters, left his petrified imprint on its flanks more permanently, visibly, and earlier than others. Petrarch's scaling of Mount Ventoux inaugurated new forms of understanding the encounter with nature and the self.[8] And what about Perrier, Pascal's brother-in-law, and his barometric experiments on top of Puy de Dôme?

Within Humboldt's programme of geological, mineralogical, and botanical observations in the Andes, to study Chimborazo meant extending the chain of data gathered and studied on Puracé, Pichincha, Antisana, Cotopaxi, and Tungurahua. If we add his later observations in New Spain (Jorullo, Nevado de Toluca, Cofre de Perote) or his ascent of Teide en route, and of Vesuvius after his return from America, we gain an idea of his breadth as an explorer. Humboldt managed to record 407 volcanoes on the earth's surface, 120 of them in America. Not only was he among the precursors of modern vulcanology, but he also scaled and recorded his own observations on many of them.

In his study of Humboldt's Andean investigations, Pere Sunyer describes how he was inspired by Horace Bénédict de Saussure, the Swiss geologist considered by many to be the father of mountaineering.[9] All the same, Humboldt's performance, his longevity and his remarkable physical exertions have served to elevate him above any rival in the history of scientific exploration, aesthetic immersion, or Olympic competition. He did not invent the science of the mountains, nor did he discover Saussure's 'laboratory of nature,' but if he was not the first, he was surely the most exhaustive. Neither Saussure nor Leopold von Buch, Ramond, Deluc, or any of the other numerous geologists or inquirers into mountains could challenge the pole position when it comes to the number of ascents.

There is a certain cumulative voracity, not only in the excursions and observations at high altitude but across all of Humboldt's science. If we follow Dettelbach in understanding Humboldtian science as 'a disciplined practice of observing and measuring a number of physical variables over continental expanses of territory, using the latest advances instrumentation and attending to all possible sources of error,'[10] it is obvious that the greater the number of observed facts, the larger the quantity of data on vegetation at high altitudes or magnetic declinations, the greater the degree of reliability of his pronouncements. Nevertheless, besides the mountainous mass of statistics, we should also pay attention to the Herculean, self-fashioning aspect of Humboldt's scientific practice. There is no small measure of Olympian competition in that massive accumulation of data, that rosary of ascents to the most remote, dangerous, and elevated peaks.

In this sense, Chimborazo was part of a strategy for a man fully aware of these kinds of considerations and, like almost everyone who becomes a public figure, concerned about everything bearing on his own reputation and fame. That Chimborazo was considered the highest peak on the planet was due not only to unfamiliarity with the mountaintops of Tibet or of the Andean peak of Aconcagua to the South, but also to the techniques of barometric and trigonometric measurement that had not yet reached perfection.[11] Until that June in 1802, as far as we know, no one had ever reached its summit, and if they did, it certainly was not with a toolkit like Humboldt's. Mountains, like intellectuals, have their reputation, and Chimborazo's was precisely that: the highest peak in the world. This is how it appeared, for example, in the *Diccionario geográfico-histórico* (1786–1789) of Antonio de Alcedo, who trusted in the measurements carried out by La Condamine and the members of the Franco-Hispanic geodesic expedition:

> It is the highest known today in the world, since its altitude, measured by the academics of science of Paris, is 3,220 toises from sea level to the top, which terminates in the figure of a cone or truncated pyramid.[12]

But let us turn for a moment to Cook and the grand voyages of circumnavigation that fired the imagination of the young Humboldt. It is well known how much Georg Forster, who accompanied Cook on his second voyage and subsequently published *A Voyage Round the World*,[13] influenced the ideas and aspirations of Humboldt. The two men met in Göttingen in 1789 and travelled together on the Rhine and through The Netherlands to England. They visited Paris on their return journey in 1790, whereupon Humboldt left to study first at the School of Commerce in Hamburg, and soon afterwards at the School of Mining in Freiburg. The man who infected him with a passion for the exotic and for what was close at hand, that is, nature and revolution, was also the one who took him on expeditions through the Harz mountains of Saxony. Before going on board with Cook, Joseph Banks, too, had boasted that his grand tour would be a voyage round the world. Towards the end of the eighteenth century, the circumnavigation had become a nautical and scientific feat par excellence, an enterprise as edifying and circular as the very *Encyclopédie*.

Humboldt too desperately wanted to sail around the world, but it proved impossible. And even though he was to achieve it in a different, metaphorical manner (his *Cosmos* embraces and encircles the universe), he did attempt to do it physically. He tells us that he had planned to meet Baudin in El Callao when he heard in Havana that his ships would pass Cape Horn and anchor at the port of Lima. However, Baudin did not sail past Cape Horn but the Cape of Good Hope. Humboldt received news of this as he was measuring the crater of Pichincha at the end of May 1802, shortly before venturing on the ascent of Chimborazo.[14] So, for the moment, he was left with one of his dearest dreams, to enter the Pacific Ocean and complete a circumnavigation of the globe, unfulfilled.

He reconstructed the sequence of the frustrated journey and the dreams of his adolescence, interweaving them in a significant way, in *Views of Nature* (*Ansichten der Natur*) (1808). After eighteen months in the interior of the mountainous country, he felt 'an ardent desire to enjoy a view of the open sea', related to the 'pleasing impressions of youth, with early predilections for particular pursuits, with the inclination for travelling, and the love of an active life.' He evokes the day when Núñez de Balboa was the first European to glimpse the Southern Sea, before himself contemplating the grand spectacle of the vast ocean: 'The view of the Pacific was solemnly impressive to one, who, like myself, was greatly indebted for the formation of his mind, and the directions given to his tastes and aspirations, to one of the companions of Captain Cook.'[15] Forster, Núñez de Balboa, and the failure to meet up with Baudin are an allusion to those who discovered and navigated the Pacific, just as in other passages Humboldt considers himself a descendant of Columbus, Fernández de Oviedo or Acosta.[16] In both cases his aim was to situate himself on the same level as these heroes and indeed to present himself as the 'second discoverer of America' (the title that Bolívar bestowed on him and that Humboldt certainly had hoped to win upon his return to Europe). Like all heroes, Baron von Humboldt needed and strove to perform a highly symbolic feat, an accomplishment that would mark and visibly represent the abstract character of his intellectual conquest. The ascent of Chimborazo was the feat that he was looking for, the accomplishment that could substitute for the curtailed circumnavigation, the unforeseen incident that could be dramatically presented as the supreme moment in which he himself reached the summit of his achievement, the meeting point of earth and sky.[17]

After his return to Europe in 1804, Humboldt embarked on a major campaign to publicise his American journey and his investigations, including his monumental, polyglot publications. Chimborazo features to a various extent in almost all these publications, indeed becoming the leading character in some of them, such as the *Essay on the Geography of Plants*. Aside from his published and unpublished writings, however, it was as the man who had almost reached the peak of the highest mountain on the planet that Humboldt was seen in the public eye. In that same year of 1804, Gay-Lussac managed to rise even higher in a hot-air balloon, reaching a height of around 7,000 m, thus breaking the Prussian's record. Hence, when making the drawing for the publication of the *Essay on the Geography of Plants* (1807), Goethe did not hesitate to include Gay-Lussac in his balloon in the central part of the image, at a level somewhat higher than that of Humboldt, who appears on the right just below the summit of Chimborazo (estimated at 6,544 m at the time). On the left of the engraving with the eloquent title *Höhen der alten und neuen Welt / Tableau comparatif des altitudes de l'Ancien et du Nouveau Monde* is Mont Blanc with Saussure on its summit (estimated 4,775 m).[18] Humboldt had the Swiss alpinist's ascent of 1787 in mind:

> The point where we stopped to observe the incline of the magnetized needle appears to be the highest that any man has reached on this mountain ridge; it is eleven hundred meters higher than the top of Mont Blanc, which that most skilled and fearless of travellers, Mr. de Saussure, had the good fortune to reach, struggling against even greater difficulties than we had to overcome near the top of Chimborazo.[19]

The real *coup de grâce* was not long in coming. In 1807 Crawford was the first Westerner to climb the mountain peaks of Tibet. Chimborazo was no longer the highest mountain in the world, much to Humboldt's regret, as he hinted years later in the prologue to the second and third editions of *Views of Nature*.[20]

Deprived of such a colossal but ephemeral distinction, Humboldt had no choice but to cling to his feat by reinterpreting and recalibrating its meaning. Once again, in *Views of the Cordillera* (1810–1813) he appears to be engaged in restoring importance to his eclipsed achievement. In the chapter on Chimborazo, 'These gruelling excursions, the results of which generally stir the public's imagination, offer only very few results that are useful to scientific progress.'[21] Nevertheless, for Humboldt Chimborazo remains the majestic peak. Although the summit of Chimborazo 'is still four hundred fifty meters below the point at which Mr. Gay-Lussac, during his memorable air voyage, conducted experiments,' and although, 'according to Colonel Crawford, the tallest peak in the Cordilleras of Tibet is over 7617 meters,' he insists that 'the absolute elevation of mountains is an unimportant phenomenon to the true geologist who, engaged in the study of rock *formations*, is accustomed to viewing nature on a large scale.' He concluded that it would not be surprising if 'at some point in the future and in some other point of the globe, someone discovers a peak that surpasses Chimborazo by as much as the highest mountain in the Alps soars above the highest point in the Pyrenees.'[22]

The final vindication was to arrive in his *magnum opus*. In the face of the numbers and measurements, the Humboldt of *Cosmos* (1845–1862), the sage who now only addressed posterity, returned to the rivalry between the mountains. That he did this in the introduction to the work is a measure of its importance in his eyes: 'But although the mountains of India greatly surpass the Cordilleras of South America by their astonishing elevation,' they cannot present 'the same inexhaustible variety of phenomena by which the latter are characterised.'[23] And 'the chain of the Himalayas is also wanting in the imposing phenomena of volcanoes.'[24]

The superiority of tropical nature and the equatorial zones lies in the fact that it is there, among the 'colossal mountains of Cundinamarca, of Quito, and of Peru,' that 'man is enabled to contemplate alike all the families of plants, and all the stars of the firmament.'[25] This is the setting not only for majestic palms, humid bamboo forests, and the other forms of the tropical world but also the place where 'a single glance embraces the constellations of the Southern Cross, the Magellanic clouds, and the guiding stars of the constellation of the Bear, as they circle round the arctic pole':

> There the depths of the earth and the vaults of heaven display all the richness of their forms and the variety of their phenomena. There the different climates are ranged the one above the other, stage by stage, like vegetable zones, whose succession they limit; and there the observer may readily trace the laws that regulate the diminution of heat, as they stand indelibly inscribed on the rocky walls and abrupt declivities of the Cordilleras.[26]

Humboldt had made the right choice after all. The ideal vantage point from which to speak about nature, from its summit, we might say, was his. He set himself outside this world, saving the supremacy of the territory from which he had constructed a large part of his science. The feat of climbing Chimborazo was transformed into the feat of gaining access to a privileged, all-encompassing point of view.

If the mountains were going to become the 'laboratories of nature' of nineteenth-century science, as the firmament had been the theatre of Newtonian mechanics,[27] then Chimborazo would be the icon of phytogeography and vulcanology, a scene from which to observe the forces of nature and their relations, a scene from which it would be possible to invent a science and its inventor. Humboldt recast Chimborazo to suit his own self-fabrication.

Let us examine the three portraits in which the scientist is featured along with his mountain. The first is the oil painting by Friederich Georg Weitsch (1810), the view from the Tapia Plateau in which Humboldt and Bonpland are represented in the foreground with an Indian to whom he is passing a sextant, while behind them we can see other Indians, mules, cacti, and clouds hovering above the majestic volcano in the background (see Figure 7.1). It is an image charged with 'imperial eyes,' the influential reading that the literary critic Mary Louise Pratt made of the Prussian scientist who otherwise 'invented' American nature.[28] The second is the portrait by Carl von Steuben (1812), in which Humboldt is resting on a rocky outcrop with his notebooks in hand and Chimborazo in the background (see Figure 7.2). Humboldt is shown in a meditative pose, ready to record the phenomena, as the volcano has recorded geological history, the ancient revolutions of the earth, indicated by the basaltic columns at the bottom right. The third portrait is the one that Julius Schrader made in the year of Humboldt's death (1859), in which the hoary head of the ninety-year-old scientist matches the snow-covered peak of Chimborazo, both

FIGURE 7.1 Alexander von Humboldt und Aime Bonpland im Tal von Tapia am Fuß des Vulkans Chimborazo, Friederich Georg Weitsch, 1810. Courtesy of the Berlin Staatliche Schlösser und Gärten.

FIGURE 7.2 Alexander von Humboldt in ganzer Figur, Carl von Steuben, 1812. Oil on
canvas. Collection Freifrau v. Heinz, Schloß Tegel. Vorstand der Deutschen
Jahrhundertausstellung (Herausgeber): Katalog zur Ausstellung deutscher
Kunst aus der Zeit von 1775–1875 in der Königlichen Nationalgalerie
Berlin. Verlag F. Bruckmann AG, München 1906 (2 Bände).

bowed down by age, to express their ruinous character in a parallelism between
nature and culture that was greatly to Humboldt's liking (see Figure 7.3).[29]

In the chapter on volcanoes in volume V of *Cosmos*, Humboldt returns to
the fame and comparison between mountains: 'Of all the volcanoes of the New
Continent, the volcanoes of Quito enjoy the most widely-spread renown' because,
thanks to the two men who measured the arc of the meridian, these mountains are
'associated with the illustrious names of Bouguer and La Condamine.' Humboldt
continues: 'Wherever intellectual tendencies prevail, wherever a rich harvest of
ideas has been excited, leading to the advancement of several sciences at the same
time, fame remains, as it were, locally attached for a long time.'[30] It is the conquests
of knowledge that confer their fame on the mountains. Hence, Puy de Dôme is
associated with Pascal and Mont Blanc owes it fame neither to its elevation, 'which
only exceeds that of Monte Rosa by about 557 feet,' nor to the dangers to be
overcome in its ascent but to 'the value and multiplicity of the physical and geological views which ennoble Saussure's name, and the scene of his untiring industry.'

FIGURE 7.3   Baron Alexander von Humboldt, Julius Schrader, 1859. Courtesy of the Metropolitan Museum of Art, New York.

The mountains were not only the scenarios of Olympian exertions but also privileged laboratories for determinate observations, inspiration, and scientific experiments. These qualities made them *lieux de mémoire*, natural monuments, places where glory and 'untiring industry' met, for 'nature appears greatest where, besides its impression on the senses, it is also reflected in the depths of thought.'[31]

Sublime Images

In the light of this conjunction of industry, intellect, and fame, if Chimborazo was not the highest mountain in the world, it could still be considered the most majestic and complete. Humboldt assigned Chimborazo a strategic place in his pages on the global distribution and arrangement of the Cordilleras, the geology, and the history of the earth. This is how the mountain appears in the corresponding passages in *Views of Nature* (1808), *Views of the Cordilleras* (1810–1813), and *Cosmos* (1845–1862). Before all these, Humboldt had already given Chimborazo a key position in his *Essay on the Geography of Plants* (1807), the first publication after the American expedition, in the famous plate designed by Schönberger and Turpin and engraved by Louis Bouquet, the *Tableau physique des Andes et Pays voisins*, the large and multidimensional colour plate in which the entire geography of plants is condensed (see Figure 7.4). The ideal Andean terrain, accompanied by all the physical and barometric variables, conditions of luminosity, humidity, and cultivation of the soils, is spread out in a single image, over which Chimborazo presides.[32]

The mountain was the ideal location to observe the tableau where all the physiognomic transformations took place, the space where the distribution of living

FIGURE 7.4 Tableau physique des Andes et Pays voisins. Engraved by Louis Bouquet in the *Essai sur la Géographie des Plantes*, 1807.

beings was situated. Moreover, the mountain made it possible to bring together all the possible views, it was the source of knowledge, the spectacle that united the whole beauty and diversity of the natural world.[33] It was a microcosm.[34] Humboldt's love of mountains found its highest expression in Chimborazo. The *Tableau physique*, a single image, is the receptacle for his entire conception of unity within diversity and a compendium of his experiments: 'This tableau contains almost the entirety of the research I carried out during my expedition in the tropics.'[35]

A lot has been written on Humboldt's originality and indebtedness in connection with phytogeography. He recognised his debt to Willdenow, Tournefort, and Georg Forster. His own observations on Teide, for example, were already heading in this direction.[36] His relations with Francisco José Caldas, studied in detail by Mauricio Nieto and Alberto Gómez in the present volume, are surely a major chapter in the controversies over priorities and borrowings between Humboldt and Creole science.[37] Phytogeography was not an isolated invention made solely by Humboldt; Humboldt was not the only one to climb Chimborazo; and the invention of Humboldt cannot be attributed to him alone. Some scholars attribute Humboldt's genius to an idea, others to a visual device suited to communicate that idea. The disagreement characterises the history of science today, torn between a discipline once preoccupied by 'what' but now increasingly interested in 'how.' That long-standing epistemic structure of the discipline, still linked to the popular notion of the pioneering discoveries of a lone genius, is now being replaced by an emphasis on the collective and social character of scientific practices. It is enough to read Humboldt's own account of his ascent of Chimborazo to realise that it was not carried out under ideal conditions for measuring, recording, and studying the distribution of plants in relation to such a range of variables.[38]

To return to the *Essay on the Geography of Plants*, it was a strategic and symbolic choice. In this first publication after the expedition, the work dedicated to Goethe in its German edition, and to Jussieu in the French, could be taken in with a single glance at the profile image of Chimborazo. In many ways, this image announced his *physics of the globe*: a science inasmuch as it expounded a way to approach the study of nature, a globe inasmuch as the great Andean volcano brought together all possible climates and habitats, arranged vertically by altitude. Besides the portraits, views, tableaux, diagrams, and other visual devices, Humboldt also wanted to construct that image through a certain literary style. After all, Humboldt was not only a compulsive reader of 'the book of nature' but also a magnificent writer.[39] His whole oeuvre, especially *Views of Nature* and *Cosmos*, is an expression of a manifest literary ambition. This alone would have earned him a place of honour in any universal history of science. The friend of Goethe who wanted to communicate to the reader the ancient communion between nature and the spiritual life of human beings, the explorer who was to fabricate an aesthetic way of discussing and presenting landscape, could not let pass the opportunity to weave a canonical image around Chimborazo, destined to last, like Humboldt himself, beyond measurements and new discoveries.

Certainly, the aesthetic of the sublime is present in the three images that Humboldt established as stereotypes of American nature: the tropical forest, the vast mesetas or plateaus of the interior, and the snow-covered peaks of the mountains.[40] Those peaks offer the grand, dynamic natural spectacle of solitude and freedom, two key concepts central to Humboldt's thought. The enumeration in the preface to the first edition of *Views of Nature* is symptomatic: the forests of the Orinoco, the steppes of Venezuela, and finally the solitude of the mountains of Peru and Mexico. The work is dedicated to oppressed spirits, whom Humboldt invites to follow him 'into the depths of the forests, over the boundless steppes and prairies, and to the lofty summits of the Andes':

> On the mountains is freedom! The breath of decay
> Never sullies the fresh flowing air;
> Oh! Nature is perfect wherever we stray;
> 'Tis man that deforms it with care.[41]

If the literary aim was to reproduce in poetic prose the grandiose aspect of the forces of nature, Chimborazo combined both lofty ideals: grandeur and liberty. In the proverbial dialectic between art and nature, a crucial one for Humboldt's project, its majestic summit will be compared with another:

> On the South Sea coasts, after the long winter rains, when the air has suddenly become more lucid, Chimborazo appears to the observer like a cloud on the horizon; it stands out from the neighbouring peaks; it looms above the entire Andes range just as that majestic dome, Michelangelo's work of genius, looms above the ancient monuments that surround Capitoline Hill.[42]

Its image had no equal. In the past, the nearby El Altar or Capac-Urcu volcano was higher than Chimborazo, but 'its destruction marks a memorable period in the physical history of the new continent,'[43] as Carihuairazo had after its eruption in 1698. Humboldt was primarily interested in the geological history of the New World, a dominion in which the volcanoes were documents and monuments. Of the three summits, the most majestic was the round shape of Chimborazo:

> Travellers who have approached the summits of Mont Blanc and Mont Rosa, are alone capable of feeling the character of the calm, majesty, and solemn scenery of these mountains of the Andes. The bulk of Chimborazo is so enormous, that the part which the eye embraces at once, near the limit of the eternal snows, is about 23,000 feet in breadth. The extreme rarity of the strata of air across which we see the tops of the Andes, contributes greatly to the splendour of the snow and the magical effect of its reflection.[44]

Its appearance before the reader is usually preceded by a moment of mystery. Enveloped in clouds, they are suddenly dispersed, or the air becomes transparent, and then its dome rises above all the others. Its view is always magnificent. In 1764, Kant had chosen the example of big mountains to illustrate the difference between the beautiful and the sublime. Like the description of a raging storm or Milton's depiction of Hell, he declared, the view of a mountain whose snow-covered peaks rise above the clouds, arouses enjoyment mingled with horror. The countryside was beautiful; the lonely shadows in the sacred grove were sublime. The day was beautiful, the night, sublime. Kant described various types of sublime scenes, often associated with the colossal.[45] Great height was as sublime as great depth, but the former aroused astonishment, the latter terror. One was the noble sublime, the other the terrifying sublime. And Kant, too, had recourse to the same example: the basilica of St Peter in Rome. Beauty occupied so much space in its vast and simple design that it conveyed the impression of the sublime, and the whole was magnificent.[46]

It is no coincidence that Humboldt chose the same example to describe the two types of sublimity. The mountain takes on a terrifying aspect in the 1837 account of the ascent of Chimborazo, in which the sheer precipices plunge the reader into the depths of terror. Seen from afar, from the Tapia Plateau or in *Views of Nature*, Chimborazo manifested itself in all its majesty and magnificence in a canonical image of the noble sublime. Perhaps that is why, in the introduction to *Cosmos*,[47] Humboldt implicitly attacked Burke's thesis on the relation between the sense of the sublime and ignorance about natural phenomena.[48] In fact, the literature of the Enlightenment is filled with examples of how the grandeur of nature arouses fear and manifest horror.[49] Humboldt's argument here is remarkable, no doubt inspired by the *Naturphilosophie*, by a new ethic and a new aesthetic of virgin nature. Scientific knowledge and the sense of the sublime could not be opposed, and if they were to be condensed into a single image, the image of Chimborazo was the obvious choice. Image, views, depictions of the sublime – while in *Cosmos* Humboldt described three ways of disseminating the study of nature (animated delineations of natural scenery, landscape painting and the cultivation of tropical flora) that connected with the image of nature,[50] it is the animated delineations of natural scenery or descriptive poetry that interest us here:

> Ought any means to be left unemployed by which an *animated picture* of a distant zone, untraversed by ourselves, may be presented to the mind with all the vividness of truth, enabling us even to enjoy some portion of the pleasure derived from the immediate contact with nature?[51]

It was a rhetorical question, for Humboldt well knew that 'the best description is that by which the ear is converted into an eye', the Arab proverb that he cites on the same page. He had risen higher than anyone else in the double exploration of the description of phenomena and the animated painting of impressive natural scenery. The specialist literature has insisted on the visual, dynamic aspects of Humboldtian science. Anne Godlewsaka, for example, emphasised more than twenty years ago that perhaps the most innovative procedure of Humboldt was his 'visual thinking', a new scientific language.[52] His unceasing experiments with thematic or proto-thematic maps, graphs, diagrams, and all kinds of scientific images are evidence of a sustained attempt to develop a visual language where the importance lay in engraving and reproducing at a single glance the unity and variety of natural phenomena. Bouquet's plate in the *Essay on the Geography of Plants* was a veritable archetype in this new language.

Ottmar Ette has made movement the leitmotiv of his studies of Humboldtian science and writing.[53] On one hand, in the face of the systematic and immobile spirit of Linnean science, Humboldt pursued an organic, lively concept of forces in movement, which also explains his interest in the migrations of living beings, geological shifts and any natural phenomenon expressing animation and movement. In Humboldt, as in Goethe, there is an echo of the continuous movement of the *physis* of the ancient Greeks. On the other hand, his very writing, his poetic prose, was designed to put things before the reader's eyes and to do so in movement (compare Ricoeur's discussion of Aristotelian poetics in his *The Rule of Metaphor*).[54] Wherever one looks, everything in Humboldt seems to lead to animated images, to a cinematographic reading.

The movement in the mountains easily gives way to vertigo, a dynamic sensation, the internal agitation triggered by the sight of grandiose natural scenery. The author of the oldest treatise on the sublime in the West, Longinus, wrote:

> Therefore it is, that for the speculation and thought which are within the scope of human endeavour not all the universe together is sufficient, our conceptions often pass beyond the bounds which limit it; and if one were to look upon life all round, and see how in all things the extraordinary, the great, the beautiful stand supreme, he will at once know, for what ends we have been born.[55]

Humboldtian science assumes that mission: man is destined to contemplate the grandeur of the universe. The ponderous, evocative style of *Views of Nature* or *Cosmos* reflects the majesty of creation. But in the last resort, what is the origin of the extraordinary, the great, the beautiful? Where does the source of the sublime lie? Longinus concluded: 'Sublimity is the note which rings from a great mind.'[56] Humboldt concurred:

> Impressions change with the varying movements of the mind, and we are led by a happy illusion to believe that we receive from the external world that with which we have ourselves invested it (…) It may seem a rash attempt to endeavour to separate, into its different elements, the magic power exercised upon our minds by the physical world, since the character of the landscape, and of every imposing scene in nature, depends so materially upon the mutual relation of the ideas and sentiments simultaneously excited in the mind of the observer.[57]

The fame of the mountains and their sublime character originate in the subject who explores and contemplates them, in this case a man who has been seen as the precursor of so many things (modern biogeography, globalisation, ecology, independence ...), but whose thought also captures a lost moment, that prior to the split theorised by Ritter and Kant between aesthetics and science, when the discovery of the landscape and the mechanisation of the world followed separate paths.[58] His universal vocation is marked by that refusal to abandon the contemplation of nature on a grand scale.[59] That is why, in speaking about Germany and the genius of Goethe in *Cosmos*, he asked himself who if not the great poet, 'has more eloquently excited his contemporaries to "solve the holy problem of the universe," and to renew the bond which in the dawn of mankind united together philosophy, physics, and poetry?'[60]

A Magical Place

The fame and sublimity of the mountains always proceeded from those who climbed, measured, and portrayed them. Some historians have wondered about his delay and reticence in describing his ascent of Chimborazo. Despite the reputation that it secured for him, and the strategic role played by the image of Chimborazo in his science, Humboldt delayed recounting the climb that he had made on June 23, 1802. The only mention he made of it at the time is in a letter from Lima to his brother Wilhelm, sent in November of that year.[61] Humboldt alluded to it in *Views of Nature*, *Views of the Cordilleras* and *Cosmos*, but he did not offer a detailed account of the ascent. It was not until 1837 that he published 'On Two Attempts to Ascend Chimborazo' in a German periodical, subsequently translated to English and published in Edinburgh in the same year.[62] In 1853 he reworked the text in 'About an Attempt to Climb to the Top of Chimborazo,'[63] eliminating the other ascent of Jean Baptiste Boussingault in 1831.

All these texts emphasise the dramatic character of the events. The ascent is an attempt, an experiment, like any journey and exploration.[64] What is put to the test is the experimenter's own body. The senses play the leading role. Scientific glory begins with the physical nature of the achievement, since man enters into relations with nature through his bodily organs. The tropics turn out to be the ideal environment for Humboldt's own body ('The tropical world is my element'). Others also attested to his tropical regeneration during the American journey.

As in his earlier works on animal electricity,[65] Humboldt experimented with his own body. The evidence starts as an autopsy of self. For example, during the ascent of Antisana in Ecuador, Humboldt and his companions Bonpland and Montúfar naturally bled from the lips and eyes. A couple of months later, at the rim of the crater of the Pichincha, 'which no one had climbed since La Condamine and which is more active now,' they felt 'more than 18 very violent earthquakes.' It was the terrifying sublime, 'the most impressive, most melancholy and most terrifying spectacle that one can imagine.'[66]

The climax comes on Chimborazo. The 1837 account of the ascent and the diary entries are full of mentions of the mountain sickness, the pain, the cold, and the suffering that they endured.[67] In these painful conditions, his ulcered foot, bleeding gums and eyes occupy centre stage. He suffers from vomiting and vertigo. His battered body makes its way amid abysses of razor-sharp rocks, through the snow, the mist, and an ever-darkening sky. Mules and horses cannot get this far. Not even his *indios* can follow him, and the *mestizo* from San Juan who accompanies them, a sturdy peasant, suffers more than they do and has to abandon the enterprise. Humboldt recalls that neither the Spanish *conquistadores* nor Gay-Lussac experienced the same lack of oxygen and the same haemorrhages as he and his (almost invisible) fellow sufferers. When a hailstorm bursts, it does not prevent them from collecting rock samples to take to Europe, 'pieces of Chimborazo.' The summit, that has been hidden from view during the whole ascent, suddenly appears to rekindle the hope of victory. Fortunately, he recognises, the climb was the last of those programmed in South America; the previous ones had given them more confidence in their strength.[68]

His suffering body fulfils the role of a scientific instrument.[69] It has been trained and acclimatised to withstand experiment. Goethe, who hated a science based on alembics, nuts and bolts, had said that man was 'the greatest and most exact physical apparatus possible.'[70] Humboldt did not go that far, trusting blindly in his *Künstliche Instrumenten* and never separated from his barometer (or, rather, from his 'Jose' who carried the barometer). But on the day of the ascent of Chimborazo, he recognised that he had brought his sextant and other measuring instruments in vain.[71] His plan to combine barometric and astronomical data was frustrated. Despite the difficulties, he managed to continue measuring and calculating. He applied the barometric formula of Laplace to fix the altitude of the point they reached at 3,016 toises (18,096 feet). He took the altitude of Chimborazo determined by La Condamine at 19,097 feet (the number that was engraved in stone in the Jesuit college in Quito) and calculated that they were 1,224 feet from the summit, a distance 'thrice the height of St Peter's church at Rome.'[72] But beyond the calculations and instruments, it was a physical exertion. The source of all the credibility of the narrative is his body.[73] The view that will later become a universal, standardised way of seeing Chimborazo is an observation made by a pair of eyes that can barely see.

Why the delay in telling the story? Is it the narrative of a failed or frustrated attempt? Why did he discount the importance of adventures of this kind on various occasions? Why are his recollections of the ascent and his wavering on whether it was a triumph, a failure, or simply another attempt, another scientific experiment, valid and insufficient like any other, so ambivalent? Was Humboldt captivated by a 'fascinating emptiness', as Lubrich argues, making him more an existentialist than a romantic hero?

Instead of looking for psychological explanations, perhaps it would be better to resort to magical ones. Long before Thomas Mann narrated the fantasies of Hans Castorp in the sanatorium of Davos, the mountains were already magic mountains that favoured hiding or revealing astonishing facts, sites of cure and redemption.

As places of experiment or natural laboratories, mountains multiplied that condition instead of losing it, since it is in laboratories too that conjuring tricks take place. There, too, things appear and disappear. As in every account of an experiment, the individual subject has to become a collective and impersonal one, the 'I' becomes 'we,' those bloodshot eyes become the eyes of all. The body as source of contamination, pain, false sensations or errors, has to disappear, as in successful conjuring tricks: seen and not seen. At the same time, more detailed accounts and trivialities have to be avoided. This is how Newton proceeded when he described the crucial experiment. Superfluous details are to be spared in the interest of objectivity. The refinements come later. The subject must be avoided. The body of the traveller must be eliminated to give the image of Chimborazo and his science the timeless, impersonal, and incorporeal aura of scientific facts and theories. Humboldt skilfully managed to convert his gaze into the panoramic vision of *Views of Nature* and *Cosmos*. His voice is no longer the dramatic first person but that of the omniscient narrator who manages to convince the reader that his voice is neutral.[74] The bodily experience vanishes. As if by magic, his body has vaporised into thin air. The terrifying sublime, where pain and terror meet, has become the noble sublime, with its grandiose spectacle of nature and the universality of its science. It is the device required to convert his own, particular observations into 'eternal bonds,' to turn his Chimborazo into Chimborazo itself.

Long before Norbert Elias spoke of the affective distance in the observation of things that is produced in the civilising process,[75] George Cuvier had extolled the armchair naturalist above the traveller with Humboldt in mind.[76] While the latter had the advantages of fieldwork (immediacy, contact, intensity), the former was the genuine naturalist who, at a distance, could reconstruct the relations, causes and links between living beings, what escaped the traveller absorbed in fragments. The only real journey of exploration in nature did not involve travel; it took place in the study. The true knowledge of the natural order came not from contact and the experience of the body on the spot, but, paradoxically, from distance, from the very fact of *not* being there: not seeing, hearing, touching. Perhaps that helps us understand the triumph of the later images of Chimborazo compared with the actual Chimborazo that Alexander von Humboldt experienced and suffered. In this respect, he was greedy too, a travelling naturalist who wanted to embody both profiles at the same time, who encountered things and also distanced himself from those things.

Other explanations for his playing down the importance of his achievement, his more or less condescending comments on adventures of that kind, oblige us to indulge in psychological interpretations. I favour two of these, which are possible but not demonstrable. First, his record altitude and the reputation of the snowy summit as the highest in the world were not destined to last long. *Sic transit*. Second, he wanted to contrast his creative genius with his excursion as a young man, his science with his experiments. It is an attitude that rather condescendingly verges on the forced modesty of those who feel themselves to be superior to ordinary mortals. As in ascetics and saints, there is a certain *contemptus mundi* in his resistance to dwelling on his youthful exploit, a contempt for the glories of this world characteristic of those who have now reached an unattainable status.

The mountain itself acquired cultural and political significance beyond phytogeography and vulcanology, even though linked in some form or other with both these sciences. As a microcosm, Chimborazo's iconic image came to stand for the great green capital that antiquity had denied the tropics. Life was now shown to be concentrated where Aristotelian and Ptolemaic geography had denied it. The biggest hotspots of biodiversity are situated in equatorial and tropical latitudes. The plate from *Essay on the Geography of Plants* has become an iconographic motif of the defence of biodiversity and ecology, the most easily recognisable image of the 'green Humboldt' who has become so popular recently.

As for Chimborazo's afterlife in vulcanology and the political discourse of liberty and revolution, to Goethe's great displeasure it was in the Andes that Humboldt abandoned Neptunism for Vulcanism. Epiphanies and conversions take place in desolate plains and in laboratories. The former pupil of Werner gave up Neptunism in favour of the plutonic origin of eruptions of rock, proving that its laws were 'indelibly inscribed on the rocky walls and abrupt declivities of the Cordilleras.'[77] While the followers of Werner defended the universal origin of all rocks from their sedimentation beneath the seas, after his observations among the Andean volcanoes Humboldt switched to defend a regional perspective and the volcanic origin of part of the masses of rock. He created a magmatic theory, linked vulcanism with the tectonic shifts and quakes of the earth and distinguished between the endogenic source of eruptions of rocks and the exogenic origin of sedimentary rocks.

This geological debate had implications for theories of New World nature.[78] To defend the generative power of volcanoes in geological history was another way of attacking the thesis of American degeneracy and to vindicate the primacy of the Andean landscape. Humboldt was interested in volcanoes as productive forces of nature, in the past and present. Goethe, for example, supposed that the American continent had no basaltic rock or medieval battlements, in other words, lacked the mineral of a feudal past and remained aloof from the violent changes that were associated with volcanoes in the language of the time.[79] 'Volcanic' was, and still is, a revolutionary term. In Humboldt's day, that revolution was the French one that formed such an important part of Humboldt's emotional and political education. It is a different version of the metaphysical pathos that for Humboldt is associated with whatever relates to enlivening dynamism. Transported to the American territory, volcanoes were associated with social as well as natural (re)generation and renewal. As 'intermittent earth-springs', the effusions of the river of hell, Humboldt here recalls the river of fire from which lava torrents originate in Plato's *Phaedo*, resonate with a revolutionary sound in the course of human history.

Humboldt did not confine himself to the Western tradition. As much interested in the moral history of the Indies as in their natural history, he transcribed the beliefs of pre-Columbian peoples on the significance of volcanic eruptions in the Kingdom of Quito as omens and portents. In his diary, he mentions a meeting with Leandro Zepla, *cacique* of Riobamba, two days after his ascent of Chimborazo.[80] It appeared that Zepla preserved the memory of a sixteenth-century book in Puruhua (a native language spoken in the region alongside Quechua), which had recently been lost and which contained important information about the history of those peoples. Among them was 'the memorable time' when the great eruption took place of Nevado del Altar, called Cápac-Urcu in Quechua at the time, 'the supreme mountain,' whose summit was once higher than that of Chimborazo. The eruption caused the collapse of the colossus. Zepla recalled having read in that book that the eruption lasted seven consecutive years, during which 'the sun was invisible on the Tapia Plateau because of the rain of ashes that fell all the time.'[81] The priests warned their ruler that the catastrophe was an omen of the end: 'The face of the universe changes, other gods drive ours out. Let us not resist the order of destiny.' It was the end of an era of prosperity, since before the eruption and the collapse of El Altar, 'the climate was healthy and the *indios* lived to the age of 160–170 years.'[82] Volcanoes as portents or divine messages and as ruins of nature are two other inexhaustible themes.

Humboldt fuses the indigenous with the classical world and geology with politics. The analogy between physical and moral forces is at work in the frontispiece of the *Atlas géographique et physique du Nouveau Continent* (1814). It features the androgynous image of an Aztec prince or princess raised by Athena/Minerva and Hermes/Mercury with Chimborazo in the background. This beautiful engraving by Barthélemy Joseph Fulcran Roger, after a drawing by François Gerard, has been widely discussed (see Figure 7.5).[83] One of the commentators has underlined the analogy between archaeological remains and volcanic eruptions that characterised the century in which excavations began at Pompeii.[84] The presence of Chimborazo seems to be a vindication of Humboldtian science, presiding over the scene of American redemption under the motto *Humanitas, Literae, Fruges* (civilisation, learning, cultivation). It is taken from a letter from Pliny the Younger to Valerius Maximus, who had been sent to Greece as a magistrate to oversee the affairs of local government.[85] It is symptomatic that two contradictory readings of the image and the motto are possible, both firmly grounded in Humboldt's attitude towards the relations of subordination and hierarchy in the Old and the New World. On one hand, it is Athena and Hermes, the deities of the Old World, who raise the New through the gift of arts and sciences. The indigenous peoples of America are backward, caught up in an infantile or primitive world. They are recipients and learners or, worse still, frozen in time, fossilised, reified as archaeological remnants. The Europeans are the collectors and investigators, the Americans the collected and investigated. However, a reading of the whole passage in the letter of Pliny the Younger offers another possible reading:

> My affection for you is such that I feel compelled not to direct you—for you have no need of a director—but to strongly advise you to keep in strict remembrance certain points that you are well aware of, and to realise their truth even more than you now do. Bear in mind that you have been sent to the province of Achaea, which is the real and genuine Greece, where the humanities, literature, and even the science of agriculture are believed to have

HUMANITAS. LITERÆ. FRUGES.

FIGURE 7.5 Frontispiece. Humanitas. Literae. Fruges. Engraved by Barthélemy Joseph
Fulcran Roger, in Alexander von Humboldt, *Atlas géographique et physique
du Nouveau Continent* (Paris: F. Schoell, 1814).

been discovered; that your mission is to regulate the status of the free cities,
or, in other words, that you will have to deal with men who are really men
and free, men who have preserved the rights, given to them by nature, by
their own virtues, merits, friendship, and by the ties of treaties and religious
observance. Pay all due respect to the gods and the names of the gods, whom
they regard as their founders; respect their ancient glory, and just that quality
of age which in a man is venerable, but in cities is hallowed. Show deference
to antiquity, to glorious deeds, and even to their legends. Do not whittle away
any man's dignity or liberties, or even humble anyone's self-conceit.[86]

The New World read here as a new Achaea offers a different aspect, a region where civilisation, humanities, literature, science, and agriculture originated and where the new Romans, the Europeans, have come to govern free men whose antiquities, legends, liberties, and dignity should be respected. Traveller and armchair naturalist, revolutionary, and courtier Humboldt bears the guise of a coloniser or a liberator, depending on the season and the eye of the beholder.

Chimborazo also experienced a post-Humboldt aftermath, though one almost always in his wake. The most famous figure here is Simón Bolívar, whose *Delirium over Chimborazo*, that monument to Chimborazo as a liberating iconography, was literally constructed in the footsteps of Humboldt. Written in 1822, the text repeats some of the themes of this chapter:

> I sought the footsteps of La Condamine, of Humboldt. Boldly I followed them, nothing could hold me back.
> I reached the glacial regions, where the air was so thin that I could scarcely breathe. Never before had human feet trodden the diamond crown placed by the Eternal Father on the lofty brow of the King of the Andes.[87]

At the foot of the abyss, delirium seized hold of him. As if set alight by a strange, supernatural fire, he felt possessed by the God of Colombia. There at the summit the apparition of Time, ruler of the past and the future, suddenly stood before him. Bolívar was seized by a holy terror. Time advised him to remember what he had seen: to trace for his fellow men the picture of the physical and moral universe, not to hide the secrets that heaven had revealed to him, and to speak the truth to mankind. It is a programme of activities that Humboldt could have signed. And now American destiny changed its course, since 'Time himself has been unable to check the march of Liberty.'

The reputation of the mountains and the sublime volcano are naturalised for a whole era beyond the scientific achievement and the geography of plants. Humboldt added another reason to include him in the pantheon of heroes of knowledge and of the history of emancipation. In projecting onto Chimborazo a whole series of fields of knowledge and desires, he had managed to link his own myth to that of the resuscitated continent: an American Vulcan, the ancient god of arts and manufactures, awaiting him slumbering among the snows of Chimborazo. That is how the authors of the coat of arms of Ecuador understood it. Designed by the Ecuadorian writer and hero José Joaquín de Olmedo in 1845 and consolidated with some modifications in 1900, it has continued down to the present. It shows Chimborazo in the background seen from the Guayaquil estuary. A steamer is passing, its mast the staff of Mercury, god of commerce, like the one in the frontispiece discussed earlier. Patricio Javier Aguirre is working on this and other images of Chimborazo, including those by not only Frederick Church but also Joaquín Pinto and Rafael Troya, often neglected by the Humboldtians of the 'great tradition.' Aguirre is also studying the ascent of Edward Whymper, the Englishman who reached the summit in 1880 with the brothers Louis and Jean-Antoine Carrel and months later with the Ecuadorians David Beltrán and Francisco Campaña. This young Ecuadorian historian has a very attractive thesis about Chimborazo as a national symbol: it is the allegory of an inexistent grandeur in the past, an image that triggers patriotic feelings and a certain nostalgia for something that never existed, a sort of 'melancholy loop.'[88]

Did liberty lie in the mountains? In 1989 an East German filmmaker, Rainer Simon, directed the first co-production of a feature film involving RDA and RDF, *Die Besteigung des Chimborazo*. It is a dramatisation of the ascent interspersed with biographical flashbacks of Humboldt, a politically charged film screened in the year in which the Berlin Wall fell. Far from Ecuador, Chimborazo dispersed its ashes on a new liberating eruption in the post–Cold War world. Like other peaks and other mountain ranges, Chimborazo and the Andes are monuments that each generation and each culture endows with new and old meanings.

Notes

1 Andrea Wulf, *The Invention of Nature. The Adventures of Alexander von Humboldt, The Lost Hero of Science* (London: John Murray, 2015).
2 Daniel Kehlmann, *Measuring the World*, trans. Carol Brown Janeway (London: Quercus, 2006).
3 Alexander von Humboldt, *Ueber einen Versuch den Gipfel des Chimborazo zu ersteigen*, Oliver Lubrich and Ottmar Ette, comps. (Frankfurt am Main: Eichborn, 2006).
4 Humboldt, *Ueber einen Versuch*, 7–77; Oliver Lubrich, 'Fascinating Voids: Alexander von Humboldt and the Myth of Chimborazo,' in Sean Ireton and Caroline Schaumann, eds., *Heights of Reflection: Mountains in the German Imagination from the Middle Ages to the Twenty-First Century* (Suffolk: Boydell & Brewer, 2012), 153–175, translated into Spanish as 'Vacíos fascinantes: Alexander von Humboldt y el mito del Chimborazo,' *Cuadernos Americanos* 159 (2017), 97–124.
5 Caroline Schaumann, *Peak Pursuits. The Emergence of Mountaineering in the Nineteenth Century* (New Haven, CT: Yale University Press, 2020), 27–75; 'Who Measures the World? Alexander von Humboldt's Chimborazo Climb in the Literary Imagination,' *The German Quarterly* 82:4 (2009), 447–468.
6 Juan Pimentel, 'El volcán sublime: Humboldt desde el Chimborazo,' in *Testigos del mundo. Ciencia, literatura y viajes en la Ilustración* (Madrid: Marcial Pons, 2003), 179–210.
7 Nicolaas A. Rupke, *Alexander von Humboldt. A Metabiography* (Chicago, IL: The University of Chicago Press, 2008), 13–27.
8 Charlotte Bigg, David Aubin and Philipp Felsch, 'Introduction: The Laboratory of Nature-Science in the Mountains,' *Science in Context* 22:3 (2009), 311–321.
9 Pere Sunyer Martín, 'Humboldt en los Andes de Ecuador. Ciencia y Romanticismo en el descubrimiento científico de la montaña,' *Scripta Nova* 58 (2000). On Humboldt and the Hispanic world, see Charles Minguet, *Alejandro de Humboldt, historiador y geógrafo de la América española* (México: UNAM, 1985), 2 vols; Estuardo Núñez and Georg Petersen, *El Perú en la obra de Alejandro de Humboldt* (Lima: Studium, 1971); Miguel Ángel Puig-Samper, ed., 'Alejandro de Humboldt y el mundo hispánico. La modernidad y la Independencia americana, in Debates y perspectivas,' *Cuadernos de Historia y Ciencias Sociales* 1 (2000), 7–28; Thomas Gomez, ed., *Humboldt et le monde hispanique* (Paris: Université Paris X, 2002).
10 Michael Dettelbach, 'Humboldtian Science,' in Nicholas Jardine, James A. Secord and Emma C. Spary, eds., *Cultures of Natural History* (Cambridge: Cambridge University Press, 1996), 287–305, here 287. Dettelbach widens the term 'Humboldtian Science,' borrowed from Susan Faye Cannon, *Science in Culture: The Early Victorian Period* (New York: Dawson, 1978), 73–111.

11 Because of the equatorial bulge, the summit of Chimborazo is the farthest point (6,364.4 km) on the earth's surface from the centre of the earth.

12 Antonio de Alcedo, *Diccionario geográfico-histórico de las Indias occidentales o América*, C. Pérez-Bustamante, ed. (Madrid: BAE, 1967), 312. 1 toise = 1.946 m.

13 Johann Georg Adam Forster, *Voyage round the World in His Britannic Majesty's Sloop Resolution in the Years 1772, 3, 4 & 5* (London, 1777). There is a modern edition edited by Nicholas Thomas and Oliver Berghof (Honolulu: University of Hawai'i Press, 2000), 2 vols.

14 Letter from Alexander von Humboldt to Domingo de Tovar y Ponte, 2/8/1802, in Charles Minguet, ed., *Alejandro de Humboldt: Cartas Americanas* (Caracas: Biblioteca Ayacucho, 1980), 90–92.

15 Alexander von Humboldt, *Views of Nature*, trans. E.C. Otté and Henry G. Bohn (London: Henry G. Bohn, 1850), 419.

16 Mary Louise Pratt, *Imperial Eyes: Travel Writing and Transculturation* (London and New York: Routledge, 1992), 111–144.

17 Most of the biographers, such as Douglas Botting, *Humboldt and the Cosmos* (New York: Harper & Row, 1973), repeat this topos. See also Wolfgang Burgmer, 'Sobre el volcán,' *Inter Nationes* 126 (1999), 46–52.

18 The engraving is included in the opusculum that follows the *Essai*, titled *Tableau physique de régions équatoriales*. All references here are to Alexander von Humboldt and Aimé Bonpland, *Essay on the Geography of Plants*, with an introduction by Stephen T. Jackson, trans. Sylvie Romanowski (Chicago, IL: University of Chicago Press, 2008). The present-day altitudes are 6,266 m for Chimborazo and 4,807 for Mont Blanc.

19 Alexander von Humboldt, *Views of the Cordilleras and Monuments of the Indigenous Peoples of the Americas. A Critical Edition*, ed. with an introduction by Vera M. Kutzinski and Ottmar Ette, trans. J. Ryan Poynter (Chicago, IL: University of Chicago Press, 2012), 126.

20 *Views of Nature*, xiii.

21 *Views of the Cordilleras*, 126.

22 *Views of the Cordilleras*, 226–227.

23 Alexander von Humboldt, 'Introduction to the Third (1849) Edition,' in *Cosmos: A Sketch of a Physical Description of the Universe*, trans. E.C. Otté (New York: Harper & Brothers, 1850), vol. 1, 29.

24 Humboldt, 'Introduction,' 30.

25 Humboldt, 'Introduction,' 33.

26 Humboldt, 'Introduction,' 33.

27 Bigg et al., 'Introduction,' 312.

28 Pratt, *Imperial Eyes*.

29 Schaumann, *Peak Pursuits*. On 'natural antiquities,' and the parallelism between nature and culture in Humboldt, see Peter Mason's contribution to this volume.

30 Humboldt, *Cosmos* V, 274–275.

31 Humboldt, *Cosmos* V, 276.

32 The full title is *Essai sur la Géographie des Plantes; accompagné d'un tableau physique des régions équinoxiales, fondé sur des mesures exécutées depuis le dixième degré de latitude boréale jusqu'au dixième degré de latitude austral pendant les années 1799–1803* (Paris, 1805).

33 Alberto Castrillón, *Alejandro de Humboldt, del catálogo al paisaje* (Medellín: Universidad de Antioquía, 2000), 90.

34 Bigg et al., 'Introduction,' and Schaumann, *Peak Pursuits*.

35 Humboldt, *Essay*, 79.

36 On Humboldt's brief stay in the Canary Islands, see Mason in this volume, and Miguel Ángel Puig-Samper and Sandra Rebok, *Sentir y medir. Alexander von Humboldt en España* (Madrid: Doce Calles, 2007).

37 See José Antonio Amaya, in this volume; Mauricio Nieto, *Americanismo y eurocentrismo. Alexander von Humboldt y su paso por el Nuevo Reino de Granada* (Bogotá: Universidad de los Andes, 2010), 49–70; Jorge Cañizares, 'How Derivative Was Humboldt? Microcosmic

Nature Narratives in Early Modern Spanish America and the (Other) Origins of Humboldt's Ecological Sensibilities,' in Londa Schiebinger and Claudia Swan, eds., *Colonial Botany: Science, Commerce, and Politics in the Early Modern World* (Philadelphia: University of Pennsylvania Press, 2005), 148–165, and Jorge Cañizares-Esguerra, *Nature, Empire, and Nation. Explorations of the History of Science in the Iberian World* (Stanford, CA: Stanford University Press, 2006), 112–128.

38 Humboldt, 'On Two Attempts.'

39 Ottmar Ette, 'Un 'espíritu de inquietud moral.' *Humboldt writing*: Alexander von Humboldt y la escritura en 'la modernidad',' *Cuadernos Americanos* XIII, 4/76 (July–August 1999), 16–43; Juan Pimentel, 'Cuadros y escrituras de la naturaleza,' *Asclepio* 56:2 (December 2004), 7–24.

40 Pratt, *Imperial Eyes*, 125; Manuel Lucena Giraldo, 'Alejandro de Humboldt y la invención del trópico,' in Gomez, ed., *Humboldt*, 43–59.

41 *Views of Nature*, x. The lines are from Schiller's *Bride of Messina*, trans. A. Lodge.

42 *Views of the Cordillera*, 127.

43 *Views of the Cordillera*, 126–127.

44 Alexander von Humboldt, *Travels and Discoveries in South America*, 2nd ed. (London: John W. Parker, 1846), 262. First published in 1840, this is an abridged translation of Alexander von Humboldt, *Voyage aux régions équinoxiales du nouveau continent*.

45 Peter Mason, *The Colossal: From Ancient Greece to Giacometti* (London: Reaktion, 2013).

46 Immanuel Kant, *Observations on the Feeling of the Beautiful and Sublime and Other Writings*, trans. Patrick Frierson and Paul Guyer (Cambridge: Cambridge University Press, 2011 [1764]), 17.

47 *Cosmos* I, 38ff.

48 Edmund Burke, *A Philosophical Inquiry into the Origins of our Ideas of the Sublime and Beautiful* (London, 1757).

49 Typical is Buffon, for whom virgin nature was feeble and tedious.

50 *Cosmos* II, 19–20.

51 *Cosmos* II, 81 (my emphasis).

52 Anne Marie Claire Godlewska, 'From Enlightenment Vision to Modern Science? Humboldt's Visual Thinking,' in David N. Livingstone and Charles W.J. Withers, eds., *Geography and Enlightenment* (Chicago, IL: University of Chicago Press, 1999), 236–281.

53 Ottmar Ette, *Alexander Von Humboldt y La Globalización: El Saber En Movimiento* (Mexico City: El Colegio de México, 2019).

54 Paul Ricoeur, *The Rule of Metaphor* (London and New York: Routledge, 1995).

55 Longinus, *On Sublimity* 35.1, trans. A.O. Prickard (Oxford: Clarendon Press, 1906).

56 Longinus, *On Sublimity*, 9.

57 *Cosmos* I, 26–27.

58 Michael Dettelbach, 'Global Physics and Aesthetic Empire: Humboldt's Physical Portrait of the Tropics,' in David Philip Miller and Peter Hanns Reill, eds., *Visions of Empire. Voyages, Botany, and Representation of Nature* (Cambridge: Cambridge University Press, 1996), 258–292.

59 Ottmar Ette, 'Hacia una conciencia universal. Ciencia y ética en Alejandro de Humboldt,' in Puig-Samper, ed., *Alejandro de Humboldt*, 29–55.

60 *Cosmos* II, 82.

61 Letter to Wilhelm, 25/11/1802, in Minguet, ed., *Alejandro de Humboldt*.

62 See note 38.

63 Published in *Kleinere Schriften*, 1853. The documents bearing on the ascent of the Chimborazo in the diaries of the American expedition can be consulted digitally at https://humboldt.staatsbibliothek-berlin.de/werk/ (Humboldt, Alexander von (1801–1802): [Diarios del Viaje Americano] VII bb u. c. Quito (Pichincha, Cotopaxi, Tungurahua, 40. Chimborazo, Altar). Staatsbibliothek zu Berlin–Preußischer Kulturbesitz (SBB–PK), Nachl. Alexander von Humboldt (Tagebücher) VII bb/c. (Staatsbibliothek

zu Berlin–Preußischer Kulturbesitz (SBB–PK), Nachl. Alexander von Humboldt (Tagebücher) VII bb/c). I am grateful to Tobias Kraft for assistance in locating the sources within this Berlin digitization project.

64 On journeys as experiments see Pimentel, *Testigos*, 25–144.

65 Alexander von Humboldt, *Versuche über die gereizte Muskel- und Nervenfaser* ... (Berlin: Poznań, 1797).

66 Letter to Clavijo y Fajardo, 12/6/1802, in Minguet, ed., *Alejandro de Humboldt*, 88.

67 Letter to Wilhelm, Minguet, ed., *Alejandro de Humboldt*, 92–100; 'On two attempts'; Humboldt, Alexander von (1801–1802): [Diarios del Viaje Americano] VII bb u. c. Quito (Pichincha, Cotopaxi, Tungurahua, p. 40. Chimborazo, Altar). Staatsbibliothek zu Berlin. On the aesthetics of suffering and the painful pleasures of mountaineering, see Maurice Chappaz, *La Haute Route* (Laussane: Bertil Galland, 1974).

68 'On Two Attempts,' 12.

69 Kapil Raj, 'When Human Travellers Become Instruments: The Indo-British Exploration of Central Asia in the 19th Century,' in Marie-Noëlle Bourguet, Christian Licoppe and H. Otto Sibum, eds., *Instruments, Travel and Science: Itineraries of Precision from the Seventeenth to the Twentieth Century* (London: Routledge, 2002), 156–188.

70 Jeremy Naydler, ed., *Goethe y la ciencia* (Madrid: Siruela, 2002), 50.

71 'On Two Attempts,' 10.

72 'On Two Attempts,' 16.

73 On the relation between the body and knowledge of nature see Cristopher Lawrence and Steven Shapin, eds., *Science Incarnate. Historical Embodiments of Natural Knowledge* (Chicago, IL: University of Chicago Press, 1998); Jean Starobinski, *Le Corps et ses raisons* (Paris: Le Seuil, 2020).

74 Pratt, *Imperial Eyes*, 125.

75 Norbert Elias, *On the Process of Civilisation*, Stephen Mennell, Eric Dunning, Johan Goudsblom and Richard Kilminster, eds. (Dublin: University College Dublin Press, 2012); Richard Kilminster and Stephen Mennell, eds., *Essays I: On the Sociology of Knowledge and the Sciences* (Dublin: University College Dublin Press, 2009), vols. 3 and 14.

76 Dorinda Outram, 'New Spaces in Natural History,' in Jardine, Secord and Spary, eds., *Cultures*, 249–266.

77 Humboldt, *Cosmos* I, 33. See above, n. 26.

78 Paolo Rossi, *The Dark Abyss of Time: The History of the Earth and the History of Nations from Hooke to Vico* (Chicago, IL: University of Chicago Press, 1984); Antonello Gerbi, *The Dispute of the New World: The History of a Polemic, 1750–1900*, rev. and enlarged ed., trans. Jeremy Moyle (Pittsburgh: University of Pittsburgh Press, 1973).

79 Umbach Maiken, 'Visual Culture, Scientific Images and German Small-States Politics in the Late Enlightenment,' *Past and Present* 158 (1998), 110–145.

80 Humboldt, *Views of Nature*, 353–379. This chapter reproduces a lecture given in Berlin in 1832.

81 'Notions données par Don Leandro Zepla' (folio 27v and following), Staatsbibliothek zu Berlin–Preußischer Kulturbesitz (SBB–PK); Nachl. Alexander von Humboldt (Tagebücher) VII bb/c.

82 'Notions données,' f. 28v.

83 Letter to Wilhelm, 25/11/1802, Mnguet, ed., *Alejandro de Humboldt*, 97.

84 Peter Mason, 'Moving Mountains and Raising the Dead,' in Maarten Jansen and Luis Reyes, eds., *Códices, caciques y comunidades. Cuadernos de Historia Latinoamericana* 5 (1997), 247–264.

85 Helga von Kügelgen-Kropfinger, 'El frontispicio de François Gérard para la obra de viaje de Humboldt y Bonpland,' *Jahrbuch für Geschichte von Staat, Wirtschaft und Gesellschaft Lateinamerikas* 20 (1983), 575–616.

86 Pliny, *Letters* VIII.24, trans. J.B. Firth, 1900.

87 Simón Bolívar, 'My Delirium on Chimborazo,' in David Bushnell, ed., *El Libertador. Writings of Simón Bolívar*, trans. Frederick H. Fornoff (Oxford: Oxford University Press, 2003), 135.

88 Patricio Javier Aguirre kindly communicated the theme of his doctoral thesis (FLACSO) titled 'El Olimpo en los Andes.' See his 'Montañas y sujetos: un acercamiento teórico a las huellas simbólicas del montañismo en el universo social y cultural,' *Antropología Cuadernos de Investigación* 14 (2014), 67–78. In Jon Juaristi, *El bucle melancólico. Historias de nacionalistas vascos* (Madrid: Espasa, 1997), the author relates the history of a humiliation that could never be righted because it never took place.

8

PERUVIAN *DESENCUENTRO*

Humboldt's Fog, Unanue's Light[1]

Mark Thurner

[En el Perú virreinal] los talentos extraordinarios, la constante aplicación, la sabiduría adquirida por uno y otro no tuvieron más premios que una dependencia inmediata de europeos orgullosos e ignorantes.[2]

– Joseph Hipólito Unanue

Humboldt estuvo algún tiempo en Lima, se paseó bastante, e hizo en un hospital dos o tres experimentos galvánicos: con lo cual se creyó autorizado para hablar ex-Catedra de todo lo relativo a América, y esto no puede ser.[3]

– Ramón Olaguer Feliú

Latin America's periodic 'Humboldt cult' bloomed on schedule in 2019 for Alexander's 250th birthday party. Although this cult surely resonates with the more celebrated German and Anglo cults, in the Americas Humboldt's moon has other suns.[4] To be sure, nearly everywhere Humboldt's name and face are objects of adulation: monuments, schools, streets, plazas, parks, glaciers, and ocean currents are named after him. In Cuba, Mexico, and Ecuador, he is part of the national pantheon of savants, although not quite in the way that he is in Germany; in Colombia, the reception of Humboldt is more nuanced, in part because rival savants such as Caldas and Mutis are also the objects of national cults (see Gómez and Amaya in this volume).[5] In Lima, where in certain regards rival savants such as Unanue have displaced him, the reception of Humboldt has been even more ambivalent, if not grey and cool like the mist of the Peruvian current that bathes its shores.[6] Nor, in the case of Peru, is the ambivalence toward Humboldt new. Suspicion and criticism of Humboldt have made their way into print since at least 1811, when the exasperated Ramón Olaguer Feliú, one of Peru's several elected representatives before the Cortes de Cádiz, who was in Lima when Humboldt visited, slammed the Baron

DOI: 10.4324/9781003231479-9

and those fawning followers who gleefully tread in his 'footsteps.' Olaguer Felíu's comment and politics (he notably defended the equality of all 'free' Americans in the Cortes, including those of native, mixed, and African descent) lead one to suspect that Humboldt's Asiatic view of the intellectual and artistic capacity of natives was being deployed at the Cortes of Cádiz by those who sought to deny citizenship and equality to them.[7] Nearly a century later, the Baron was pelted once again in print when Ricardo Palma, one of Peru's leading intellectual figures at the time, published in the *Ateneo de Lima* a now infamous private letter in which Humboldt trashed Lima. The wry Palma quipped that the Baron's vision must have gotten fogged up by Lima's famous mist, a natural product of what, before Humboldtians named it after their hero, was known as the Peruvian Current.[8] There is perhaps some sliver of justice in the current's modern misnaming after the Baron: in Peru Humboldt probably spread more fog than light. This irony becomes particularly clear when we contrast the Baron's views on Peru and Peruvian civilisation with those of his contemporary, the Peruvian polymath José Hipólito Unanue, who, incidentally, has a better claim on the current.

Humboldt apologists have been quick to point out that Alexander also praised certain residents of Lima. But the Baron's praise of his American counterparts is limited to a handful of individuals, several of them foreigners, and in any case the praise rings hollow, as if it were an apologetic afterthought. More importantly, it is rarely remembered. As the late Teodoro Hampe noted, the tone of Humboldt's private letter to Checa, and published posthumously by Palma, contrasts in sharp fashion with the public statement that appeared in print in 1804.[9] Writing to Checa, Humboldt wrote, uncharacteristically in the Spanish language, these words:

> En Lima no he aprendido nada del Perú. Allí nunca se trata de ningún objeto relativo a la felicidad pública del reino. Más separada del Perú está Lima que Londres, y, aunque en ninguna parte de la América española se peca de un patriotismo excesivo, no conozco otra ciudad en la cual este sentimiento esté más apagado.[10]

So, will the real Humboldt please stand up? Did Alexander learn something about Peru while in Lima? Or did he learn nothing? It is tempting to conclude that both are true, although not in the sense that Humboldt would have us believe. Rather than choose one or another of these rhetorical opposites, it makes more sense to read in the double talk of having it both ways an authorial strategy. Indeed, the attentive reader encounters this rhetorical strategy throughout Humboldt's opus. There is no question, of course, that although one could (and can) learn something of Peru in London one could (and can) learn a great deal more about Peru in Lima, if you know where to look and who to talk to. But what the reader remembers, and what is most often quoted, is the quip that 'in Lima I have learned nothing of Peru.' It is that damning, private statement that rings true, and it is this sentence that has lived long in literary, political, and popular discourse. But the undeniable fact is that thanks to Lima's libraries, archives, presses, and intellectual communities,

Humboldt harvested significant sources on Peru while in Lima. On the other hand, given the short length and nature of his stay, his preference for foreign company, his Prussian disdain for the 'lujo vicioso' of Lima's social and intellectual elite, and the stunted extent of his travels in the country, Humboldt clearly failed to learn many things about Lima and greater Peru that he might have. In short, if he 'learned nothing of Peru in Lima,' it was in part his own fault. In any case, the relative brevity of his stay and the consequent lack of expertise, coupled with his ambivalent views of Peruvians and Peruvian civilisation and his tendency to range wildly over subjects he knew little about, opened Humboldt up to criticism. In many cases, the criticism was sound.[11] Humboldt wrote rather less, and rather less convincingly, about Peru than he did, say, about Cuba, Venezuela, Mexico, New Granada, and Quito where he spent more time and expended greater effort.[12] In the end, the mixed Peruvian reception of Humboldt appears to reflect Humboldt's own ambivalence toward Peru, expressed not only in his writing but in his relatively short, and not altogether happy stay there. Surely Humboldt betrays this ambivalence toward 'the locals' in other American locales as well, but it was in 'grey Lima' that its consequences were more readily displayed and called out. For this reason, and because it is the exception that may not only prove but be the rule, the Peruvian *desencuentro* is particularly enlightening.

The 'Second Columbus' Belatedly Discovers America

That Bolivar regarded the Prussian a 'second Columbus' was in an ironic sense apt since neither Columbus nor Humboldt could possibly have 'discovered America.' Columbus was too early, and Humboldt was too late.[13] Neither 'encountered America.'

Oliver Lubrich has noted that Columbus and Humboldt were alike in the sense that both mistook 'America' for an imagined Orient. Columbus saw 'the Indies' everywhere while Humboldt could see nothing but reflections of Napoleonic 'Egypt.'[14] In both cases, 'America' was an accidental find, a collateral damage, the ironic source of unending fame. But the vague parallel ends there. The Admiral of the Ocean Sea was not, and could not possibly be, an Orientalist. He sailed with the winds into a late fifteenth-century horizon, opening a southwestern route via the Canary Islands toward what existing cartographic knowledge predicted should be the *Mar Indicum*. The appellative name of 'Indies' ('Indias' in Spanish) was the enduring mark of his encounter with what could only be islands (and, later, Tierra Firme) in that vast 'Indian' ocean. As Iberian navigators pushed on, the ancient, Hellenic name and legend of 'Indias' came to gloss the global 'Torrid Zone.' Only later was the name and concept of 'America' invented and applied to the 'New World.'[15] As it were, the coins of 'Indias' and, later, 'America,' were the enduring marks of the first globalisation, established two centuries before Humboldt stumbled onto a 'New Continent.' The young Alexander inhabited and sailed into an entirely different, late eighteenth-century conceptual horizon that, in the Franco-Anglo-Germanic world of letters, was rapidly being colonised by an enlightened scientific and increasingly 'Orientalist' imagination unknown in Columbus's time.[16] Humboldt was not an admiral but instead an ambitious, relatively unexperienced young man with an inheritance to burn and a reputation to gain. He captained no expedition other than the imaginary, paper one of a would-be cosmopolitan natural science that, for the moment, was precariously based in, or exiled from, Paris. This second Columbus sailed as a passenger on military and commercial vessels plying well-known routes, with a passport authorised by the Emperor of the Indies.

Despite his very un-Columbian voyage, means, and times, there is nevertheless a telling, inventive sense in which Humboldt did indeed 'second' the misread example of the much abused and many-times refashioned Columbus. Precisely because of the historical chasm separating the two, Humboldt self-fashioned himself after a Columbus made in France and the United States. This invented Columbus was conjured up by a list of distinguished characters including Raynal, Diderot, Miranda, Bolivar, and Jefferson. As Lina del Castillo has shown, the sculpted 'Discoverer of America' that many of us would recognise, celebrate, or topple today, and that Humboldt surely imagined then, had little to do with the beleaguered admiral of the late fifteenth century. This enlightenment's Columbus was a shining mirror of its own cause of natural liberty and reason, a rhetorical weapon in the polemic against 'despotic' Spain and its Indies. For them, Columbus was a martyr of science and progress persecuted by an 'ignorant and tyrannical' Hispanic Monarchy. Like his Colombian contemporary Francisco de Miranda, Humboldt found it useful to embody this invention.[17] This second Columbus did indeed discover Peru. What did he find there?

Humboldt's Dark and Melancholy Peru

Humboldt's most indelible footsteps were tread in Parisian ink, not on volcanic Andean mountainsides. To 'follow in his footsteps' is to follow a paper trail of scribbles, letters, and images that leads to his apotheosis as a secular saint of globalisation. Humboldt discovered an ancient Oriental or 'Asiatic' Peru that persisted in the faces and spirits of contemporary Peruvians, and he brought this Peru back with him to Paris, where it would leave a deep and lasting impression. In his lavishly illustrated and widely consulted parlour album, *Vuès des cordillères et monumens des peuples indigènes de l'Amerique*, published in Paris in 1810, Humboldt speculated not only that Peru's 'semi-barbarian civilization' was Oriental 'in spirit' and 'style.' He suggested that its mythical founder, the first Inca Manco Capac, was likely of 'Asiatic' origin. Although Humboldt was surely not the first European to propose the Asiatic origins of civilisation in the New World, he was the first to do so in a philosophical fashion, in the process influencing enlightened debates on aesthetics, civilisation, nature, and law.

Humboldt's 'yellow' Incas challenged the 'white' Incas conjured by French philosophes. Thanks in part to courtesan French translations of *The Royal Commentaries of the Incas*, these philosophes had found in the Incas historical inspiration for a universal civilising mission or project to be led by enlightened 'legislators' who, like the Inca, respected 'the natural law.'[18] The wise Incas of eighteenth-century Parisian literature and drama were, as Raynal or Diderot speculated, white or, rather, almost white. In the best-selling, and widely censored, *History of the Two Indies*, thousands of readers learned that Manco Capac and Mama Ocllo were 'whiter than the natives of the land' and so likely 'descendants de quelques navigateurs d'Europe ou des Canaries, jettes para la tempête sus les côtes du Brésil.'[19] In contrast, Humboldt now speculated that Manco Capac had brought benign but nevertheless despotic 'Asiatic laws' to Peru. Gleaning passages from *The Royal Commentaries of the Incas* and other works he likely read while in Lima, Alexander noted (in French) that

> men with beards, and with clearer complexions than the natives of Anahuac [Mexico], Cundinamarca [Colombia], and the elevated plain of Cuzco [Peru], make their appearance without any indication of the place of their birth; and, bearing the title of high-priests, legislators, and friends of peace and the arts which flourish under their auspices, they operate a sudden change in the policy of these nations, who hail their arrival with veneration. Quetzalcoatl, Bochica, and Manco Capac are the sacred names of these mysterious beings.

For Humboldt, 'the history of these legislators, which I have endeavoured to unfold in this work, is intermixed with miracles, religious fictions, and with those characters that imply an allegorical meaning.' In a critical reference to the authors of the *History of the Two Indies*, Humboldt notes that 'some learned men have pretended to discover that these strangers were shipwrecked Europeans,' but, he reasons,

> a slight reflection on the period of the Toltec migrations, on the monastic institutions, the symbols of worship, the calendar, and the form of the monuments of Cholula, of Sogamozo, and of Cuzco leads us to conclude that it was not in the north of Europe that Quetzalcoatl, Bochica, and Manco Capac framed their code of laws. Every consideration leads us rather towards Eastern Asia, to those nations who have been in contact with the inhabitants of Tibet, to the Shamanist Tartars, and the bearded Ainos of the isles of Jesso and Sachalin.[20]

In an oft-cited passage, Humboldt is quoted (again, in English translations) as noting that

> nothing is more difficult than a comparison between nations who have followed different roads in their progress towards social perfection. The Mexicans and Peruvians must not be judged according to the principles laid down in the history of those nations which are the unceasing object of our studies.[21]

This rhetorical manner of, at least on the surface, having it both ways, is, as we have already noted, vintage Humboldt. In this case, as in many others, the gesture obscures his meaning, for Humboldt did not hesitate to make such comparisons in print. But no matter: that part of the rhetorical equation is forgotten. When, as in this case, Humboldt momentarily hesitates or sows a seed of doubt, it is not because he was, as some scholars have argued, a convinced cultural relativist. It was, rather, that he considered ancient Mexico and Peru to be 'Oriental' in 'spirit' and 'style,' that is, to be radically 'other' in these respects to 'those nations which are the unceasing object of our studies' (those who were presumed to form the West), precisely for having travelled 'another road.' The sentence immediately following the above, oft-quoted line makes the distinction clear: 'They are as remote from the Greeks and the Romans as they bear a near affinity to the Etruscans and the people of Tibet.' In short, the reason one should not compare Mexican and Peruvian civilisations to ancient Greece and Rome was because they surely followed different roads 'toward their own perfection.' The objection obviously did not apply to 'Eastern' nations nor to those American civilisations that Humboldt considered to be Oriental in origin or 'spirit.'[22]

The historicist concept of diverging East/West lines of civilisational development expressed here was hardly a novel insight that may be attributed to Humboldt. It is found in German historicist thought since at least Leibniz and was further cultivated by Herder and Goethe, and traces of it are also found in Kant, whose thought exercised a strong influence on Alexander's generation. But historicist veins of thought were by no means the sole property of the Germanic mind. Unheralded varieties of historicism flourished across the polycentric, eighteenth-century Hispanic world as well.[23] This pervious world, it should be remembered, included swaths of southern, central, and western continental Europe plus most of 'the rest of the world,' then generally shorthanded as 'the Indies.' The historians of historicism rarely mention the fact that Giambattista Vico (1668–1744) was appointed Royal Historiographer of Charles III, King of Naples and the Two Sicilies until 1759, then King of the Spains and Emperor of the Indies until 1788. In *New Science* (1725) the Neapolitan savant argued that Eastern and Western nations and civilisations followed different, spiral paths of development guided by their own poetics and culture heroes. Based on his investigations in classical and early modern sources, Vico noted that Eastern nations tended to be founded by 'Zoroastrian' sages, while Western ones claimed origins in 'Herculean' heroes (notably, 'Spain' had both kinds of founders). Similar historicist arguments about the deep history of Spain and Peru, based also on the close reading of classical and early modern texts and oral traditions, are found in the work of Vico's Peruvian contemporary, Pedro de Peralta Barnuevo (1663–1743). Unanue was also a historicist thinker, but his historicism differed in critical ways from Humboldt's now canonical Germanic variety.[24]

Humboldt's influential view, whose boot prints treaded heavily on the nineteenth-century minds of many an admiring or needy savant, was that 'among the Peruvians, a theocratic government, while it favoured the growth of industry, the construction of public works, and whatever might be called general civilisation, presented obstacles to the display of the faculties of the individual.' For Humboldt, 'the empire of the Incas may be compared to some great monastic establishment, in which each member of the congregation was prescribed the duties he had to perform for the general good.' With reference to the founding Inca, Manco Capac, Humboldt argued that

> the most complicated political institutions recorded in the history of mankind had crushed the germe of personal liberty, and the founder of the empire of Cuzco, in flattering himself with the power of forcing men to be happy, reduced them to the state of mere machines.

Indeed, Humboldt compared the 'theocracy' of the Incas with slavery in the American South. He concluded that 'the Peruvian theocracy was, no doubt, less oppressive than the government of the Mexican kings' but that both 'contributed to give the monuments, the rites, and the mythology of the two nations *that dark and melancholy aspect which forms a striking contrast with the elegant arts and soothing fictions of the people of Greece.*'[25]

The enlightened Prussian's bifocal vision, Occidentalist when it came to 'our' soothing arts and sciences, Orientalist when the subject was the 'Asiatic laws' of Incas, Aztecs, and Toltecs, found its iconic, visual expression in the striking engraving by Barthélemy Joseph Fulcran Roger after a drawing by François Gerard, commissioned by Humboldt as the frontispiece for *Atlas geographique et physique du Nouveau Continent* (Figure 8.1).

As Juan Pimentel notes in his chapter in the present volume, 'it is symptomatic' that at least 'two contradictory readings of this image and its motto are possible.' Humboldt extracted the legend or motto, *Humanitas, Literae, Fruges* from a letter from Pliny the Younger to Maximus, written on the eve of his departure on a mission to the Roman province of Achaea, where from his base in the city of Athens he would be charged with governing Sparta. In the letter, Pliny bids Maximus to

> [b]ear in mind that you have been sent to the province of Achaia, which is the real and genuine Greece, where the humanities, literature, and even the science of agriculture are believed to have been discovered; that your mission is to regulate the status of the free cities, or, in other words, that you will have to deal with men who are really men and free, men who have preserved the rights, given to them by nature, by their own virtues, merits, friendship, and by the ties of treaties and religious observance. Pay all due respect to the gods and the names of the gods, whom they regard as their founders; respect their ancient glory, and just that quality of age which in a man is venerable, but in cities is hallowed. Show deference to antiquity, to glorious deeds, and even to their legends. Do not whittle away any man's dignity or liberties, or even humble anyone's self-conceit. Keep constantly before you the thought that this is the land which sent us our constitutional rights, and gave us our laws, not as a conqueror, but in answer to our request.[26]

Perhaps the most obvious and oft-repeated reading of the image is that it expresses a 'colonialist' and perhaps Black Legend view of 'the New Continent,' in which

HUMANITAS, LITERÆ, FRUGES.

FIGURE 8.1 Frontispiece. Humanitas. Literae. Fruges. Engraved by Barthélemy Joseph Fulcran Roger, in Alexander von Humboldt, *Atlas geographique et physique du Nouveau Continent* (Paris: F. Schoell, 1814).

modern European commerce and knowledge redeem a ruined and hapless native America. Pimentel entertains another possible reading, derived in part from the contents of Pliny's letter: that for Humboldt the 'New Continent' is a new Achaea. Still, Humboldt's views on ancient civilisation in the Americas cause us to hesitate at such a reading. Humboldt's New Continent was an Oriental, anti-Achaea that, as we have seen, for the Baron formed 'a striking contrast with the elegant arts and soothing fictions of the people of Greece.' It was those 'soothing fictions' (of an ancient Greece conjured by enlightened Germans) that would uplift this anti-Achaea.

Beyond the ready, 'symptomatic' readings of Humboldt as 'coloniser' or 'liberator' lies a third, more encompassing possibility: Humboldt as 'fashioner' or, in today's social media lingo, 'influencer.' That is, the engraving may suggest as much about Humboldt's self-fashioning and self-promotion as it does about his composite vision of 'the New Continent.' The two were always hard to separate, in part because Humboldt appears to have cultivated a certain complicity or ambivalence in this regard. In the image, Humboldt's greatest conquest and most enduring source of fame, oversees the scene. Below this mythical, Humboldtian Chimborazo, a fallen Aztec king (presumably Moctezuma but later identified with the Inca Atahualpa[27]) is being lifted from his pyramidal tomb, ruined by the Spaniards. The lifting is done by Hermes/Mercury, the god of commerce and patron of travellers, while Athena/Minerva, the god of letters and wisdom, extends an olive branch. Below Moctezuma's right knee yet another fallen Humboldtian icon adorns the scene: the 'sacerdotisa azteca' or Xilonen that Humboldt himself uplifted and delivered, Mercury-like, in print, to European readers as the very first plate in his *Vùes des Cordillères et monumens des peuples indigènes de l'Amérique*, no doubt selected for its Oriental resonances.

Pliny the Younger's famous letter to Maximus closes with lines that must have resonated with the Humboldt who had recently returned to the new Rome named Paris:

Quo magis nitendum est ne in longinqua provincia quam suburbana, ne inter servientes quam liberos, ne sorte quam iudicio missus, ne rudis et incognitus quam exploratus probatusque humanior melior peritior fuisse videaris, cum sit alioqui, ut saepe audisti saepe legisti, multo deformius amittere quam non assequi laudem.[28]

Humboldt would not have missed that Maximus had travelled to Achaea 'by lot' (a sort of honorary post, granted by the Roman emperor, it seems) and not by choice; he had travelled to America 'by lot' as well, and with a passport issued by the emperor. Both men travelled under the auspices of Mercury/Minerva to an unanticipated outpost, but there the comparison ends. Maximus travelled to the old homeland of Hermes/Athena, Humboldt to the new land of 'Quetzalcoatl, Bohica, and Manco Capac.' This sad and melancholy province to which Humboldt had come by a twist of fate and from which he now returned triumphant, was very, very far from ancient Greece.

It cannot be missed that the 'individual' faculties that Humboldt found lacking in Peruvians were precisely those that Winckelmann had identified as the fountainhead of ancient Greek aesthetics. In Winckelmann's evolutionary scheme, this aesthetic represented the pinnacle of artistic sensibility. In this new land of Peru, however, Humboldt found a 'downtrodden race' utterly lacking in imagination, aesthetics, and individuality.

I know no race of men who appeal more destitute of imagination. When an Indian attains a certain degree of civilization, he displays a great facility of apprehension, a judicious mind, a natural logic, and a particular disposition to subtilize or seize the finest differences in the comparison of objects. He reasons coolly and orderly, but he never manifests that versatility of imagination, that glow of sentiment, and that creative and animating art which characterize the nations of the south of Europe, and several tribes of African negros. ... Many Indian children educated in the college of the capital or instructed at the academy of painting founded by the king, have no doubt distinguished themselves; but it is much less by their genius than their application.

This passage is immediately followed by Humboldt's authorial double-speak, his rhetorical acumen for having it both ways, in which he first delivers memorable, gross accusations and then mawkishly withdraws them in a subsequent passage that practically no one remembers, and only 'revisionist' scholars, reference. As the epigraph to this chapter makes clear, some Peruvians understood that this strategy reflected Humboldt's own ambivalence; they also noted with alarm the damage that such a strategy could inflict on its victims.

I deliver this opinion, however, with great reserve. We ought to be infinitely circumspect in pronouncing on the moral or intellectual dispositions of nations from which we are separated by the multiplied obstacles which result from a difference in language and a difference of manners and customs. A philosophical observer finds what has been primed in the centre of Europe on the national character of the French, Italians, and Germans, inaccurate. How, then, should a traveller, after merely landing in an island, or remaining only a short time in a distant country, arrogate to himself the right of deciding on the different faculties of the soul, on the preponderance of reason, wit, or imagination among nations?[29]

Unanue's Brilliant and Ingenious Peru

Joseph Hipólito Unanue's (1755–1833) 'physiological' historical science of the civilisation, imagination, environment, and aesthetics of Peruvians stands in stark contrast to the Prussian's. Where Humboldt saw dark, melancholy, 'Asiatic' automatons, Unanue beheld ingenious, beautiful, and universal men well-adapted to their spectacular 'clime.'

The Lima-based polymath's medical and historical science both anticipated and notably diverged from that of the more celebrated Humboldt. The anticipation followed from the historical fact that the sciences of both men were products of a polycentric enlightenment characterised less by a unified 'European project' than a methodological and epistemological pluralism that reflected the distinct vantage points of intellectual communities and cultural milieus unevenly connected via faulty networks and idioms of exchange. This Babelian, transoceanic Enlightenment nevertheless mattered.

Another reason that Humboldt's work was anticipated by Creoles like Unanue was a certain historical resemblance between, on one hand, 'cameralist' or statist science in the protestant principalities and provinces of northern and central Europe where von Humboldt was weened and, on the other hand, the so-called bureaucratic or court and council paperwork of knowledge production in the kingdoms and provinces of the polyglot Iberian world, which provided outlets for Unanue. Both these traditions privileged the gainful production of useful knowledge at the behest of state, church, and commercial or company interests. In both cases, these interests could be defined as 'national' and/or 'imperial,' albeit on vastly different scales. Humboldt's *Political Essay on the Kingdom of New Spain* (1811) is a case in point. Humboldt himself notes that his *Political Essay* is modelled in part on the brilliant precedent of Unanue's annual statistical essays on the Kingdom of Peru. In addition, and by his own admission, the Baron's statistical essay merely 'digests' data culled from Mexican archives, savants, and local informants. Finally, this derivative work is dedicated to his patron, the Bourbon king of Spain and the Indies, Charles IV, with these words: 'How can we displease a good king,' Humboldt writes, 'when we speak to him of the national interest, of the improvement of social institutions, and the eternal principles on which the prosperity of nations is founded?'[30] Notably, this tardy dedication was fruitless, for it was printed only days before Charles IV was deposed at the behest of Napoleon Bonaparte.

Our two enlightened men of science and letters could have, but apparently did not meet in Lima, as neither man mentions any such meeting in their letters. If, in fact, they did not meet, as seems to be the case, then surely it was a huge, missed opportunity for Humboldt, a monumental *desencuentro*. If they did meet and there is no record of it, the result is the same. Unanue was probably the most influential intellectual figure in Lima at the time and for decades to come. His library rivalled any in Lima.[31] We know that, by the time of his death, Unanue's library held several of Humboldt's works. Eventually, each took advantage of the other's reputation to advance their respective careers, but it cannot be said that the reciprocal influences were strong. As many scholars have noted, Humboldt at one point dutifully praised Unanue's *Observaciones* (and Unanue made use of the recognition), but he does not seem to have engaged or absorbed its more critical insights.[32] He also cast doubt on Unanue's population estimates, and Unanue, in turn, defended himself. Did Humboldt learn anything of Peru in Lima? Not much from one of its leading lights, it seems.

Undoubtedly, the two enlightened 'physiologists' would have had many things in common, but a theory of the origins of the Incas and 'American civilisation' was not one of them. It was not merely a matter of Unanue's 'patriotic epistemology' or 'amor propio' getting the better of him, as some have implied. Despite his cosmopolitan airs and Parisian fashions, Humboldt, too, was a 'patriotic epistemologist' of the Germanic variety. He served vested, courtly interests in Prussia; moreover, his thought was marked by the German classicist fantasy of Greek origins.

Both men were liberal courtesans who sought to create useful knowledge and make that knowledge available to academies, periodicals, universities, princes, viceroys, kings, governors, and 'legislators.' And when the time came, both found themselves siding with anti-colonial revolutions, Unanue in the flesh, Humboldt only, and then hesitatingly, in a spirit of solidarity. What separated the otherwise kindred souls, then, was not epistemology or patriotism but contrasting visions of global history and aesthetics. The contrasting visions were in part a product of the distinct milieus from which, and for whom, they wrote.

Following Marie-Noëlle Bourguet, Miruna Achim has suggested that Humboldt pioneered not only a 'universal science of man' but also an 'anthropology of diversity.'[33] If this is so, then one would have to venture that Unanue also did so in his own way, and indeed that he was quite possibly ahead of the Prussian on both counts. How could this be? How could a provincial Peruvian anatomy professor in the gloomy, vainglorious 'City of Kings' be 'ahead' of a globetrotting, Prussian aristocrat who resided in the fashionable City of Lights? Why, in short, might it be better to think about 'man' from Lima or Madrid rather than Paris or Berlin?

To entertain seriously such risky if not foolhardy questions, one must begin by recalling that both men built their respective 'physiologies' upon an imperial Hispanic tradition of 'natural and moral history' and 'cosmography' that, with varying degrees of success, had long sought to apply the sciences to 'the arts of rule' in the Indies. In this regard, the Jesuit polymath Joseph de Acosta was greatly admired and frequently cited by both men.[34] As one might expect, Unanue readily recognised his debt to Jesuit and Peruvian cosmographers. Humboldt also recognises this debt, if sporadically, as several contributions to the present volume attest. For his part, Unanue sought to extend the Hispano-Peruvian cosmographic tradition with a 'physiological' or natural-historical science of 'clime' and 'man,' based primarily on 'observations,' or cases, and informed by key recent works of enlightenment science, medicine, and philosophy, including those of Montesquieu, Condillac, Newton, Hume, Cullen, and von Haller. In addition, Unanue's approach to the study of 'man' is supported by historical research in archives and early chronicles, supplemented by archaeological, linguistic, and ethnographic forms of evidence. Humboldt sought to combine mineralogical, botanical, natural-historical and mathematical methods, but in the Americas, he also ventured into antiquarian, linguistic, political, and historical research. Although the wingspan of Humboldt's career and science eventually stretched over the entire planet, Unanue's vision is, in a Peruvian way, also global. But unlike Humboldt's vision and much to its credit, Unanue's historical science sidesteps the East/West divide that dogged enlightenment thought in Europe.

Unanue's wide-ranging 'observations' on the influences of Peru's clime, published first in Lima in late 1805 (one year after Humboldt's return to Europe) and then in an expanded and revised edition in Madrid in 1815, is a good example of the global reach of his thought and method.[35] As one would expect of any enlightened natural-historical work of the period, in *Observaciones* it is possible to trace many textual influences, including those that we now call, anachronistically, 'French,' 'Dutch,' 'German,' 'Italian,' 'Spanish,' or 'British.' But as in so many works of the period on the elusive subject of 'clime,' Montesquieu could not be avoided. In chapters 2 and 3 of Book XIV of *The Spirit of the Laws*, titled 'On the Laws in Their Relation to the Nature of the Climate,' the Baron of Breda had declared:

> Cold air shrinks the extremities of the exterior fibres of our bodies, and that increases their elasticity and favours the return of blood from the extremities towards the heart. It decreases the length of these same fibres; thereby further increasing their strength. Warm air on the contrary relaxes the extremities of the fibres and lengthens them; it thus reduces their strength and elasticity. People therefore have more vigour in cold climates. ... The peoples of warm countries are timid, like the aged; those of cold countries are courageous, like the young. ... The Indians naturally lack courage; even the children of Europeans born in the Indies lose the courage of their own climate. But how can we reconcile this with their atrocious acts, their customs, their barbaric forms of penitence? Men subject themselves there to unbelievable sufferings, the women set fire to themselves: that is a lot of strength for so much weakness. Nature, which has given those peoples a weakness that makes them timid, has also given them such a lively imagination that everything impresses them to excess. That very organ delicacy that makes them fear death also serves to make them dread a thousand things more than death; it is the same sensitivity that makes them flee all perils and yet defy them all. As a good education is more necessary to children than to those whose mind has reached maturity, so the peoples of these climates have greater need of a wise legislator than the peoples of our own. The more easily and strongly one is impressed, the more important it is that it be in the right way, that it not receive prejudices, and that it be led by reason.[36]

For Unanue, the political and cultural consequences of hypersensitive fibres or nervous tissue in equatorial American humanity differ notably from those that Montesquieu imaginatively ascribes to India and, indirectly, to all those 'children of Europeans born in the Indies' who 'lose the courage of their own climate.' Unanue was one such child, and he would have taken offence. But it was more than a question of 'patriotic epistemology.' In the Indies or 'Indias Occidentales' of which Peru was part, the equatorial sun similarly exercised elastic nerve endings but in such a way that rapid pulsations of the images of observed objects are transmitted to the brain, producing powers of imagination unknown among men confined to temperate zones. Here, Monte's 'exalted imagination' becomes the 'generous imagination,' producing not only the enervating passions imagined by Europeans but also rapid advances in the arts and sciences. To back up his observations-based argument about the American imagination, Unanue draws on the relevant historical, philosophical and travel literature, including the 'sensationism' of Condillac, the 'neural physiology' of von Haller, the 'topographic medicine' of British and French physicians, and the Hispano-Peruvian cosmographic tradition (Acosta, Peralta, Bueno, etc.) to which he is heir.

Based on his experiments and observations, Unanue concluded that 'to those who are born in this New World is granted the privilege of exercising a superior imagination and of discovering all that flows from comparison.' In clear contrast to Montesquieu, and drawing on Condillac and von Hallen, by 'imagination' Unanue did not mean 'those strong and tumultuous impressions, excited by objects analogous or contrary to our passions, and profoundly engraved upon our sensory organs, recurring perpetually and involuntarily, almost forcing us to act like brutes, without deliberation or reflection,' but instead, the physiological 'power to rapidly perceive the images of objects, their proportions, and qualities, from which arises a facility to compare and energetically exploit those images.' In this manner, Unanue continued, 'are our thoughts illuminated, our sensations augmented, and our sentiments painted with vigour.' This acute tropical imagination was surely

> the source of that astonishing eloquence with which the savages of America often express themselves: the natural and strong comparisons of their discourse, and the vitality of their sentiments. After hearing the stirring speeches of Araucanian warriors, we are convinced that Colocolo was no less deserving of the consideration of Ercilla, than was Nestor of Homer. ... From this same precious source flows skill and dexterity in sculpture and painting, *without any other master than one's own ingenuity*.[37]

The clear implication of Unanue's 'physiology' of the American tropical imagination was that Manco Capac's wise laws were the product of 'his own ingenuity,' that is, of those comparative powers of reason that blessed the native-born 'legislator.' Nor was this keen ingenuity a thing of the remote past. Burning under the equatorial sun but sweetened by an unparalleled, Andean diversity of climes, Unanue argued that Peru's homegrown ingenuity had guided her history from the Inca age down to the present.

The human diversity of Peru paralleled its natural diversity and, for this reason, defied and exceeded European classification schemes such as those developed by Montesquieu, Buffon, or Camper. There was nothing simple-mindedly 'torrid' about Peru, however, and nothing 'Asiatic' about its laws or faces. Peru embraced all the climes of Europe, Africa, and Asia, from snow-capped volcanos with permanent glaciers down to high pasture and temperate valleys to subtropical slopes followed by cool deserts and ocean on the western slopes of the Andes and rain forests and the Amazon River system down its eastern flank.[38] All manner of humanity, and with the modern arrival of Old World peoples, every race and mixture of races, gainfully inhabited Peru's altitudinal niches and teemed in her cities. In short, Peru was a microcosm of the globe in a way that no European country could ever be. The Inca's sun spanned the globe without abandoning Peru. The same was true of Unanue's science. It did not need to map Asia, Europe, and Africa because the climatic and human elements of all the continents were found in Peru.

Ironically, Humboldt's 'cosmopolitan' view of Peru and Peruvians was the more provincial of the two. After Winckelmann, he was convinced that ancient Greece was, for its individual freedoms and aesthetics, the pinnacle and fountainhead of the arts and sciences, patriotism, and beauty. This wildly influential view was notably underwritten by the notion that, burdened by despotism, Oriental civilisations yielded only monotony and decadence. Unanue did not share these novel 'European' prejudices. For the Peruvian, Asia and Africa and not ancient Greece were the original sources of beauty, ingenuity, and civilisation in the Old World. These cultural gifts found their way to the New World in the sixteenth century via the Arabs and Iberians, where they met with civilisations that, 'by their own ingenuity,' had developed without Old World assistance. Moreover, it was Spain's Arabic or Andalusian and Jewish intellectual culture that transmitted these gifts northward to a once barbarous and now 'ingrate' Europe.

Unanue's primary target in *Observations* is not Humboldt nor Montesquieu but instead the prestigious Dutch physician, Petrus Camper (1722–1789), author of *Über den natürlichen Unterschied der Gesichtszüge in Menschen verschiedener Gegenden und verschiedenen Alters.*[39] Camper had studied at Leiden, was a member of the Royal Society at London and his acquaintances included Buffon and Goethe. In *Über den natürlichen Unterschied* Camper advances a very precise thesis on the distribution of intelligence and beauty, based on a comparison of facial angles. Unanue does not single out Camper but instead directs his critique broadly at what he calls the European 'tribunal.' Although

> all the nations on earth dispute the favour of ingenuity, that precious gift that distinguishes man from beast, the Europeans – who today triumph over the other three-fourths of the globe, not less by the energy of their pens than by the force of their victorious arms – have erected themselves as tribunal and have passed sentence in their own favour.[40]

This self-appointed European 'tribunal' now proclaimed, 'that the external features of the body are a certain sign of the excellence of the soul that inhabits it' and that

> all the world's beings are linked together in a single chain tied to the foot of God's throne, descending from angels to man, who loses the beautiful dispositions of his body in accordance with the degradation of the privileges of his soul, until he meets with the beasts.

The principal indicator of talent was now 'the raised brow, such that, even among irrationals, for having this distinction the elephant is the wisest; at the same time, the raised brow that is proportionate to the rest of the facial features constitutes beauty.' In short, Europe's 'tribunal' (Camper) on intellect and beauty had reduced the Great Chain of Being to the curvature of the brow.

This European 'tribunal' now decreed that the 'ancient statuaries of Greece, which have left us the models of beauty, measure the most perfect angle of 100 degrees' and that 'any wider angle supposes an imperfect face.' The statuaries of Rome were a close second, with an angle of 95 degrees. Under this European scheme, if these two models are applied to the 'faces of nations,' they

> will reveal that the Europeans occupy the first slot, with a facial measurement of around 90 to 80 degrees; the Asians are in second place, with angles of from 80 to 75 degrees.... The Americans ... with facial angles of 75 to 70 degrees are third. Finally, these proportions decrease in Africans, among whose Blacks the facial angles are only 70 to 60 degrees, which is arriving at the measurement of quadrupeds.

Of course, it is 'for that reason the Negro is the last on the chain, the one who links man with beast.' With the same scheme of degradation, 'the talents' are said to

> descend from the celestial and sublime of the European to the stupid and rude of the Negro. That is why men born in that happy part of the globe are thoughtful men, and only among them may laws, arts, sciences, and valour flower.

Under this scheme, 'the Asian is without talent to reform his pleasures and despotism'; the American 'cannot leave his ignorance behind'; and 'the Negro' is incapable of conquering 'his brutality.' Consequently, the 'other three quarters of the globe' may not present 'any other advantage relative to the European except that of their bodily senses, supposing that the acuity of those senses increases in proportion to the decrease in the privileges of the spirit.'[41]

A superficial and not uncommon reading would conclude that Unanue agrees with Europe's self-appointed 'tribunal.' Americans did after all possess an 'acuity of the senses.' But in the American case the relation between 'bodily senses' and 'privileges of the spirit' was, in fact, the inverse. Americans were privileged spiritually because their sensual acuity produced a heightened ingenuity based on 'quick comparison.' Unanue's anatomical investigations in Peru demonstrated the existence of what he took to be rapid pulsations from sensory organs to the brain; these were read as evidence of an accelerated capacity for comparison and thus imagination. In short, South Americans were naturally agile thinkers, well-adjusted to the challenges of their own, circum-equatorial milieu.

On the related question of aesthetics and civilisation, Humboldt and Unanue could not be more distant. In his characteristically ironic tone, Unanue wrote:

> Let us congratulate the Europeans for their scheme, in which handsome features are the most certain sign of the nobility of spirit; for if such is the case, then ... there are many peoples capable of competing with, and even exceeding Venus.

Unanue's sources on the stunning beauty of men and women in Asia and Africa were drawn from the same travel literature that had inspired European philosophes. In a reference to Louis Antoine de Bougainville's *Voyage autor du monde* (1771), Unanue notes that

> there is no place, says Mr. Bougainville, where one may encounter models more bizarre than Hercules or Mars than in Tahiti. The women exhibit features no less gratifying than those of European women, and in the symmetry and fine proportions of their members, they compete with the well-endowed.

But Unanue is also an eyewitness to a rare beauty beyond Europe.

'Here in South America big black eyes animated with fire are common. The Greek artists paid much homage to this aspect of beauty in both sexes, so that in their busts and medals the eyes are always larger than in those of the ancient Romans.'[42] Unanue observes that the eyes of Andean llamas are bigger, wider lashed, and at least as beautiful as those of the fabled antelope, with whom beautiful African women were so often compared. Who could deny the seductive beauty of the limeña? Did it not beckon from her big, dark, llama-sized eyes and eyelashes? Judging by the letter he wrote in Spanish and dispatched from Guayaquil to the Governor of Jaen, Humboldt turned a blind eye to this seductive beauty.

Reclamations of the key role of Arabic learning in the history of European civilisation were not uncommon in eighteenth-century Spanish and Creole historical discourse, when Hispanic intellectuals battled to deflect 'the ever-louder northern European criticism of the Spanish mind' and its empire.[43] From Montesquieu in the 1740s to Hegel in the 1820s, 'Europe' or at least its 'head' had been shrunk to cramped, Franco-Germano-centric contours, where it was refashioned as a 'fortress of reason' with ancient Hellenic roots. In effect, this fortress 'Europe' now marginalised the southern and eastern expanses of the subcontinent from which most of what was now called 'European civilisation,' had flowed. The peninsula of Iberia was semantically dislodged and set adrift, as 'Europe' was fancied to end at the Pyrenees (and the Italian Alps), as indeed it does in Hegel's *Philosophy of History*.

Responding critically to this Francophile fashioning of Europe, many intellectuals based in Spain, Italy, Portugal, and their Indies, lands considered by French critics (Montesquieu was especially influential) to be in decisive ways more 'Oriental' and 'African' than European, moved to quickly provincialise this new 'Europe' as an ingrate and *arriviste* conceit. The vital intellectual defence of Spain and its Empire of the Indies as modern heirs to the science and civilisation of ancient Mediterranean and Near Eastern empires found a resounding echo in New World cities like Lima where Creoles, Mestizos, and Indians, many of whom fancied themselves to be heirs to the fabled Inca Empire, had long commingled with peoples of African and Asian origins. For Unanue, the follies of the modern European mind were exposed by 'the vicissitudes of human affairs,' and the global history of ingenuity and beauty. Asia and Africa were the cradle of the arts and sciences, and since to found sciences takes more talent than it does to advance them, I am not at all sure whether the souls that animate the bodies of those with high brows have any clear advantage over those with low ones.

Far from 'aping Europe' as so many have claimed, Creoles like Unanue saw modern Europe not as the fountainhead and tribunal of history but instead as the new home of a self-interested amnesia. As Unanue noted:

> Empires fall under the weight of their own splendour and culture, leaving behind only imperfect traces of their existence; at the same time others arise amidst rustic nations and, in their newfound felicity, forget the origins of their lights, and like ungrateful children destroy the breast that suckled them.

Those rustic nations of northern Europe had forgotten that it was in the sixth century that 'the lights that Asia and Africa had carried to Greece, and in Europe to Spain, were eclipsed.' During this extended eclipse,

> two peoples emerged to subjugate the lovely provinces of the Roman Empire. One came from the North of Europe, the other from Arabia. The first introduced extreme barbarism; the second began to dissipate that barbarism and to elevate Europe by degrees to the heights in which it now basks.

Paris was not then the centre of learning and culture that it was now. Instead, Unanue noted, 'Baghdad was then the centre of politics and culture, as also were Cordova and Seville, colonies which had acquired victorious arms.' The French, Italians, and Germans had found it 'necessary to come [to Al-Andalus, Spain] to obtain any knowledge of the natural sciences.' After imbibing the knowledge imparted by the celebrated Spanish and Arab schools and returning to their homelands, Unanue noted sardonically, 'they were taken for witches and magicians.' Finally, Unanue asked rhetorically: 'What do you think they would have made of that comparison of faces as an indicator that, in the pursuit of the sciences, certain souls were more capable than others?'[44]

Humboldt's Face

In Humboldt's face and vision, we may behold not only the handsome Prussian aristocrat surrounded by 'sublime nature' but the inverse image of those 'Asiatic' faces and 'laws' that haunted him not only in the Americas but also in Paris and even in Berlin. It is true that Humboldt's 'soft' Orientalism was less preoccupied with the precise measurement of skulls and brows and more with the emotional, visual, or stylistic traces of 'Asiatic' despotism. These imputed traces, which included submission, indifference, and small-mindedness, had been established earlier, mostly by French and German writers. For Humboldt, the enduring traces of the 'Asiatic laws' of Manco Capac could be readily observed on the contemporary faces of Peruvians, since

> when on the spot we study those Peruvians who, through the lapse of ages, have preserved their national physiognomy, we learn to estimate at its true value, the code of laws framed by Manco Capac, and the effects produced on morals and public happiness.

Written on those contemporary faces was

> a more submissive resignation to the decrees of the sovereign than patriotic love for the country; passive obedience, without courage for bold enterprises; a spirit of order, which regulated with minute precision the most indifferent actions, while no general views enlarged the mind, and no elevation of thought ennobled the character.

Humboldt celebrated, and was celebrated for, the most 'general views' of nature and man. These 'general views' that 'enlarged the mind' were manifested in the ascent of summits like Chimborazo, and in the pictorial representation of 'monuments,' and vertical profiles of entire countries, mountain ranges, and, later, the globe. In his *Political Essay on the Kingdom of New Spain*, Humboldt attributes the presence of 'enlarged mind' among Europeans to the cultivation of 'individual liberty' and 'character,' a practice that for him was traceable to the ancient Greeks. This aspect was otherwise lacking in Orientals, or the Oriental-spirited, common Peruvians and Mexicans he says he observed during his travels. Humboldt's comments on these matters are telling, for they contrast sharply with Unanue's observations on the very same subjects:

> As the Indians almost all of them belong to the class of peasantry and low people, it is not so easy to judge of their aptitude for the arts which embellish life. I know no race of men who appeal more destitute of imagination. When an Indian attains a certain degree of civilization, he displays a great facility of apprehension, a judicious mind, a natural logic, and a particular disposition to subtilize or seize the finest differences in the comparison of objects. He reasons coolly and orderly, but he never manifests that versatility of imagination, that glow of sentiment, and that creative and animating art which characterize the nations of the south of Europe, and several tribes of African negros.

In characteristic Humboldtian double-speak, the 'circumspect' Alexander attempts to withdraw the brunt of the comment, but it is to no avail:

> I deliver this opinion, however, with great reserve. We ought to be infinitely circumspect in pronouncing on the moral or intellectual dispositions of nations from which we are separated by the multiplied obstacles which result

from a difference in language and a difference of manners and customs. A philosophical observer finds what has been primed in the centre of Europe on the national character of the French, Italians, and Germans, inaccurate. How, then, should a traveller, after merely landing in an island, or remaining only a short time in a distant country, arrogate to himself the right of deciding on the different faculties of the soul, on the preponderance of reason, wit, or imagination among nations? Many Indian children educated in the college of the capital or instructed at the academy of painting founded by the king, have no doubt distinguished themselves; but it is much less by their genius than their application.[45]

The killer line is the last: 'it is much less by their genius than their application.' Here again, Unanue and Humboldt could not disagree more.

Humboldt read similar Oriental traits into the faces of ancient Mesoamerican sculpture. His Orientalist or 'Humboldtian' heirs in Peru inevitably found such traits everywhere as well, including, as we shall see below, in fake artefacts. Humboldt's career-launching album, *Vues des cordillères et monuments des peuples indigènes de l'Amérique* (1810–1813), opens with an image of 'an Aztec priestess' drawn from the antiquarian collection of Guillaume Joseph Dupaix (see Figure 8.2). The 'priestess' reminds Humboldt of a Greek sculpture of Isis he saw in Rome, and of sculpted heads encrusted in the columns of the Temple of Hathor in Egypt, which he apparently saw in Paris in print. Hathor was a sun goddess and wife of Ra, the sun god, and over time Isis eventually absorbed traits once associated with Hathor, including the association with the sun. The comparison likely came to Humboldt's mind less for coincidences of 'style' and more for the deeper, solar deity associations, which were taken to support his conviction that New World civilisations were 'Asiatic' in spirit if not origins. In short, the 'Aztec priestess' was a New World iteration of Hathor and Isis. Achim suggests that Humboldt's Oriental speculations and comparisons followed Winckelmann's evolutionary notion of 'style,'[46] but in this case, Humboldt correlated Aztec aesthetics (or rather, the lack thereof) with America's 'perennially savage and agitated nature.' One may imagine that Unanue would have been puzzled by such an odd projection onto American nature of Prussian classicism, French Orientalism, and Vulcanism.

In Peru, Unanue's sensory theory of the American imagination combined with his global history of the routes of ingenuity and beauty undercut Humboldt's Orientalising speculations and his Occidentalist aesthetics, restoring Manco Capac and Inca civilisation to their native or 'American' roots, and elevating in the eyes of science the aesthetic quality of her people and her arts. The Creole from Arica further recognised that modern Peru was not merely the debased product of Spanish 'divide and rule,' as Humboldt had argued, but instead the historical product of the conjuncture of the native ingenuity of its civilisation and currents of knowledge flowing from Asia and Africa via Andalusia to Europe and the Americas. Although Unanue's anthropology was surely not relativist,[47] it reached beyond Peru and well beyond Europe, granting 'the other three quarters of the world' a strong claim

Buste d'une Prêtresse Aztéque.

FIGURE 8.2 Xilonen. Humboldt, *Vuès des cordillères et monuments des peuples indigènes de l'Amérique* (Paris, 1810–1813), Plate 1.

on 'the glory to which man aspires.' Humboldt's 'global anthropology' pales in comparison.

Unanue's views on the origins and nature of American civilisation did not pass with him. They found a resounding echo and enduring legacy in nineteenth-century republican Peru. To be sure, Humboldt's 'footsteps' continued to attract many followers as well, particularly among European savants who settled in Peru. The contrasting 'footsteps' of Unanue and Humboldt are worth tracing, if only to demonstrate that the Baron's footprints were buried in the sands of the Peruvian coast. The respective legacies may be traced most clearly in debates about the origins of Manco Capac, a subject to which we now turn.

Burying the Baron's Footprints

In one of the first general histories of Peru to be published in independent republican Peru, José María de Córdova y Urrutia protested that prestigious foreign speculations on the Incas had 'reached the extreme of depriving us of the glory of Manco Capac's being born in our country, persuading [us] instead [to believe] that he came from some foreign land.' Following Unanue and Ulloa, Córdova's *Las tres épocas del Perú o compendio de su historia* (Lima, 1843) dismissed Humboldt's view that Manco Capac's laws were 'Asiatic,' and he ridiculed John Ranking's claim, based in part on earlier British speculations as well as new crania despoiled at Tiahuanaco, that Manco Capac had been a 'son of the Gran Kublai Khan.' Instead, the Peruvian historian returned to Inca Garcilaso de la Vega's account, which had in turn been informed by Inca oral tradition and Cieza de Leon, insisting that the best evidence suggested that Manco Capac 'came from a small island in Lake Titicaca.'

As the debate raged on, the altiplano region south of Cuzco around Lake Titicaca became a Mecca for savants seeking to clarify the true origins and nature of Inca civilisation and its fabled founder, Manco Capac. In *Vues des Cordillères* the frustrated Humboldt, who never made the voyage south, put his stamp on Pedro de Cieza de Leon's pioneering call, made in his *Crónica del Perú* (1575), issuing a resounding plea for 'some learned traveller [to] visit the borders of Lake Titicaca, in the district of Collao, and the high plains of Tiahuanaco, the theatre of the ancient American civilization' to verify Cieza's claims. Léonce Angrand, Alcide d'Orbigny, Jacob von Tschudi, and Ephraim George Squier were among those 'learned travellers' who heeded Humboldt's echoing of Cieza de Leon. Each of these nineteenth-century explorers published travel accounts or scientific reports with drawings of the ruins. In all these drawings, one monument stands out: the Puerta del Sol or Sungate at Tiahuanaco. Squier appears to have been the first to take photographs of the Sungate, which served as the basis for an illustration that appeared in his travelogue, *Peru: Incidents of Travel and Exploration in the Land of the Incas* (1877; see Figure 8.3). Squier had been an aspiring student of the Yankee historian William Prescott, the celebrated author of popular histories of the conquests of Mexico and Peru. Squier had authored notable studies of Mississippian Indian mounds and Nicaraguan antiquities. Following his Peruvian investigations, he christened Tiahuanaco 'the Baalbec of the New World.' Ancient 'Baalbec' (Heliopolis under the Romans) was a Phoenician city devoted to a solar cult.

Humboldt was not alone in doubting the Peruvian credentials of Manco Capac. Prestigious European explorers, chroniclers and natural scientists, from Walter Raleigh to Paul Ricaut, the Abbe Raynal, and Alcide d'Orbigny, had all suspected that Manco must have been of foreign origin. As late as 1901, the question was still hotly debated in learned circles, as is clear from W. Golden Mortimer's notable *Secret History of Coca*. Mortimer wondered: Were the first Incas white, yellow, or red? Were they of the same race or 'trunk' as that mass of 'Peruvians' or 'indigenes' over whom they ruled? After a review of the literature, our author concluded that the consensus of scholarly opinion 'now is that these people in some prehistoric

FRONT OF GREAT MONOLITHIC GATE-WAY.

FIGURE 8.3 Front of Great Monolithic Gate-Way. Ephraim George Squier, *Peru: Incidents of Travel and Exploration in the Land of the Incas* (New York, 1877).

time found their way to the shores of South America from China and other parts of Eastern Asia.' But Mortimer also noted, finally, that 'whatever opinions and traditions there may be on the early origin of the Peruvians, all coincide on one point: that the first appearance of the progenitors of the Incan race was in the Titicaca region.' Indeed, 'the most frequently cited legend of Inca origins describes a pair of white people – Manco Capac and Mama Ocllo – as mysteriously appearing on the shore of Lake Titicaca.'

The nineteenth-century pursuit of 'Asiatic' Inca origins produced curious 'artefacts' in Peru that resonated with the images publicised by Humboldt in *Vues* and elsewhere. Cross-legged 'buddhas' now appeared not only in Uxmal, Mexico, but in Ica, Peru as well. What these two regions had in common was Chinese coolie labour. In 1890, a handful of previously published archaeological field reports was collated and reprinted under the title of *Antigüedades nacionales* (National Antiquities). The compilation included an 1866 report on a dig in the south-central coastal region of Ica. The Ica report sported a crude drawing of an unearthed *huaca* or ceramic figurine, evidently inscribed with Chinese characters (see Figure 8.4). The characters inscribed on the Buddha-like figurine evoked a notion dear to European theories of 'Oriental despotism': centralised control of irrigation works.[48]

This fabulous artefact attracted the attention not only of scores of charlatans but also of serious 'learned travellers,' including the fervent Humboldtian, Marcos

Idolo sacado de las antiguas huacas peruanas

FIGURE 8.4 Ídolo sacado de las antiguas huacas peruanas. García y Merino, Manuel and Teodorico Olaechea, *Antigüedades nacionales* (Lima: Escuela de Ingenieros, 1890).

Jiménez de la Espada (1831–1898). Jiménez de la Espada's archive, held at the *Consejo Superior de Investigaciones Científicas* or Spanish Council for Scientific Research (CSIC) in Madrid, contains several photographs taken by J. Laurent of the very same artefact described in the 1866 report (Figure 8.5), and there is also a notable drawing in the same archive made by Diaz Carreño.[49] In short, the forgery existed and was deemed authentic or curious enough to be photographed and recorded by the Scientific Commission of the Pacific led by Jiménez de la Espada. The whereabouts of this curious object today is unknown.

Reports of ancient Chinese figurines in Ica resonated not only in learned circles. More urgently, they played, or were intended to play, roles in contemporary policy debates. At the time, Peru was engaged in a national polemic over immigration policy. One faction of the urban elite, based primarily in Lima, would restrict immigration to white Europeans only, so to 'improve the race' and foster 'civilisation' and colonisation in the frontier regions of the Republic to the east. On the other hand, Peru's coastal sugar planters desperately needed 'brazos' or field hands to cut cane. Facing the decline of African slavery, abolished in Peru in 1854, and given the reluctance of highland Indians to work in the plantations for more than short periods of time, the only ready source of labour was the Emperor of China, who, via British merchants, supplied Chinese 'coolies' or bonded labourers, as well as rice and opium to sustain them. Ica was the first plantation region in Peru to import

FIGURE 8.5 Ídolo de cerámica. J. Laurent y Cia., Biblioteca General de Humanidades y Ciencias Sociales, Centro Marcelino Menéndez Pelayo / Centro de Humanidades, Marcos Jiménez de la Espada, Iconografía.

Chinese bonded laborers on a vast scale. Although further research is needed, the archaeological 'discovery' may well have been intended to lend support to planter arguments that the Chinese were not the 'undesirable' and 'uncivilised' immigrants painted by the urban elite who sought German immigration.

Yet another curious archaeological report, this one of blonde mummies disturbed in the cloud forests of northern Peru, provided fuel for the Whites-only side of the debate on immigration policy. The 1843 report by Chachapoyas judge Juan Crisóstomos Nieto was first printed in the official daily, *El Peruano*. It would have been forgotten if not for the fact that it was reprinted in 1892 in the highly respected *Boletín de la Sociedad Geográfica de Lima*. The report described seven mummies with fine, blonde hair 'unlike that of today's indigenes.' Nieto speculated that the fortress of Quelap had been a 'Tower of Babel in Peru.' Its finely crafted artifacts had been 'transferred by an enlightened and great nation that occupied this territory, but that later came into decadence … until it was encountered, in a state of isolation, by the great Manco [Capac].' It is quite possible that Nieto had in mind the learned French legends (Voltaire, Raynal) about Manco Capac and Mama Ocllo being wayward 'whites.' The dark age of barbarism that Manco Capac had encountered (and that Inca Garcilaso de la Vega had described) had been preceded, it now seemed, by an earlier age of high, white civilisation in Peru. The judge concluded that the ruins of Babel at Quelap demonstrated that 'America is the Old World with respect to the other four parts of the globe.' The editors of the *Boletín*, who sought to attract German immigrants to the region, added a learned commentary on Nieto's report, authored by Modesto Basadre Chocano. Although admitting that additional research was required, Basadre confirmed Nieto's view that

> the crania covered with fine, trim blonde hair proved, without need of additional evidence, that the men who built that edifice [at Quelap], and who have long since disappeared, were of a completely distinct race … indeed, a very ancient and very superior race, compared to that of the Indians.[50]

In an equally imaginative racialist line of speculation, the Philadelphian phrenologist and patriotic epistemologist Samuel G. Morton argued that

> the most natural division of the American race is into two families, one of which, the Toltecan family, bears evidence of centuries of demi-civilization, while the other, under the collective title of the American family, embraces all the barbarous nations of the New World excepting the Polar tribes or Mongol-Americans.

This 'Toltecan family' included 'the civilised nations of Mexico, Peru and Bogotá.' In the chapter of *Crania Americana* dedicated to 'the Ancient Peruvians,' Morton claimed to have examined

> nearly one hundred Peruvian crania: and the result is, that Peru appears to have been at different times peopled by two nations of differently formed crania, one of which is perhaps extinct, or at least exist only as blended by adventitious circumstances, in various remote and scattered tribes of the present Indian race.

Morton assigned pre-Inca status to one of these two 'nations' or 'families,' calling it the 'Ancient Peruvian [nation], the remains of which have hitherto been found only in Peru, and especially in that division of it now called Bolivia.' These remains had been dug up in tombs that

> abound on the shores and islands of the great Lake Titicaca, in the inter-alpine valley of the Desaguadero, and in the elevated valleys of the Peruvian Andes. … The country around this inland sea was called Collao, and the site of what appears to have been their chief city, bears the name of Tiaguanaco.

Like everyone else, Morton cited Cieza de León's early observations at Tiahuanaco, to the effect that the buildings were pre-Inca, that the first Incas had held court there, and that they had modelled the great walls of Cuzco upon the stonework at Tiahuanaco. Nor did Morton fail to cite Inca Garcilaso de la Vega's *Los Comentarios Reales* and 'an unnamed author of an article appearing in the *Mercurio Peruano*' (namely, Unanue), to lend what he thought was support for his speculations. But to back his thesis that these were 'Ancient Peruvians,' Morton relied on the published observations of Mr. Pentland,

> who has recently visited the upper provinces of Peru. This gentlemen states that in the vicinity of Titicaca he has discovered innumerable tombs, hundreds of which he entered and examined. These monuments are of a grand species of design and architecture, resembling Cyclopean remains, and not unworthy of the arts of ancient Greece or Rome. They therefore betokened a high condition of civilization.[51]

For Morton 'the most extraordinary fact belonging to them is their invariably containing the mortal remains of … an extinct race of natives who inhabited Peru above a thousand years ago, and differing from any mortals now inhabiting our globe.' The Philadelphian thus divides the Peruvians into 'Ancient Peruvians' (the Collas) and 'Inca or Modern Peruvians.' This 'modern' family is also further subdivided in two races, one called 'Toltecan' and the other, 'American.' As with Humboldt, for Morton the founders of the Inca dynasty were Asiatic 'Toltecans' who had migrated south from Mexico in the twelfth century.

In words that anticipated those of Mortimer decades later, Morton lamented that 'the origin of the Incas of Peru is shrouded in fable.' These origins

> are represented in their traditions as two celestial personages, a son and daughter of the sun himself, who were sent from heaven to instruct and civilize a favoured people … Manco Capac, the first Inca, and Coya Mama, who was both his sister and his wife. They appeared first on an island in the Lake Titicaca, and taking the people under their jurisdiction, began at once a reform of all the institutions of the country.

He concludes that 'this preference for [an origin in two rulers] was calculated to render the account more marvellous, and the descendants of the individuals more respected.' Doubting this account and unwittingly echoing the views of Antonio de Ulloa, Morton asks: 'Can it be credited that this total revolution in social and civil government was the result of moral causes, operating on nations who were as strongly devoted to their own institutions as any other people? Certainly not.'

Given the impasse, Morton turns to a foreign origin hypothesis of a racialist nature. He writes: 'On the contrary we are compelled to attribute this change to an influx of foreigners, whose number and intelligence enabled them to overcome every obstacle that arose in their path. Who could these strangers be?' For Morton, the likely candidates were 'the Toltecans, the most civilized nation of ancient Mexico.' After governing that country for four centuries, he noted, they 'suddenly abandoned it about the year 1050 of our era.' There is, he continued, 'a coincidence in the squared and conical form of the head in the Toltecans and Peruvians that is very striking, and which will be more particularly adverted to in a future part of this work.' Morton concludes:

> whether the preceding inference, which is by no means new, be correct of not, there can be little doubt that the Inca family was an intruding nation, led perhaps by a few individuals of the sacerdotal class; and having conquered Peru, much the same political relations appear to have subsisted between them and the pre-existing inhabitants, as we at present observe between the modern Greeks and the Turks.

Morton's phrenological investigations furthered two notions: first, that the Incas were a 'superior race' not native to Peru, that an 'ancient' and now 'extinct' race probably founded Tiahuanaco, and that the 'Toltecan' Incas drew upon the precedent of this ancient civilisation when they founded their own. Notably, Morton was a 'patriotic epistemologist' who, like Humboldt, defended 'American civilization' against European cynics who doubted its very existence. After 'a review of the … facts,' Morton exclaims: 'How idle is the assertion of Dr. Robertson, that America contained no monuments older than the conquest! How replete with ignorance are also the aspersions of Pinkerton and De Pauw!' Morton declares that

> it is in vain any longer to contend against facts; for however difficult it may be to explain them, they are nevertheless incontrovertible. Whence the Peruvians derived their civilization, may long remain a mooted question; that they possessed it, cannot be denied.

'Soft' Orientalist defences of American civilisation such as those of Morton and Humboldt were prestigious but unsatisfying to serious Americanists based in Peru. For many of them, to argue that civilisation existed in the Americas only because 'a superior race' had brought it from Asia, smacked of charlatanism if not racism. In general, Peru-based antiquarians would seek firmer ground.

Humboldt acolyte and Peru's founding National Museum director, Mariano de Rivero y Ustariz, drunk from both cups, straddling the competing views of Humboldt and Unanue on the question of Inca origins. Rivero favoured the arguments of the Swiss anthropological linguist and long-time Peruvian resident, Jacob von Tschudi, with whom he collaborated, to those of the distant Morton. Von Tschudi developed a firm command of the Quechua language, carrying out extensive field and linguistic research in Peru. He was among those who made the journey to Tiahuanaco. Von Tschudi argued that skeletal and lexicographic evidence suggested that three, equally endowed 'races' (which, being the linguist he was, he named 'Aymara,' 'Quechua,' and 'Chincha') had populated Peru in ancient times. Tschudi argued that 'race' in Peru was the product of cultural phenomena and geography and that it had little to do with 'civilisation' per se, since all three of his linguistic 'races' achieved civilisation, albeit in different regions of Peru. In *Antigüedades Peruanas* (1851), Rivero guardedly accepted Humboldt's speculation that Manco Capac was most likely an Oriental sage, noting that he could have been a Brahmin or a Buddhist priest 'who, by [virtue of his] superior and civilizing doctrine [was] able to dominate the animus of the indigenous peoples and elevate them to political supremacy.' Yet Rivero arrives at this confirmation of Humboldt's speculations not because he merely wished to follow in his mentor's footsteps (Humboldt was instrumental in furthering his career, and both had studied at Freiburg) but because, like the intrepid Jesuit missionaries in Peru and China, he saw Buddhism and Christianity to be analogous religious traditions. Notably, Rivero's nod toward Humboldt's thesis did not stop him from asserting that the Inca dynasty was 'indigenous' to Peru. Rivero surmised that Manco Capac had been a sagacious, priestly 'reformer' and therefore could not have been an actual 'Inca' of the royal bloodline of the Sun. Rather, the enterprising Manco had placed Inca-Rocca (in Inca Garcilaso de la Vega's canonical account, Inca-Rocca or 'Sinchi-Rocca' is the second Inca dynast), an able member of an illustrious local family, on the Peruvian throne. Rivero reasoned that his royal name of 'Inca' was strong lexical evidence that he was the likely founder of the lineage. Thus, for Rivero, Inca-Rocca was 'the first Indian autocrat, the stump of the tree of the Peruvian Monarchs.' In coming to this conclusion, Rivero followed the lexicographic methods of British antiquarian John Ranking, who had argued that the name 'Manco' was likely to be of Oriental origin.

Rivero and von Tschudi's *Antigüedades peruanas* (1851) was dedicated not to princes or royal academies of science but to 'the cause of National Sovereignty' and 'of memory against ruin.' As founding director of Peru's National Museum (appointed by Bolivar at Humboldt's behest; he also founded Colombia's National Museum in Bogotá), Rivero had secured funding from the Peruvian Congress to subsidise the printing of lithographic plates in imperial Vienna, thanks to his Swiss colleague von Tschudi, a member of Vienna's Royal Academy of Sciences. In the preface, Rivero sounds Unanue's refrain, lamenting the sorry colonial legacy of destruction and neglect for, he writes, 'centuries have passed before Peru possessed a collection [of artefacts] drawn from her ancient archaeological monuments.' Today, however, 'these mute yet eloquent witnesses reveal the history of past events and they demonstrate to us the intelligence, power and greatness of the nation ruled by our Incas.' This material demonstration was of the utmost importance, for 'the 'history of nations … is not of interest merely to know what stage of power and culture was attained … but rather to instruct us in their progress … and to prepare the people for the enjoyment of national liberty.… Babylon, Egypt, Greece and Rome are not the only empires worthy to serve as nourishment for a generous imagination.'[52]

The frontispiece to the second volume of *Antigüedades Peruanas* offered as 'nourishment' for the 'generous imagination' an iconographic representation of Peru's

FIGURE 8.6 Frontispiece. Mariano Eduardo de Rivero and Juan Diego de Tschudi, *Antiguedades Peruanas*, t. 2 (Vienna: Imprenta Real, 1851). Courtesy of the John Carter Brown Library, Brown University.

deeply promising history (see Figure 8.6). The ancient ruin of the Sungate (Puerta del Sol) at Tiahuanaco now appears not as a solar cult of despotic, Oriental inspiration but as the triumphant arch of the Republic, as the ancient native threshold of the Peruvian national future. The pastoral Indian family and the native flora and fauna 'animate' the bounty of the landscaped native soil, while the ancient glory of Inca kings and the monumental stone portend the even greater glories to come. Rivero's republican Sungate would soon become a logo, serving as an architectural motif for the façade of Peru's national museum of archaeology and turning up in reproduction in national museums in Bolivia and Argentina. Here the 'country' or landscape is represented in Humboldtian terms, with signature volcanoes and native flora and fauna; the portraits of the Inca kings appear as the pillars of 'Peruvian civilisation,' framing Peru's bountiful, republican future. The common Indian or Peruvian dressed 'in 'national costume' animates the scene. The man points upward to the title of the work held aloft by the condor, at once 'sovereign of the avian kingdom' and 'sovereign of these regions.'

The distance that separated Humboldt from Rivero (and which marked his affinity with Unanue) may be appreciated in these lines, written by Rivero after he visited the haunting ruins at Chavin de Huantar. 'Fatigued and at the same time satisfied with my arduous investigations,' at Chavin, Rivero writes, 'I took rest upon a slab of granite some three meters long, engraved with certain signs or designs that I could not decipher, and which I had stumbled upon as I emerged from a subterranean passage near the river.' In that moment of repose, and in seeming exercise of those powers of imagination that Unanue had theorised and defended in his work on the Peruvian clime and genius,

> my imagination took flight with the speed of lightning, ranging over all the ancient sites I had visited and all the great events that had taken place in the time of the conquest. Saddened, I raised my eyes toward the ruins of this silent site and saw the deplorable images of destruction committed by our ancient oppressors.

Three centuries of pillage, Rivero noted,

> had not been enough to erase from memory those infinite maladies suffered by the pacific and simple inhabitants of the Andes, and I thought I saw in the narrow current water stained with the blood of victims, and that the rubble along the bank was a mountain of cadavers upon which fanaticism had erected its throne of tyranny, and from which it had given thanks to Heaven for having achieved its work of destruction.

Absorbed by such melancholy meditations, 'and sympathizing with the unhappy luck of such a laborious and sagacious nation,' Rivero records,

> I thought I heard a voice from the depths of the subterranean passageway, a voice that spoke to me thus: What motives have you to wander in these places of eternal rest, to remove rubble and tread on the ashes that time has respected, now that men are content to deprecate them? Are the data of histories not sufficient proof of our greatness, our simplicity, our hospitality, and love of work? Are the monumental remains that escaped the bloody sword of the inhuman conqueror not better testimonies of the opulence of our ancestors? Are they not better still than the robbery of our treasures, the sacking of our cities, the treacheries, the death of our adored Inca, of our wise men [amautas] and of our nobility?[53]

This ancient Peruvian portal of this lightning-fast, 'generous' republican imagination was largely invisible to foreign historians and critics. William Prescott, for example, held that Manco Capac was a mere 'figment of the vain imagination of Peruvian monarchs.' Although he never visited Cusco, and if he had could not have seen it, he nevertheless could not resist speculating on the meaning of the Inca ruins at Sacsayhuaman outside Cuzco. The ruins were material proof of a grinding 'despotism' that was ultimately of Oriental origin. Transporting himself in prose to the site, Prescott wrote that

> we are filled with astonishment when we consider that these enormous mass-
> es were hewn from their native bed and fashioned into shape by people igno-
> rant of the use of iron … we see in it the workings of a despotism which had
> the lives and fortunes of its vassals at its absolute disposal, and which, however
> mild in its general character, esteemed these vassals, when employed in its
> service, as lightly as the brute animals for which they served as a substitute.

Citing Morton (incorrectly, as it turns out) Prescott claimed that the Incas nevertheless were a 'superior race' that lorded over primitive tribal natives. Prescott
added that he was uninterested in the origins of this 'superior race' since that was
a matter for 'speculative antiquarians,' not real historians. In the end, for the docu-
ment-bound and blind Yankee historian, Inca origins lie in 'a land of darkness …
far beyond the domain of history.'

In contrast, Sebastián Lorente's Manco Capac bore the bright light of the 'national
spirit.' In Lorente's account Manco is also not a dynast but instead an enlightened
'reformer' imbued with the national spirit, and who 'in his native wisdom knew
how to amalgamate all those elements of civilisation that already existed in Peru.'
Lorente's considered view thus drew both upon Rivero and Unanue, among others,
and, in the end, it buried the Oriental view of the origins of Peruvian civilisation
fomented by Humboldt and Morton.

For Lorente, 'anyone who impartially interrogates history' will find that 'the
'origin of Manco Capac' is 'not in doubt.' That man

> who so perfectly knew the lay of the land and its people – who was so inun-
> dated with the national spirit that with its knowledge he could amalgamate
> all of the elements of the anterior civilization – that man was without doubt
> born in Peru.

Manco Capac's 'works bear the seal of the national race and that of the land; it is
the expression of his epoch as a man of genius would comprehend it.' To support his
view, Lorente turned, as eighteenth-century Creole historians had, to non-literate,
native forms of memory as well as the Inca oral testimony registered in the early
colonial archive. However, unlike Humboldt, Morton, or Mortimer, Lorente did
not read these sources as mere 'fables.' In the fashion of Giambattista Vico, whose
work he much admired, Lorente examined oral history sources as cultural evidence
that could support other kinds of data concerning distant historical events. The
'nationality of Manco [Capac] is deduced from other conclusive evidence and is
also revealed to a certain point by legends and direct oral testimonies,' Lorente concluded. Although in Lorente's view Manco Capac was not an Inca dynast but 'only
a reformer of institutions,' this does not diminish his place in Peruvian history, since
he 'secured the unity of Peru, the basis of its future greatness.' Lorente's reading of
Manco Capac was a critical step in the republican rewriting of Peruvian history.
'Peruvian civilisation' would now be re-founded, not on 'the Incas, kings of Peru
that were' of colonial dynastic history or enlightened, philosophical history but on
the primitive 'soul' and living 'communal spirit' of villagers.

As Unanue had argued, for Lorente it was upon the soil of the country that the ruins of ancient Peru stood, and it was from the Peruvian clime that the genius of Manco Capac and the Incas sprang. Moreover, this wellspring of ingenuity was available to all who exercised the inherently 'generous imagination' afforded by Peru's clime. Rivero's arch is precisely an allegory of and for that 'generous imagi- nation.' That ancient republican threshold, heralded by the Incas, would now open onto the national territory and future of Peru, where Sebastián Lorente would deliver his republican history lessons on 'Peruvian civilisation.' Those history lessons would begin not with Manco Capac, but instead with Manco Capac's 'Peruvian' cultural origins. Humboldt's Orientalist footprints were buried in Peruvian soil.

Or not. Championed primarily by the French but with notable British, German, and Dutch contributions and variations, European Orientalism continued to res- onate in Peruvian intellectual circles. An influential work in this vein was Louis Baudin's *L'empire socialiste des Inka* (1928). Its Peruvian traces may be seen in the writings of the Peruvian socialist José Carlos Mariátegui, among others of his gener- ation. Like Montesquieu, Humboldt, Hegel, and Marx, Mariátegui considered the Inca's 'ancient law' to be 'Asiatic' in spirit. Unlike them, the Peruvian socialist con- sidered this spirit to hold revolutionary potential as a mobilising myth for 'the new Indian.' In the end, however, Peruvian historicist thought carried the day. Thanks to Unanue and others, the threat posed by European Orientalism was success- fully resisted, and the Incas became firmly Peruvian once again. With Humboldt's cultural legacy retreating to the fringes, only Humboldt the naturalist remained, fittingly commemorated by a curious monument donated by the German settler community or 'colony' circa 1935 (see Figure 8.7), and now ensconced in Lima's *Parque de Exposiciones*, a species of graveyard for demoted monuments.[54]

FIGURE 8.7   Humboldt Monument donated by German Colony, Expositions Park, Lima. Photograph by Juan Antonio Lan.

The cultural and antiquarian branch of 'Humboldtian science' was now, in Peru at least, a dead letter. This dead letter, I have suggested, was the product of a Peruvian *desencuentro*, but it may very well reflect a larger *desencuentro* with Hispanic America that lies at the very heart of the Humboldtian enterprise, then and now.

But our story does not begin or end with Humboldt. Natural and human history has had a much longer and brighter, indeed pioneering history in Peru.[55] Humboldt was a small, if distracting blip on its wide screen. Monuments in Lima dedicated to Inca Garcilaso de la Vega, Pedro de Peralta Barnuevo, José Hipólito Unanue (see Figure 8.8), Mariano de Rivero, or Sebastián Lorente, confirm the

FIGURE 8.8 Unanue Bronze. Faculty of Medicine, University of San Marcos, Lima. Public domain.

living legacies of these towering figures of Peruvian thought. All should be recognised beyond Peru for their contributions to the history of knowledge but apart from Garcilaso, none are.

Notes

1 The Spanish concept of 'desencuentro' is not readily translated to English. *La Real Academia Española* gives two meanings. The first, more literal meaning is 'encuentro fallido o decepcionante,' while the second, more expansive sense is 'discordia,' which may be translated as 'failed or disappointing encounter' and 'discord,' respectively. I suggest that both senses of the word apply to Humboldt's encounter with Peru.

2 'Discurso de Don Hipólito Unanue al ser elegido presidente del Congreso Constituyente, Sesión del 20 de diciembre de 1822,' *Colección documental de la independencia del Perú*, t. I, v. 8 (Lima, 1974), 839–842.

3 'Discurso del Señor Feliú en que hace la apología de los indios contra las imputaciones del Barón de Humboldt,' *Noticias del Perú*, n. 8, 1–12, no date [¿1811?]. Consulted at the John Carter Brown Library.

4 Nicolas A. Rupke, *Alexander von Humboldt: A Metabiography* (Chicago, IL: University of Chicago Press, 2008).

5 Although more research is needed, German cultural diplomacy and funding combined with the initiatives of academic associations, German schools, 'Humboldt clubs' and the like may account for much of the 'Humboldt cult' in Latin America. In any case, these elements were certainly visible in 2019.

6 Most explanations for the mixed Peruvian reception of Humboldt tend to recycle the Baron's own views, while others reflect more recent prejudices. Among these, the most stubborn is the provincial cliché that Lima's elite was and is vice-ridden, 'aristocratic,' 'colonial,' unpatriotic, 'rancid,' and otherwise divorced from and wholly ignorant of 'the deep Peru' or 'real Peru' of the Andean and Amazonian interior. In such a place, the enlightened, cosmopolitan Humboldt could hardly expect a warm welcome. This long-lived trope predates Humboldt, but, as with much else that concerns the Baron's legacy, it surely received a strong boost from Alexander's pen, and it is still commonplace. But the claim notably ignores the fact that more aristocrats and colonials inhabited Paris and Berlin and that most of them were entirely ignorant of the interior provinces of their own regions, never mind the New World. For a mixed compilation of Humboldt's writings on Peru in translation, see Estuardo Núñez and Georg Petersen, eds., *Alexander von Humboldt en el Perú: Diario de viaje y otros escritos* (Lima: Banco Central de Reserva, 2002).

7 Ramón Olaguer Feliú (1784–1831) was apparently born in Ceuta to an enlightened military family from Cádiz, posted for duty in Peru and Chile. Ramón arrived in Lima in 1793. There he studied law at the prestigious Real Convictorio de San Carlos, part of the University of San Marcos, graduating with a degree in 1802. He was in Lima when Humboldt made his short visit. Ramón was elected representative of Peru to the Cortes de Cádiz, where he defended broad American rights of equal representation, and served as secretary.

8 *El Ateneo*, v. II, n. 40 (Lima, 1906), 116–120.

9 See Teodoro Hampe Martínez, 'El Virreinato del Perú en los ojos de Humboldt (1802): Una visión crítica de la realidad social,' *Ibero-Amerikanisches Archiv*, v. 26, n. 1/2 (2000), 191–208, and, by the same author, 'Humboldt y el mar peruano. Una exploración de su travesía de Lima a Guayaquil,' *HiN*, v. VIII, n. 15 (2007), 13–22.

10 See Alexander von Humboldt to Ignacio Checa, Guayaquil, 18 enero de 1803, in Charles Minguet, *Alejandro de Humboldt: Cartas americanas* (Caracas: Biblioteca Ayacucho, 1980). For an American English translation of the infamous letter, see Carlos Aguirre and Charles Walker, eds., *The Lima Reader* (Durham: Duke University Press, 2017).

11 As Rizo-Patron notes:

> No matter how brilliant the observations of the German scientist in the field of natural phenomena, it seems evident that, despite the elements of truth they contained, his impressions on the human and social field were incomplete and tainted with subjectivity.

Peruvians took note. See Paul Rizo-Patrón, 'Arrogance and Squalor? Lima's Elite,' in *Alexander von Humboldt: From the Americas to Cosmos* (Online Publication, New York: Bildner Center for Western Hemisphere Studies, 2004), 69–82. https://www.gc.cuny.edu/CUNY_GC/media/CUNY-Graduate-Center/PDF/Centers/Bildner%20Center%20for%20Western%20Hemisphere%20Studies/Publications/humboldt.pdf.

12 Humboldt's claim that Lima fortunes were relatively small and squandered reflected both his limited contacts in Peru and his more sustained and intimate associations with wealthy, slaveholding elites in Cuba and Venezuela. On Humboldt's fabulously wealthy, 'second slavery' hosts in Cuba and Venezuela, see Michael Zeuske, 'Humboldt in Venezuela and Cuba: The "Second Slavery,"' *German Life and Letters*, v. 74 (2021), 1468–0483 (online). On Humboldt's limited contacts in Lima, see Rizo-Patrón, 'Arrogance and Squalor?'

13 The 'second Columbus' trope is still very much alive. As Rupke notes, in 1992 Humboldtian literary scholar Ottmar Ette revived the comparison, once again in Humboldt's favour. See Rupke, *Alexander*, 183–184.

14 Oliver Lubrich, 'Egipcios por doquier: Alejandro de Humboldt y su visión orientalista de América,' *HiN*, v. III, n. 2 (2002), 3–20.

15 Edmundo O'Gorman, *La invención de América: Investigación acerca de la estructura histórica del Nuevo Mundo y del sentido de su devenir* (Mexico City: Fondo de Cultura Económica, 1995). On the poetics of naming 'America' and 'Peru,' see Mark Thurner, 'The Founding Abyss of Colonial History, Or "The Origin and Principle of the Name of Peru,"' *History and Theory*, v. 48 (February 2009), 44–62.

16 French, German, and British variants made enlightened European Orientalist writing at this time a far more heterogenous and contradictory affair than Edward Said's *Orientalism* would suggest. In addition, Hispanic and Italian writing on the 'Oriental Indias' differed notably from French, Anglo and German discourse.

17 Lina del Castillo, 'Inventing Colombia/Columbia' in Mark Thurner, ed., *The First Wave of Decolonization* (New York: Routledge, 2019), 48–76.

18 Mark Thurner, 'El sol de la Ilustración y el espejo del Inca en la imaginación ilustrada,' in Ramón Mujica Pinilla, ed., *Forjando la nación peruana: el incaismo y los idearios políticos de la Republica en los siglos XVIII–XX* (Lima: Banco de Crédito del Perú, 2021), 41–80.

19 Guillaume Thomas François Raynal, *Histoire philosophique et politique des établissements et du commerce des Européens dans les deux Indes*, t. 4, libro 7, (Ginebra, 1783) 19-20.

20 Humboldt appears to have taken the idea of southward Toltec migrations from Francisco Clavigero, combining the Mexican Jesuit's thesis with observations made while passing through New Granada, Quito, and Peru. See Alexander von Humboldt to Wilhelm von Humboldt, Lima, 25 November 1802, in Minguet, *Alejandro de Humboldt*.

21 Alexander von Humboldt, *Vües des cordillères et monuments des peuples indigènes de l'Amerique* (Paris: Schoell, 1810), 18.

22 The origins and nature of Etruscan civilisation were subject to debate in Humboldt's time, but the Oriental thesis appears to have been the dominant view.

23 See Mark Thurner, 'An Old New World for the History of Historiography,' *Storia della Storiografia*, v. 67, n. 1 (July 2015), 29–50.

24 See Mark Thurner, 'Historical Theory through a Peruvian Looking-Glass,' *History and Theory*, v. 53, n. 4 (December 2015), 27–45.

25 My emphasis.

26 Pliny the Younger to Maximus VIII, Ep. 8. 24. Web source: Attalus, Greek and Latin Authors on the Web. http://www.attalus.org/old/pliny8.html#24 accessed 27 July 2021.

27 Fulcran Roger's iconic engraving of 'the New Continent' found an anonymous Inca echo in nineteenth-century Ecuador, where 'Moctezuma' becomes 'Atahualpa' arising from his tomb, albeit now with a French rather than Latin, inscription, and with national and republican overtones. See Alexandra Kennedy, *Construir la Nación: Imágenes y espacios del Ecuador en el siglo*, v. XIX (Quito, 2018), 19–42.

28 'So you will have to do your best to prevent people from thinking that you have shown greater humanity, integrity, and tact in a far-off province than in one nearer Rome, among slaves than among freemen, and when you were chosen for the mission by lot rather than by deliberate choice, and that you were an untried and unknown man, and not one of tried and proved experience. Moreover, as you have often heard and read, it is much more disgraceful to lose a good reputation than to fail to win one.' Web source: Attalus, Greek and Latin Authors on the Web. http://www.attalus.org/old/pliny8. html#24 accessed 27 July 2021.

29 Alexander von Humboldt, *Political Essay on the Kingdom of New Spain*, trans. John Black, v. 1 (London: Longman, 1811), 170–171.

30 Humboldt, *Political Essay*, v. 1, xvii.

31 Tragically, part of Unanue's library was torched in the disturbances surrounding Lima's independence. His testament provides a record of the contents of his library in 1833, but we may suspect that circa 1800 his collection was considerably larger. See Luis Alayza y Paz Soldan, 'Biblioteca de Hipólito Unanue según el inventario judicial, practicado después de su muerte,' *Anales de la Facultad de Medicina*, v. 38, n. 3 (1955), 702–708. Also see Oswaldo Salaverry, 'Los orígenes del pensamiento médico de Hipólito Unanue,' *Anales de la Facultad de Medicina*, v. 66, n. 4 (2005), 357–370.

32 In *Political Essay* Humboldt refers to 'the author of an excellent physiological treatise on the climate of Peru' (115). The reference is to the 1805 Lima edition of *Observaciones*.

33 Miruna Achim, 'Los objetos de Humboldt: una colección para un mundo global,' cultura.nexos.com.mx (14 September 2019). https://cultura.nexos.com.mx/los-objetos-de-humboldt-una-coleccion-para-un-mundo-global/.

34 'Estudio introductorio,' José Carlos Ballón and Lucas Lavado, in *Aproximación a Unanue y la Ilustración peruana* (Lima: Fondo Editorial de la Universidad Nacional Mayor de San Marcos, 2006), 17–44.

35 The first edition was published in Lima in 1805 and the second, expanded and revised edition in Madrid in 1815. Citations are to the Madrid edition, consulted at the John Carter Brown Library, Providence, Rhode Island.

36 Charles Louis de Secondat, Baron von Montesquieu, *The Spirit of the Laws*, Book XIV, Chapters 2–3.

37 José Hipólito Unanue, *Observaciones sobre el clima de Lima y sus influencias en los seres organizados, en especial el hombre* (Madrid: 1815), 97–98.

38 José Hipólito Unanue, 'Geografía física del Perú,' *Mercurio Peruano*, v. 106 (1792), 11.

39 Unanue likely consulted the French translation by H. J. Jansen, *Dissertation sur les variétés naturelles qui caractérisent la physionomie des hommes des divers climats et des différents âges. Suivie de Réflexions sur la beauté …* (Paris and The Hague, 1791).

40 Unanue, *Observaciones*, 90–91.

41 Unanue, *Observaciones*, 87–89.

42 Unanue, *Observaciones*, 93.

43 Jorge Cañizares-Esguerra, *How to Write the History of the New World: Histories, Epistemologies, and Identities in the Eighteenth-Century Atlantic World* (Stanford, CA: Stanford University Press, 2001).

44 Unanue, *Observaciones*, 90–93.

45 Humboldt, *Political Essay*, 170–171.

46 Achim, 'Los objetos.'

47 For Unanue, the mixed castes (castas) and Afro-Peruvians (negros) were handicapped in their ability to take full advantage of the benefits afforded by Peru's clime.

48 Incidentally, archaeological and historical research in the Andes generally supports a pattern of decentralised irrigation works under the control of local polities and kin groups.

49 See Sara Badia, Carmen María Pérez-Montes y Leoncio López-Ocón, 'Una Galería Iconográfica,' in Leoncio López-Ocón and Carmen María Pérez-Montes, eds., *Marcos Jiménez de la Espada (1831-1898): Tras la senda de un explorador* (Madrid: Consejo Superior de Investigaciones Científicas), figs. 52–53 and p. 149.

50 *Boletín de la Sociedad Geográfica de Lima*, t. 1, n. 10–12 (31 March 1892), 444–448.

51 Samuel G. Morton, *Crania Americana* (Philadelphia, 1838–1839), 63.

52 Mariano Eduardo de Rivero and Juan Diego de Tschudi, *Antigüedades Peruanas*, v. 2 (Vienna: Imprenta Imperial, 1851), iii.

53 Rivero and Tschudi, *Antigüedades Peruanas*, v. 1, 286–287.

54 The monument was originally hewn in granite. It was relocated from the Avenida de la República to the Parque de Exposiciones and coated in bronze.

55 On the pioneering history of Peruvian historical thought, natural and human, see Mark Thurner, *History's Peru: The Poetics of Colonial and Postcolonial Historiography* (Gainesville: University Press of Florida, 2011).

9

AIR IN A FLASK

The Mexican Making of Humboldt's Objects of Knowledge

Miruna Achim and Gabriela Goldin Marcovich

> We were five and climbed in zigzag because if one happened to find oneself behind another, he would have risked being buried under the mass of stones and soil that crumbled behind the person ahead. We feared a thousand times for the barometer's safety. … We finally arrived at the summit, at the highest slope of the edge, at 8 am. … We wanted to collect air, but the Indian had forgotten the flask. He had to go back to get it. This took a long time given the difficulty of the climb and allowed us to examine the inside of the crater and descend almost to the bottom of it. [Note in the margin]: Compare with the description in the *Gazette du Mexique*, by Mr Riaño, who examined the crater with Mr Fischer.[1]
>
> – Alexander von Humboldt

The general plot lines of Alexander von Humboldt's travels in Mexico are well known. He arrived in Acapulco in February 1803, from Lima, and travelled northward across the Sierra Madre, making stops along the way at the silver mines at Taxco and Cuernavaca, before arriving in Mexico City by May. He would spend a few vibrant months in the city of palaces, as the capital of New Spain was known, visiting its celebrated institutions of learning, including the School of Mines, the Botanical Garden, and the San Carlos Royal Academy of Fine Arts, and socialising with Mexico's elite.[2] He admired the rich cabinets of the likes of Fausto de Elhuyar (1755–1833), the director of the School of Mines, and Ciriaco González de Carvajal (1745–1828), a judge at the Real Audiencia, whose collection included 'notable' geological specimens, antiquities, and shells from the Philippines. At the School of Mines and at the archives of the Viceroyalty, he spent days reading and copying manuscripts: geological and topographical maps, demographical statistics and export tables, indigenous histories, surveys of archaeological sites.[3] On other days, and with the guidance of his scholar hosts, he organised excursions to the

DOI: 10.4324/9781003231479-10

nearby ruins of Teotihuacan and to places of geological interest around the city, such as the Xitle volcano. In August he began a tour of the mining districts in present-day Hidalgo, Guanajuato, and Querétaro. In Michoacán, he was especially curious about the Jorullo, a volcano that had sprung out of nowhere over the course of a couple of days in late September of 1759. It is a fragment of his description of his ascent of the Jorullo, undertaken exactly forty-four years later, almost to the day of the eruption, that we transcribe in the epigraph that opens this chapter. Humboldt left Mexico for good in February of 1804; on his way to Veracruz, he made stops at the archaeological site at Cholula and at various volcanoes en route, including Popocatépetl, Iztaccíhuatl, and Orizaba. He had spent over a year in Mexico, which was rather longer than he had originally planned.

The more concrete products of Humboldt's travels through Mexico are his diaries, specifically books VIII and IX of his *Tagebücher der Amerikanischen Reise*, now held at the Berlin Staatsbibliothek; his letters to a wide network of correspondents across the Atlantic; and rich collections of plants and minerals, pre-Hispanic antiquities, books, manuscripts, and journals. Installed in Paris by August 1804, Humboldt spent the following years organising his Mexican (and, more broadly, American) materials into print and museum displays and, along with them, carefully fashioning his own image. This self-image continues to hold a powerful grip on his admirers and biographers, who often depict him as the inventor of the disciplines that make sense of the New World's unruly natural and human history. On Mexico specifically, Humboldt's most substantial works are the *Essai politique du Royaume de la Nouvelle Espagne* (1811) and the *Vues des cordillères et monuments des peuples indigènes de l'Amérique* (1810–1813), the latter being an ambitious attempt at theorising the vestiges of America's ancient civilisations in relation to the natural history of the 'New Continent.'[4] Both books, quickly translated into English, German, and Spanish, and re-edited numerous times in the course of the nineteenth century, became obligatory references on Mexico both in Mexico and abroad. These works have contributed strongly to establishing the image of Humboldt as the sage in his cabinet, diligently working to systematise fragments of an unknown and unexplored reality.[5]

There is little trace in this new, Humboldtian order of things of the fragility and precariousness of the things themselves or of the human relations articulated by those things, nor do we get a good sense of the kinds of work involved in moving air, rocks, antiquities, and manuscripts from Mexico to Paris. There is no trace, for instance, in the published chemical analysis of volcanic gases from the Jorullo, of the story of the hapless Indian guide which Humboldt recounts in the passage from his diary that opens this chapter. This guide, who had to climb the Jorullo twice because he (or the illustrious travellers he was accompanying) forgot the stopper flask. To fill a flask with air is, it turns out, hard work: somebody must remember the flask or go back for it. To convert that air into a formula on a page involves a long series of steps: once the air is inside the flask, its stopper tightly shut, the flask must be carried gingerly back down the steep slope by an Indian from Michoacán, and then on horseback back to Mexico City, packed unto a trunk and embarked on a ship to Europe, where it is analysed in a laboratory, probably by Humboldt's friend, the chemist Joseph Louis Gay-Lussac (1778–1850), who compared it with analyses of air samples from the Jorullo made a decade earlier by Mexican scientists. The more precarious or unstable an object (think airy or volatile) the more collective work it takes to stabilise it, as Humboldt's diaries, with their cut-out slips of paper glued to the pages, cross-outs, and reminders to compare his to others' descriptions, show.

The densely social and material dimensions of Humboldt's knowledge-making in the Americas, mostly erased from Humboldt's printed work and often ignored by his biographers, are at the centre of this present volume. This chapter follows the trajectories of three objects Humboldt took back with him from Mexico: erythronium, a metal; cochineal, the dye made from the body of an insect; and Xochicalco, an archaeological site to the south of Mexico City. These objects travelled in the form of mineral samples, descriptions, and drawings. Our purpose is not to tell three anecdotes about Humboldt's Mexican collections. Rather, we propose to look behind the vitrines of the curated displays in Humboldt's printed texts and in the cabinets in Paris and Berlin, where many of his collections were stored and classified, to ask, more broadly, about the kind of work involved in ferrying things physically, geographically, and conceptually, from one place to another, in this case, from Hispanic America to Europe, from field to cabinet, from cabinet to cabinet, from a Mexican archive to a text designed for a European readership, at the turn of the nineteenth century. We also seek to reveal the limits of translation and to reconstruct the types of material conditions and epistemological configurations that made possible (or impossible) the movement of objects or their accumulation at certain destinations. Erasure, fragmentation, loss, and mistranslation are inherent to Humboldt's collecting practices, as they are to any collection. As a result, the collections exhibit political and ideological assumptions about the place of American nature and history, both past and present, in Humboldt's scholarly project and the place (or non-place) of Hispanic American science and scientists in it. In short, we ask: What does a story like the suppressed anecdote of the misplaced flask tell us about the ways in which Humboldt produced knowledge about the Americas?

Erythronium: The (Un)making of a Metal

In June 1803, Humboldt shipped a boxful of minerals he collected during his Mexican travels to the 'citoyens' of the Institut National de France.[6] At the time, the Institut brought together some of the more celebrated scientists in France, with correspondents all over Europe and beyond. For his part, Humboldt sought to keep his name and services on their minds and thereby eventually secure a place for himself in the prestigious body by regaling them with specimens from across the Atlantic. From Peru he sent guano, which was analysed by chemist Louis Nicolas Vauquelin (1763–1829);[7] from New Granada, he shipped plant specimens and a superb collection of botanical drawings entrusted to him by José Celestino Mutis.[8] On this occasion, he sent minerals from Mexico of rare shine, colour, and composition.

Among the green, brown, grey, 'silky,' 'shimmering,' and 'striated' obsidians still being mined in the shafts Humboldt visited at the Cerro de las Navajas, and which he selected for their uncommon features, and because of their historical value as ancient tools and ritual objects, Humboldt included a 'new' kind of obsidian from Zináparo known locally as *plata encantada* (enchanted silver, named so, he explains, because it shone like silver). There was also a 'new and unknown' mineral from Zinapécuaro which, Humboldt explained, Andrés Manuel del Río (1764–1849), professor of mineralogy at the School of Mines, had classified among the werner-ites for its prismatic arrangement and specific gravity. He also sent a 'new kind' of prismatic quartz, as well as hyalites from Zimapán and fibrous tin from Guanajuato. In his description of the box's contents, Humboldt recommended his samples to the members of the Institut on account of their novelty: the minerals were 'new' to the chemists and mineralogists in Paris who were at the time compiling standardised tables and nomenclatures of the mineral kingdom.[9] Humboldt's implicit expec-tation was that his correspondents would find his samples sufficiently unusual or appealing to study them, much in the same way 'newly discovered' plant specimens were studied at the Jardin des Plantes.

The specimen numbered '14' in Humboldt's box merited longer explanation. Humboldt identified it as 'brown lead' (*plomo pardo*) from Zimapán, analogous, he suggested, to a mineral found in Zehoppan in Saxony, Hoff in Hungary, and Pollawen in Britain. He informed his correspondents that del Río had isolated a purportedly new metal from this sample and named it erythronium because it pro-duced beautiful red (*eruthrós* in Greek) salts in reaction with acids.[10] Besides 14.80% erythronium, the plomo pardo from Zimapán also contained 80.72% yellow lead oxide, a bit of arsenic, and a bit of iron oxide. Hyppolite-Victor Collet-Descotils (1773–1815) at the École des Mines in Paris received and analysed the sample and decided that what Del Río had supposed to be a new metal was nothing more than chromium, which had been recently isolated by Vauquelin.[11] Thirty years later, in 1831, Swedish chemist Nils Gabriel Sefström (1787–1845) re-discovered a new metal in the sample of plomo pardo from Zimapán and called it vanadium, the name by which the element is known today.[12] The general plot lines of this deferred discovery are not unknown and are often used to reclaim del Río as the 'real' dis-coverer. Indeed, a Wikipedia entry does as much by establishing the discovery date of vanadium in 1801 by del Río.[13] Here, we reflect on the kinds of material config-urations, epistemological assumptions, and social relations that shaped the reception and (mis)translation of the new metal first identified by del Río in Mexico City into the chrome oxide obtained by Collet-Descotils in Paris, before it eventually became vanadium in Sweden. Why did Humboldt's correspondents at the Institut fail to produce del Río's erythronium out of the sample of plomo pardo Humboldt sent along? The reasons, as we shall see, are not limited strictly to chemical misiden-tifications but lie at the heart of the friendship, and its betrayal, between Humboldt and del Río, and, more broadly, reflect on the kinds of assumptions that were being made about the place of American minerals and Hispanic American scientists in the context of early nineteenth-century knowledge-making.

Humboldt's relationship with Spanish-born mineralogist Andrés Manuel del Río goes back to the early 1790s, when both studied at the Freiberg Mining Academy under Abraham Gottlob Werner (1749–1817). Werner's theory of Neptunism explained the formation of the earth's crust from an all-encompassing ocean that had receded and precipitated, leading to the emergence of mountains and the generation of minerals.[14] From Freiberg, del Río went on to study metallurgical techniques and chemical analyses in the mines and industrial complexes of Central Europe, including at Schemnitz, Prague, Transylvania, and Dresden. In 1791 he travelled to Cornwall to study developments in iron metallurgy, then to Paris to study under Antoine-Laurent de Lavoisier (1743–1791). This stage of his career was cut short by the revolutionary events that put an end to Lavoisier's life. Del Río returned to Spain. In 1794, his training under some of Europe's most prestigious chemists and mineralogists earned him the chair of mineralogy at the School of Mines in Mexico City, at the invitation of its director, Fausto de Elhuyar (1755–1833), another colleague at Freiberg. Like the Botanical Garden and the Royal School of Surgery, the School of Mines was a product of the Bourbon Reforms and was founded in 1793 with the mandate to systematise knowledge about New Spain's mineral resources and thereby contribute to a more efficient exploitation of its famed silver mines. Del Río was contracted with the expressed mission to organise the school's collection of minerals and instruments and to contribute to the inventory, classification, and extraction of useful minerals. His *Elementos de Orictognosia* (the first volume, dedicated to 'earths, stones and salts' was published in 1795 while the second, focusing on 'combustibles, metals, and rocks,' appeared in 1802), the first Mexican textbook of mineralogy, followed Werner's system to classify and incorporate the mineral species found in New Spain.[15] Del Río's unpublished treatise *Arte de minas* (1795) was devoted to describing and proposing improvements for machines, instruments, and procedures used in mining, and it represents the more practical side of his thinking.[16] His water-column machine, used to drain flooded mines, was the first of its kind in Mexico.

During his stay in Mexico City, Humboldt renewed his friendship with del Río and with Elhuyar, who introduced the traveller to colleagues at the School of Mines and granted him access to its laboratories and archives. Humboldt later spoke highly of the institution, praising its architecture and its collections of minerals and machines[17] which del Río had organised. Upon his departure from Mexico, he sold his instruments to the school and purveyed requests for other instruments which he promised to procure and remit once in Europe.[18] It was on del Río's recommendations that Humboldt journeyed afield to the mining districts at Taxco, Querétaro, and Guanajuato. He visited the famous Valenciana mine, the richest and deepest in the world at the time, and the Morán mine, where he had a chance to see del Río's water-column machine at work.[19] Closer to Mexico City, and often in the company of del Río himself, Humboldt visited such sites as Chapultepec, Xitle, and Peñon de los Baños, on account of their topographical and geological interest.[20] On these trips, he collected minerals and exchanged observations and samples with del Rio's colleagues at the School of Mines. Humboldt certainly benefited from their knowledge and generosity. Rafael Dávalos, for example, a mining engineer, gave Humboldt his geological tables.[21] Del Río granted him access to documents and manuscripts at the school, including the geological reports by Federico Traugott Sonneschmidt, another graduate of Freiberg who had been commissioned by the viceroy to study mining districts across New Spain. Sonneschmidt's explorations served as the basis for Del Río's *Elementos de Orictognosia* and provided guidance to Humboldt's own work.[22] Del Río also provided Humboldt with a manuscript copy of his Spanish translation of Dietrich Ludwig Gustav Karsten's mineralogical tables. Del Río's translation, *Tablas mineralógicas dispuestas según los descubrimientos más recientes e ilustradas con notas por D.L. G. Karsten*, was published in 1804, after Humboldt left Mexico. It included del Río's annotations to Karsten's tables, with information about minerals and their location in New Spain.[23]

Based on the evidence of their exchanges as documented by Humboldt in his diaries, Humboldt and del Río enjoyed a professional friendship of mutual admiration.[24] Humboldt profited enormously from this friendship, gaining access to experienced experts, confidential documents, and rare collections, and probably to private cabinets as well. He also found a receptive audience for his own theories there, as del Río included Humboldt's introduction to 'geological pasigraphy' in his second volume of the *Elementos de Orictognosia*. The 'pasigraphy' was Humboldt's proposal for a universal language, based on Werner's theories, to represent rock layers in geological profiles. Obviously, del Río would have expected Humboldt to treat him with the same generosity when he entrusted his friend with a sample of plomo pardo from Zimapán collected some years earlier by Sonneschmidt. As we saw, Humboldt shipped this sample to the Institut with del Río's report on the isolation of erythronium. Given the Prussian's access to a wide network of correspondents, including the powerful body of chemists and mineralogists in Paris, del Rio likely expected him to act as a broker, ensuring not only that the new metal reached its destination but that del Río's report be examined and validated by scientists at the Institut. Del Rio knew that they held the power to call discoveries, to ratify a lump of matter as a new chemical element.[25] He might even have allowed himself to hope that his discovery would place him back on the transatlantic circuits for the production and exchange of scientific knowledge, that it would provide him with an opportunity to re-activate his relations with scientists in Europe. Humboldt fell short of del Río's expectations. While the sample of plomo pardo, the raw material which purportedly contained erythronium, did reach scientists in Paris, Humboldt did nothing to promote del Río's chemical report on the sample. In fact, there is no extant copy of that report. To gain a sense of its contents, we must piece together references and fragments of analyses which del Río published at different moments.

Del Río had first provided a description of the 'mineral de plomo pardo' of Zimapán in 1795 in the first volume of his *Elementos de Orictognosia*, although at the time he did not suggest that it might contain a new metal.[26] Some years later, in 1803, Spanish mineralogist Ramón Gil de la Quadra (1774–1860) published a comparative table of metallic substances in the Spanish journal *Anales de las ciencias naturales*, where he included a first mention of *pancromo*, a 'new metallic substance announced by D. Manuel del Río in a report directed to Sr. D. Antonio Cavanilles, dated 26 of September, 1802.'[27] Del Río's report, however, was not published by de la Quadra, which might explain why del Río entrusted Humboldt with his report when the latter visited Mexico in 1803. In 1804, del Río published another reference to *pancromo* in a note to his translation of Karsten's mineralogical tables, although, by then, he had a change of heart and no longer believed the sample of plomo pardo contained a new element. He wrote:

> Because it seemed to me this was a new substance, I called it panchromium [pancromo] because of the universality of the colors of its oxides, dissolutions, salts, and precipitates, and then erythronium for forming, with alkalis and earths, salts that turned red with fire and acids. But knowing that chrome also produces, through evaporation, red and yellow salts, I think *plomo pardo* is a yellow chrome oxide.[28]

Del Río retracted again his earlier claim of having discovered a new substance in similar terms in a text written a year later and devoted to the mining district of Zimapán and which appeared in *Anales de las ciencias naturales*.[29] There, he explained that since his first announcement of the discovery of a new element he had had the opportunity to review Fourcroy's latest work (*Système des connaissances chimiques*, 1801–1802). He became convinced that what he had taken to be a new element was a chromate of lead. In other words, the arrival of Fourcroy's book, which compiled the latest discoveries and chemical procedures, in New Spain provided del Río with the opportunity to compare his own findings against Fourcroy's system of substances. The sample of plomo pardo from Zimapán, which Humboldt sent from Mexico together with del Río's report of erythronium, must have reached France before del Río's detractions did, that is, if indeed del Río's detraction ever reached France. There, Collet-Descotils published a detailed report of his own analyses of the sample, which led him to conclude that the 'new element M. Del Río claimed to have discovered' was chrome oxide.[30]

We could at this point let this case of missed opportunities rest and conclude that the experiment ended. But for del Río the case was hardly settled. It is hard to know when Collet-Descotils's report reached him, but in 1811, in an article first published in the *Gaceta de México* then reprinted in 1819 in the *Annales des mines*, Del Río bristled at his being caricatured as a preposterous fool. The Frenchman seemed to convey by insisting, in the very title of the report, that del Río was claiming to have discovered some new element. Del Rio now reminded the scholarly public that by the time Collet-Descotils published his findings he had himself identified the unknown substance as chrome.[31] But del Río's bitterest words were saved for Humboldt, whom he addressed that same year in an open letter published in the *Mercurio de España*, accusing the Baron of nothing short of betrayal in his failure to make del Río's report known to the members of the Institut. After all, there was merit in the soundness of his procedures and in his ability to isolate the various components of plomo pardo regardless of whether those components contained new elements or not. Del Río had, after all, isolated arsenic in the sample. Humboldt could have made sure the scientific community in Europe engaged del Río's experiments directly in the true fashion of scientific dialogue and debate. Instead, del Río complained to Humboldt, 'You saw fit to give [the sample] to your friend, doubtless for the reason that we Spaniards should not make any discoveries, no matter how small, either in chemistry or mineralogy, these being a foreign monopoly.'[32] Collet-Descotils made and published his own experiments on plomo pardo, while del Río's remained unknown in France, although he had published them in Spain before Collet-Descotils did so in France. Del Río added bitterly that nobody read Spanish journals even though the *Anales de las ciencias naturales*, where he published his studies, was a leading scientific journal in Spain.[33]

It would be difficult to overestimate del Río's frustration when the contentious metallic substance in plomo pardo was eventually recognised to be the new metal vanadium. Del Río took up the issue again in the 1832 edition of the *Elementos de la Orictognosia*, clarifying to his readers that

> when Humboldt left Mexico, I gave him a copy in French of my experiments, in order that he might publish them. If he had judged them worthy of public attention, *the discovery of a new metal would not have been delayed for thirty years*, which is the critique I receive now for no fault of mine. He did not even show a copy of my experiments to Descotils, who, because he was a chemist, would have appreciated them more, would have repeated them, and, because he had knowledge of chrome which I did not, would have easily decided this was a different metal.[34]

What del Río bitterly perceived to be a missed opportunity for himself and the larger community of scholars and scientists is an opportunity for us to reflect on the stakes of 'discovery.' What, in other words, determines that the mineral sample from Zimapán, Mexico, is recognised to contain a new metal? Or, rather, what does the failure to recognise the new element in this sample suggest about how scientific 'discoveries' are validated?

On a very basic level, the sample had to be translated both physically and then conceptually, that is, from Del Río's report into a standardised system of classification and chemical nomenclature that could be internationally recognised. To claim the discovery of a new element, it was not enough for Del Río to claim to have seen or isolated that substance. He had to prove it was new by comparing it against known substances, that is, to insert it into an ever-growing taxonomy. Chemistry and mineralogy are, to use a phrase coined by Lorraine Daston, 'sciences of the archive,' that is, what is known and how it is known are shaped not only by what happens in the laboratory but by the archive or collective repository of knowledge in manuscripts, books, instruction manuals, taxonomies, and so on. As Daston explains, 'the stability of the objects of inquiry depends crucially on a long disciplinary memory'; likewise, all confrontation with a new object or unusual phenomena requires the scientist or scholar to return to the archive to verify if and how it had been studied in the past.[35] At the turn of the nineteenth century when del Río performed his experiments on plomo pardo, the archive of chemistry was undergoing rapid change and Paris more than any other place had become the repository of this archive. For del Río, access to the archive of chemistry was measured in transatlantic time and distance. Chemical knowledge, in the form of not only books, magazines, and reports, but also instruments and people, reached New Spain with some delay, their traffic dictated by and intricately bound with the schedule of ships that left Europe via Spain for Mexico. It was a moment of increasing censorship of French material related to the French Revolution, the Napoleonic invasion of Spain, and the American wars of independence. These factors help explain del Río's hesitations about laying claims to his discovery before he could determine if the new element he had named erythronium had not been discovered elsewhere.

Distance happens in both directions, however. For while news of the latest chemical developments took time to reach New Spain from Paris, news from Mexico, such as del Río's report of his experiments, did not reach Europe any faster. Physical distance, in this direction, was compounded by other kinds of distancing, which foreclosed any possibility of translating del Río's discovery into a new element on the standardised table of elements. It had to do with a lack of faith that del Río's experiments were worth repeating or made known. Before entrusting Humboldt with the sample of plomo pardo from Zimapán and with his reports, del Río had already sent his analyses to Antonio José Cavanilles (1745–1894), the doyen of natural sciences in Spain. But the report was not published in the *Anales de las ciencias naturales* in Madrid, only news of the discovery. In this context a prestigious friend's intervention could have made a big difference. Del Río had introduced Humboldt to the vibrant world of Mexico's School of Mines and its associates. Surely, Humboldt would return the favor and introduce Del Río's report to the chemists and mineralogists at the Institut.

Humboldt did not return del Rio's favours. He treated del Rio's plomo pardo as an inert lump of matter, another piece of the tropics, and passed it on to Collet-Descotils at the Écoles des Mines in Paris, where the latter inserted it into the taxonomies that were taking shape there. Humboldt, del Río would remark, preferred to give it to his new friend in Paris, breaching the trust of his old friend in Mexico. And Humboldt forgot about the report completely. Without del Río's report and, more pointedly, without Humboldt's support and conviction that del Río's science was worth knowing, erythronium never became a metal.

Cochineal: 'The dye which exists as long as Indians tend to it'

Unlike erythronium, many of the species Humboldt brought to Europe, drugs, foodstuffs, or dyes, were known and widely used in the Americas. Humboldt was to an extent aware of this. It was the ground for his hope that he could systematise Amerindian knowledge for his European public. Such is the case with *Essai politique sur le royaume de la Nouvelle Espagne*, which showcased Humboldt's project of both inventorying Mexico's 'natural' resources and assessing the possibility that, like chocolate and tobacco, they could be globalised. Humboldt's plan rested on the principle that, given similar conditions of cultivation, plant and animal species could be transplanted economically. In this conviction Humboldt followed in the paradigm created by Adam Smith and added to by David Ricardo, which proposed that the cultivated products of nations were substitutable, which allowed nations to privilege those products that they produced at the lowest cost and to import others.[36] This classical liberal premise depended on the free flow of information. Middle persons or brokers, who discovered products and translated information into the most common European languages, were thus invaluable.

Still, even when information about things travelled relatively easily, not all plant or animal species could be easily transplanted from one place to another. Unlike a species name or place of origin, or its encyclopaedic description of appearance and use, the tacit, local knowledge needed for cultivating certain commodities could not be incorporated into universal systems of abstract knowledge. A good example of a tacit-knowledge heavy product was cochineal, the deep red carmine dye made from the pulverised body of the cochineal insect, *Dactylopius coccus*. Even after cochineal was stabilised as an object of science when the royal secrecy that shrouded it from public view was finally lifted, no amount of translating the insect and its uses into the language of nineteenth-century zoology proved sufficient to allow it to thrive beyond New Spain and a handful of other places.

Originally harvested in Tlaxcala, cochineal had been a luxury item since pre-Hispanic times. Its earliest visual representation appears in tribute rolls, records of the taxes imposed by the Mexicas or Aztecs on the conquered peoples of their vast empire. Cochineal arrived in Europe with the first shipment of gifts sent by Hernán Cortés to Emperor Charles V. It was soon taken up by manufacturers of luxury vestments, where it was employed to dye the robes of princes and cardinals. In the second half of the eighteenth century, the viceroyalty of New Spain commissioned studies meant to boost cochineal production and expand its cultivation into the central valleys of Oaxaca and around Guadalajara.[37] By then it was also harvested in South America, especially around Quito.[38] As a monopoly of the Crown, by the late eighteenth century, cochineal constituted an extraordinary source of revenue, surpassed only by silver.[39]

Its immense economic and aesthetic appeal explains why, while cochineal dye was widely trafficked, the habitual secrecy of the Spanish imperial bureaucracy and the guarded reserve that surrounded the cultivation of the cochineal among indigenous groups made for the mysteries regarding its cultivation and manufacture well into the eighteenth century. Those European countries that most envied Spain's monopoly were uncertain of its biological identity into the eighteenth century, as suggested by the name itself, *grana cochinilla*, or cochineal grain.[40] In *Histoire philosophique et politique des establissemens et du commerce des Européens dans les deux Indes*, abbé Guillaume-Thomas-François Raynal (1713–1796) gave a sense of the rampant ignorance around cochineal:

The nature of cochineal, without which one cannot make scarlet or carmine, and which can only be found in Mexico, has been unknown even
to the nations that most use it. The Spanish, naturally reserved, who turn
mysterious on the topic of their colonies, keep a secret when they believe it
is important.[41]

Although Raynal correctly identified cochineal as an insect the size and shape of a
bedbug, he agreed that knowing its nature was not enough. To obtain the dye, it
was necessary to have first the plant on which it grew and to learn to cultivate and
treat it. There were various attempts to do so by enterprising foreigners and spies,
the most notorious being Nicolas-Joseph Thiéry de Menonville (1739–1780), who
led an espionage mission to Oaxaca in 1776.[42] The Frenchman managed to penetrate into the zone where it was cultivated and, through an elaborate hoax worthy
of an eighteenth-century adventure novel, managed to smuggle out *nopales* (the
prickly pear cactus, *Opuntia*, on which the insect thrived), together with vanilla
pods and indigo seeds. He took them to Saint Domingue and planted them in the
Jardin du Roi at Port-au-Prince. But the experiment was short-lived, and the cochineal harvest collapsed soon after Thiéry de Menonville's death in 1780. His book
detailing his travels, contraband activities, and methods for harvesting cochineal was
published posthumously in 1787 as *Traité de la culture du nopal et de l'éducation de la
cochenille dans les colonies françaises de l'Amérique.*[43]

Humboldt begins his entry on cochineal, in book IV of his *Essai politique*, with a
critique of Thiéry de Menonville's work. The French spy 'ignored the language and
was wary to call attention to himself and generate mistrust by seeming too active,
and therefore was unable to gather together more than very imperfect notions
of the Mexican nopaleras' or nopal fields.[44] Humboldt seeks to redress Thiéry de
Menonville's 'imperfect notions' by availing himself of first-hand observations, his
own but mostly those of others. He presents relatively detailed descriptions on a
variety of topics related to cochineal, from the study of insect species (*fina* and *silvestre*) and their habitats to the kinds of dye produced by each species, with attention
paid to the care and breeding of the insects, the production of different categories of
dyes, and the commercial value assigned to each category. Besides his limited observations on grana *silvestre* in Peru, Quito, and Mexico, Humboldt did not have the
opportunity to see *grana fina*, but he did have direct access to people engaged in the
cochineal industry and to documents on the topic. His access to knowledge about
cochineal was typical of his access to most kinds of information he gathered in New
Spain, including those relative to mining, minting, cartography, and demography,
all of which formed the basis of his *Essai politique*. Unlike Thiéry de Menonville,
Humboldt was not, in Mexico at least, under suspicion as a lone spy sent by a rival
power. In short, the cochineal case underscores the collective nature of Humboldt's
project of gathering knowledge about Mexican nature. Like chemistry and mineralogy, zoology and dye manufacturing were and are sciences of the archive, where
knowledge is built on and shaped by the repositories of information on a subject.
In this case, the archive was Mexican.

By the time Humboldt arrived in Mexico, the cochineal archive was vast, consisting of numerous detailed reports that implicated a wide network of observers and informants. Of central importance to eighteenth-century manufacture of the dye were those scientific reports commissioned between 1772 and 1779 by Viceroy Bucareli.[45] Among these reports, the best known is the 'Memoria sobre la naturaleza, cultivo y beneficio de la grana' (1777) produced by José Antonio Alzate y Ramírez (1737–1799).[46] Other reports and memoranda were equally bent on standardising production and increasing profits. Such was the case of the report of the conde de Tepa (1730–1804), who served as judge in the Marquesado of Oaxaca and was a member of the *Sociedad económica de los Amigos del País* (an enlightened circle devoted to the production of useful knowledge about Mexico), and that commissioned by Pantaleón Ruiz de Montoya, magistrate of the cochineal-producing region of Nexapa.[47] In addition to these reports and documents, Humboldt had at his disposal export tables of cochineal dye and other products which documented yearly revenues.[48] A little over two decades after Thiéry de Menonville barely escaped the vigilance of port authorities in Veracruz, Humboldt was able to traffic information about cochineal freely across the Americas and the Atlantic.

What had changed in twenty-five years? If cochineal production had been kept secret for so long, why was Humboldt given access to the cochineal archive? Part of the reason was that, by the turn of the nineteenth century, others had already published detailed descriptions, including Thiéry de Menonville's in 1787. In 1794, Alzate issued his 'Memoria en que se trata del insecto grana o cochinilla,' which was based on the 'Memoria' he had written for Bucareli, as a supplement to his periodical *Gazeta de literatura de México*. A year later, the text was published again in Madrid. In addition, the British had already transplanted a Brazilian variety to India. It now made sense from the Spanish point of view to promote the product even at the cost of revealing the secret, since the quality of the Mexican dye was superior to that of the Madras cochineal.[49]

Furthermore, cochineal production no longer needed to be kept secret because it was such a labour-intensive activity that it was not considered to be cost-effective in the European market. In Europe, the cost of labour was higher, the time to establish a consistent product was long, and the quality was inferior. As a result, cochineal production remained associated almost exclusively with the indigenous communities of New Spain. Building on Alzate's detailed description, Humboldt describes how cochineal production depended on an intimate knowledge of seasonal cycles of cold and rain that involved complex regimens of care, such as building nests for the insects before releasing them on the cactus, and protecting them from predators like lizards, birds, rats, and other insects. 'Infinite care,' writes Humboldt, was taken to keep the nopales clean of dust. The plants were brushed gingerly with squirrel and deer tail hairs, a task mostly performed by women who spent entire days bent over each single plant.[50] To protect the insects from the cold, their indigenous caretakers covered the nopales with woven mats (see Figure 9.1).[51] Others, such as

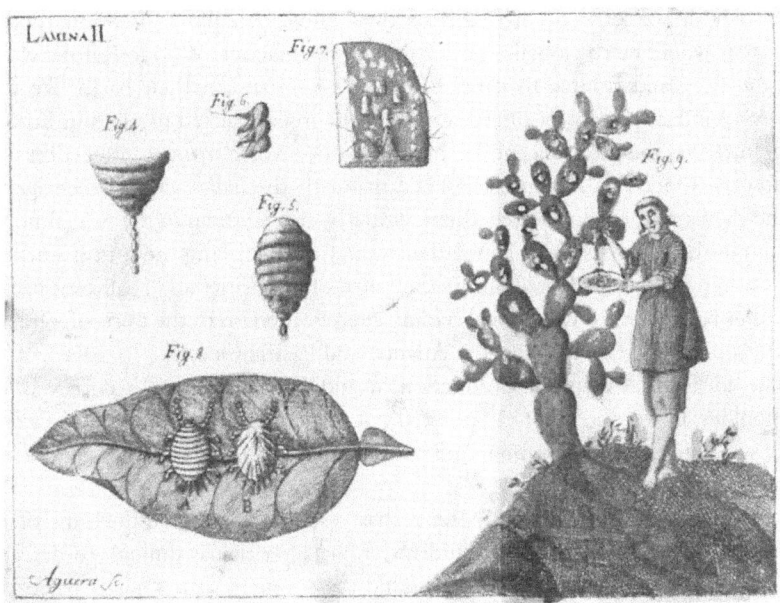

FIGURE 9.1 Harvesting cochineal. José Antonio Alzate y Ramírez, 'Memoria sobre la naturaleza, cultivo y beneficio de la grana' [1777], *Gazeta de Literatura*, 12 May 1794. Courtesy of the National Library of Medicine.

the cochineal growers in the mountainous regions around Oaxaca, kept an ancient tradition, which involved moving the insects about:

> Instead of conserving the insect in their huts during the rainy season, the Indians layer cochineal 'mothers' one on top of each other, inside baskets made with very flexible reeds, cover them with palm leaves, and carry them on their backs, as quickly as possible, to the Sierra of Istepeje, above the village of Santa Catalina, nine leagues from Oaxaca. The cochineal mothers procreate on the way and, upon opening their baskets, [the Indians] find them full of little insects and distribute them among the nopal fields in the mountains; [the insects] remain there until October when the rains end in the lower regions; then the Indians come back for the cochineal and 'plant' them on *nopaleras* in Oaxaca. This way, Mexicans move the insects about to spare them the pernicious effects of humidity, just like the Spanish move Merinos to avoid the cold.[52]

Alzate never mentioned merinos in his writings on the cochineal. The fact that Humboldt does so reflects his wider enterprise to translate American nature into European practices his readers were familiar with. At the same time, the comparisons

Humboldt draws serve to highlight its own limits by calling attention to how unlikely it would be to globalise or universalise the practice. Who in Europe would take the time and trouble to carry basketfuls of insects on their backs like this? The Amerindian pattern of interaction with the insects lacked parallels in Europe. Apiculture was the great exception, but it did not involve manual interaction with the insects. Once the cochineals reached maturity, the *indios* scraped them gently off the cacti and then 'suffocated' them, either by drying them in the sun (although this diminished their weight, hence their value) or by placing them in steam-bath ovens or temazcales. The final product, cleaned of impurities and chaff, was packed in leather bags, shipped to regional trading centres, then on to the ports of Veracruz and Acapulco, and finally across the Atlantic and Pacific Oceans.

The idea that cochineal production would be difficult to export was not Humboldt's, however. Alzate had pointed out that cochineal production was exclusively suited to indigenous labour and temperament:

> It seems like divine omnipotence has reserved certain productions of nature to the character of the Indians, a character that is difficult to describe, a patience that makes them withstand the strongest sun. Their sobriety and constancy in everything they undertake makes it possible for them to dedicate themselves to the cultivation of the grana, a very weak insect, prey to enemies and which, when dried, is reduced to nothing. Grana will exist only while the Indians tend to it. The other castes do not possess the phlegmatic character necessary for this unremitting and daily occupation.[53]

In Alzate's ethnography, the complex regimen of care required for production of the dye was a barrier to the transplantation of cochineal cultivation beyond New Spain. Discussing Thiéry de Menonville's 'theft' of cacti and insects and his attempt to 'cultivate' cochineal in Saint Domingue, Alzate writes: 'the French colony … expected great profit, but their hopes have vanished because the trade in cochineal will continue only as long as it is cultivated by the indios, who are a phlegmatic people and astute artisans; it is not a trade that can be of any utility for other castes.'[54] In any case, Alzate was strongly opposed to making cochineal viable in other places because its price would fall and so the communities directly involved with its care would lose interest in caring for it. Cochineal constituted a way of life, an intricate pattern of relations between humans and non-humans, the coming together, in dense and complex ways, of plants, insects, weather, soil, and temperaments. Cochineal could not survive without its human caretakers, and the Indians could not make a living without it. Alzate's study of cochineal production is inflected by his economic patriotism: divine providence had made Spain and its overseas kingdoms the caretaker of this great treasure, and no effort should be spared to ensure that this gift kept on giving.[55] For Alzate, the best way to do so was for the local and metropolitan authorities to promote studies of Mexico's 'generous' nature and of its inhabitants, building on their character to improve local agricultural practices.

In contrast, Humboldt was certain and anxious that New Spain's agricultural resources be exported not only via commerce but also by expanding production. With Alzate, however, he also thought that it would be very difficult, indeed unsustainable, to expand cochineal production beyond New Spain. Even as he made references to incipient cochineal industries in places as diverse as Rio de Janeiro, Calcutta, Chittagong, and Madras, he was not very sure they would succeed.[56] The reason was, once again, the Indian's inclination for laborious and time-consuming work. This explained why, 'despite the excessive price of cochineal, [Humboldt] doubted it would be profitable in countries where one knows how to take advantage of time and work.'[57] For Alzate, cochineal thrived locally because Mexico's native artisans had the phlegmatic temperament suited to produce it. A Galenic understanding of the temperaments, still widely influential in the late eighteenth-century Atlantic world, undergirded his argument. Humboldt also thought that the kind of temperament described by Alzate remained a factor in confining cochineal production to the Spanish Empire. In his description of the potato, for example, Humboldt wrote that 'the predilection of certain tribes for the cultivation of certain plants … indicate either the identity of a race, or ancient communications between peoples who lived in diverse climates. Vegetables, like languages and national physiognomies, function as historical monuments.'[58] Like cochineal, the Indians who raised and harvested it were also a 'historical monument,' allochronic vestiges of a pre-modern time.[59] For Humboldt, the European colony as a unit of production and extraction materialized the potential of people and plants:

> the smallest corner of the world, if it may come to be the property of European colonists …, will become witness to the activities that have engaged our species in the last centuries. A colony brings together in a small space all the precious things discovered by man on the surface of the globe.[60]

Nature is available as resources, that is, as things that can be counted and standardised, and these are in theory transferrable from one place to another and are thus open to speculation and financial projections. Cochineal, despite its preciousness, was an exception in this system of exchange because it could not be translated into the routines of modern European capitalism, even if the insects and the nopal on which they fed could be cultivated elsewhere. There were limits to turning colonies into laboratories for transregional industries, inasmuch as traditional routines of manufacture could not always be adjusted to a monetized system of free labour. Cochineal, Humboldt has to admit, was not just an insect on a nopal transplanted across borders but a congeries of tacit knowledges, gestures, rituals, traditions, regimens of care, social organisations, and political opportunities. Despite all attempts at standardisation, including all those that Humboldt and Bonpland exercised on the thousands of plants and minerals they collected, it was impossible to extricate cochineal from the racialised bodies that produced it and the local memories of its uses. Europeans would continue to import the dye until the invention of cheaper synthetic dyes.

Xochicalco: Fragments for a Universal History of Civilisation

'From the heights of Cuernavaca, we saw the ruins of the great pyramid at Xochicalco,' wrote Humboldt in the margins of his travel manuscript.[61] Just as he did not observe first-hand cochineal production, Humboldt did not visit the archaeological site itself. Instead, he made a note to himself in the margin to use the description of Xochicalco published over a decade earlier by Alzate: 'Decrivez-la d'après la Description d'Alzate de 1791.'[62] Humboldt collected a few choice antiquities, of which the most famous is the so-called Humboldt Celt, actually a gift from del Río.[63] In short, Humboldt relied, as he did in the case of many mountains and volcanoes, on images and descriptions made by other scholars. Alzate's *Descripción de las antigüedades de Xochicalco* forms the basis for Humboldt's treatment of the site in his *Essai politique de la Nouvelle Espagne* and in *Vues des Cordillères et monuments des peoples indigènes de l'Amérique*. Humboldt did not transcribe Alzate's description, however. Instead, he transported Xochicalco to his European readership through a process of selection and re-interpretation. The differences between the respective descriptions of Alzate and Humboldt are key to understanding Humboldt's larger antiquarian enterprise. That enterprise was to write America's ancient past into the philosophical histories of the origins of humankind then in vogue in Europe.

Although the *Vues des Cordillères* became a fundamental reference for the study and conceptualisation of American antiquities in the early nineteenth century in Europe and across the Atlantic, Humboldt did not 'discover' antiquities nor were antiquities his primary interest when he travelled to the Americas. Humboldt's antiquarianism was an acquired passion, as he suggested in a lecture he gave at the Philomatic Society in Berlin in 1806. He had set out to study 'the extraordinary natural phenomena of tropical countries,' but, he explained, 'the truth is that [he] also took every spare moment to follow the traces of the slow and sometimes mysterious march of the civilizing progress of American indigenous peoples.'[64] His own project was informed by contemporary debates about the history of the American continent and by the emergence of antiquarianism as a subject of study across the Republic of Letters.[65] Mexican savants had played an especially vocal role in these debates on both sides of the Atlantic: Mexican creole Jesuits exiled in Italy had penned studies about the natural and ancient history of their homeland. Their books, which travelled back to Mexico, shaped late-eighteenth-century debates about the vestiges left behind by Mexico's ancient peoples. Accidental discoveries, as was the case with the statue of the Coatlicue or the Aztec calendar stone (better known as the Piedra del Sol) unearthed in Mexico City in 1790, pitted Alzate and Antonio de León y Gama (1735–1802), both alumni of the exiled Jesuits, against each other, as they debated the meanings of ancient vestiges and sought to establish the methods for studying them.[66]

By the time Humboldt arrived in Mexico, both Alzate and León y Gama were dead. But the Flemish antiquarian Guillermo Dupaix (1746–1818), who later went on to lead the Royal Antiquarian Expeditions (1805–1808) across New Spain, was then studying antiquities and ruins around Mexico City.[67] Humboldt met with

Dupaix; indeed, his *Vues des Cordillères* opens with a plate of a statue in Dupaix's possession. Humboldt also visited private cabinets of antiquities and was granted the special favour to have the Coatlicue unearthed (it had been buried again in the courtyard of the university since it was first discovered) so he could study it for a short time before it was buried again. Humboldt dedicated a chapter of *Vues des Cordillères* to the Coatlicue.[68] To interpret these vestiges, Humboldt had access to manuscripts on indigenous history and linguistics held at the viceroyalty's archives, as well as to antiquarian writings such as those produced by Alzate and León y Gama in previous decades.

Alzate had first visited Xochicalco in 1777 but he did not publish his *Descripción de las antigüedades de Xochicalco* until 1791. It appeared as a supplement to his periodical, the *Gaceta de literatura de México*. His 'frail essay' was an answer to the call, issued earlier by the exiled Jesuit historian Francisco Javier Clavijero (1731–1787), for the study and preservation of Mexican antiquities.[69] Alzate also hoped to intervene in transatlantic debates on the nature of pre-Hispanic civilisations in the Americas.[70] Mexico's ancient inhabitants, writes Alzate, were looked upon with 'excessive contempt' by some Spaniards and painted in 'black and vile colors' by 'foreign authors.' It was this injustice that moved him 'to investigate their origin, their uses and customs, and all that concerns their arts and sciences.'[71] In the absence of writing, material vestiges offered a privileged way to 'know the character of the people that fabricated them.'[72] For Alzate, Mexico's ancient peoples were very different from contemporary Indians, whom he saw as the product of the Spanish conquest and whose rusticity, he believed, should not be taken as proof of the level of civilisation reached by their ancestors.[73]

In the *Descripción*, Alzate constructs and reinforces the epistemological posture of an eyewitness who produces knowledge directly on site. He takes measurements whenever he can, although he acknowledges that he did not bring all his instruments, having been wrongly 'persuaded, by experience, that people tend to exaggerate in their reports.'[74] What he found clearly exceeded his expectations. The ruins, consisting of various pyramids and the remains of other structures, sit on the top of a hill which had been carved into five receding terraces contained by stone walls and surrounded at the base by a trench.[75] Alzate compares the effectiveness of these ancient building techniques to those used in his day in Europe to build fortified cities 'where the defense goes from the circumference to the center.'[76] He concludes that 'these manmade constructions show how intelligent the Indians were in the military art.'[77] He acclaims the technical skills required to make the carved reliefs, especially those representing large figures that covered the surface of several blocks of stones. Notably, Alzate is uninterested in deciphering the meanings of the 'Mexican hieroglyphs' carved in the stones because, as he made clear in the course of his debates with León y Gama on the Coatlicue and the calendar stone, he believed any attempt to decipher the glyphs was bound to fail.[78] He does, however, venture hypotheses about the uses of the structures at Xochicalco suggesting, for instance, that a stone seat known as the *chimotlate* could have served as 'a throne whence Moctezuma or some prior monarch would have manifested his power.'[79]

Notably, Alzate never saw the seat because it was no longer in place; the stone used to carve it had been removed by local *hacendados* who had been pillaging the site to construct furnaces for sugar production.[80] What he knew of Moctezuma's throne was oral tradition, related to him by local Indian guides.

To gather information on the site, Alzate relied on the material vestiges and on the skills and knowledge of his local Indian guides. He recounts, for instance, how Indians took him to the site, showed him some of its more curious features, and informed him of the existence of a vast network of subterranean tunnels that bore through the hill. But when he pleaded with his informants to lead him to the entrance to the tunnels, they staunchly refused. This Alzate attributes to their ignorance, as exemplified by his guides' purported belief in spirits that made the earth tremble when treasure hunters ventured into the tunnels. Alzate suspects that his guides' refusal had to do with a more general reluctance to reveal their antiquities to 'Spaniards.'[81] In their willingness or refusal to reveal their vestiges, Alzate depicts Indians as active agents in the process of knowledge-making about their past.

Having been brought down by the ravages of time and by destruction at the hands of sugar cane planters, Alzate's *Descripción* offers itself as a way of preserving on paper that which was in danger of disappearing forever. The text ends with a testimonial addendum, where Alzate reports Xochicalco's geographical coordinates in the event, as he fears, the site is completely destroyed.[82] The addendum is accompanied by five plates, most of which are signed by the engraver Francisco Agüera. Two represent the site as seen from Cuernavaca and from above; another depicts the façade of the 'castillo' or castle (which today we identify as the Pyramid of the Feathered Serpent); the others are speculative, showing the tunnels Alzate was not allowed to examine, as well as reconstructions, on the basis of hearsay, of carved stones and structures that were no longer present at the time he explored the site.[83]

Alzate's *Descripción* formed the basis for the site's treatment by Pedro José Márquez (1741–1820) in his *Due antichi monumenti di architettura messicana* (1804). A Mexican Jesuit exiled in Rome and an expert on classical Roman architecture, Márquez finally made Xochicalco accessible to his European reading public by stripping Alzate's description of its more narrative and anecdotal passages and by systematising the information provided by Alzate to allow for comparisons and associations with classical art. Márquez's book opens with a definition of cosmopolitanism as a state of openness that takes 'all men for compatriots' and considers all peoples to be as 'polite and learned' as the ancient Greeks.[84] He sets out to write the achievements of the ancient Mexicans in the conceptual categories used to study the material vestiges of ancient Hebrews and Romans. Citing historians of ancient Mexico and its institutions, Márquez suggests that Xochicalco was a palace 'probably constructed by the Toltecs,' who 'made palaces of stone worked with figures and characters wherein all their hardships, wars, and defeats were signified, as well as their triumphs, good successes and prosperity.' Alternately, Márquez proposes it could have been a temple comparable to that of Janus, as suggested by the similar uses of sacrifice among the ancient Mexicans and Romans.[85]

Humboldt most likely had access to Márquez's *Due antichi monumenti* during his visit to Rome in 1804; in any case, he adjusted Alzate's study of Xochicalco for his European readers in similar ways. In *Essai politique*, Humboldt presents a succinct description of the site stripped, like Márquez's description, of Alzate's narrative interest, as part of a 'tableau rapide des antiquités aztèques':

> It is an isolated hill, 117 meters high, surrounded by ditches, and divided into five manmade platforms or terraces that are covered with masonry. The whole forms a truncated pyramid, whose four faces are exactly oriented according to the four cardinal points. The porphyritic stones with their basaltic base show very regular cuts and are decorated with hieroglyphic figures, among which one distinguishes some crocodiles throwing water and, what is very curious, men sitting with their legs crossed in the Asian way. The platform of this extraordinary monument is nearly 9000 square meters and presents the ruins of a small square building which undoubtedly served as last retreat to the besieged.[86]

In *Vuès des Cordilleres*, Humboldt provides a more detailed description of the ruins, offering interpretations of the site's primary function based on comparisons with similar structures from the classical world. 'In [Mexico], Xochicalco is regarded as a military monument attributed to the Toltecs,' he writes, noting that '[the Toltecs] are, for the Mexican antiquarians, what the Pelasgian colons were a long time for the antiquarians of Italy,' that is, an original civilisation credited with all the accomplishments for which there is no precise author. 'All that is lost in the mists of time is regarded as the work of a people among whom one believes to find the first terms of civilization.'[87] Drawing on both the work of Alzate and Marquez, Humboldt suggests Xochicalco was neither a temple nor a fortress but a combination of the two: a 'fortified temple' like those built by the ancient Hebrews and some Asian civilisations. As Humboldt put it, 'there is nothing more natural for man than to fortify the places where they keep their tutelary gods.'[88]

Humboldt's most visible modification of earlier studies of Xochicalco has less to do with what he adds to the texts than with what he omits, however. Alzate's *Descripción* presented his readers with several views of the site. Márquez reprinted copies of Alzate's engravings albeit modifying their perspective. In *Vuès des Cordilleres*, Humboldt chose to reproduce only the lateral view of the structure that Alzate had referred to as a 'castillo' or castle. It depicts, according to Humboldt, 'men sitting with their legs crossed in the Asian way' amidst what the Prussian took to be water-spewing crocodiles. This latter detail he found most 'surprising,' given that the pyramid is high on a mountaintop, above the 'torrid zone.'[89] That he had not chosen to include images depicting the ancient builders' sophisticated terracing techniques – which had especially interested Alzate – is important. In other chapters in his *Vuès des Cordillères*, Humboldt does offer illustrations and descriptions of the building techniques used by the ancient Peruvians and Mexicans. Here, however, he had the opportunity to represent a different kind of vestige, one that could lend itself to iconographic or semiotic interpretation on account of its glyphs and figures. In the larger context of his book, he opts thus to include examples of a variety of "monuments." His decision is borne out by the way in which he prepares the illustration for better legibility. If we compare Humboldt's plate to Alzate's, it becomes immediately obvious that Humboldt's offers a 'cleaner' view. The plants and debris that appear in the original plate have been erased so that the frieze of men and 'crocodiles' looks like it is floating on the page, thereby creating an object of knowledge that is commensurable with objects from other times and places (see Figures 9.2 and 9.3). Stripped of narrative details, Humboldt's description achieves a similar effect. It removes Xochicalco from the immediate context in which it continued to be used by indigenous peoples and sugar cane hacendados to offer it more readily for inspection by students of antiquities elsewhere.

FIGURE 9.2   View of Xochicalco Castle. José Antonio Alzate y Ramírez, 'Descripción de las antigüedades de Xochicalco,' *Gazeta de Literatura de México*, 1791. Courtesy of John Carter Brown Library.

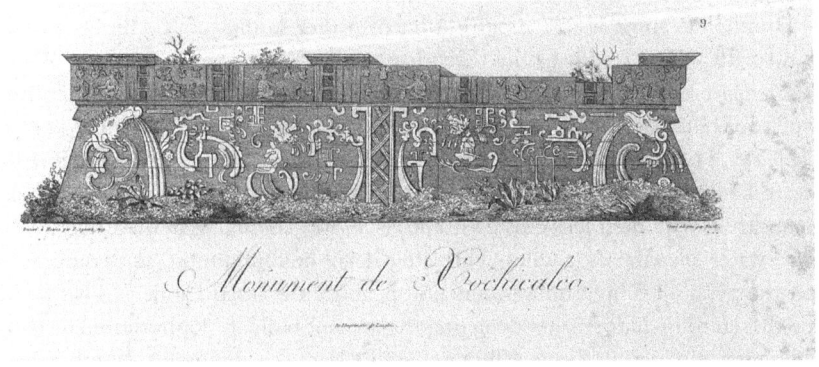

FIGURE 9.3   Xochicalco Monument. Alexander von Humboldt, *Vues de cordillères et monuments des peuples indigènes de l'Amérique* (Paris, Schoell, 1810–1813). Courtesy of Beinecke Rare Book and Manuscript Library, Yale University.

Although Humboldt vows in *Vues des Cordillères* to gather 'everything that is related to the origin and first progresses of the arts among indigenous peoples of the Americas', he is in fact very selective in his collecting, displaying special interest in those objects that 'may shed some light on the analogies observed between the inhabitants of the two hemispheres.'[90] As he confessed to his audience in a lecture delivered to the Philomatic Society in Berlin, it was his Italian journey in the summer of 1804 that allowed him a rare opportunity to compare, in the space of less than a year, not only 'the colossal volcanoes of the Andean Cordillera with the fire-spewing mountains of Europe, but also the colossal and perfect monuments of Roman art with the crude vestiges of the developing Mexican culture.'[91] In short, the Italian journey had 'motivated him in no small measure to undertake his research' on 'the monuments of the primitive American peoples.'[92] Comparison and analogy between the vestiges of ancient peoples across the Atlantic world (and beyond it) were thus at the heart of his study of prehispanic antiquities, which sought to discover common architectural, religious, and political structures, propose hypotheses about the progress of humanity, and 'fill in the blanks' where the historical record was missing.

The similarities he perceived between pyramids in Mexico and the temple of Belus in Babylon, for instance, led Humboldt to ask questions about the origins of the Toltecs, the purported builders of Mexico's ancient past. 'From where had they borrowed the design of their buildings? Were they of the Mongolian race? Were they descended from a common stock ... with the Chinese, the Hiong-nu and the Japanese?'[93] In the context of his wider antiquarian project, Humboldt's transformation of Xochicalco makes manifest the stakes of the Prussian's antiquarianism. Ancient monuments are transported, visually and conceptually, from Mexico to Europe so that they may be examined side by side with antiquities from the Old World in the cabinets of European cognoscendi. Humboldt's *Vues des Cordillères* thus became an obligatory reference for these cabinets, much in the tradition of earlier antiquarian albums, such as Anne Claude de Caylus's *Recueil d'antiquités égyptiennes, étrusques, grecques et romaines* (1762) or Dominique-Vivant Denon's *Voyage dans la Basse et la Haute Égypte* (1802).

Concluding Reflections

Since 2011, dressed in breeches, ruffled shirt, and vest, Ecuadorian artist and filmmaker Fabiano Kueva has been retracing Humboldt's footsteps from Taxco to Paris, Quito to Berlin, from the prismatic basalts at Santa María de Regla to Tenerife, stopping along the way to collect minerals and nopales, soil samples, archaeological souvenirs, and vistas. His growing collection has been displayed in Mexico City, Paris, and at the Humboldt Forum in Berlin. It will be deposited in the 'Archivo Alexander von Humboldt' after the project comes to an end in 2022. There is obvious parody in the proposal of this twenty-first-century Ecuadorian to follow in Humboldt's footsteps in reverse, activate his networks of mineralogists, botanists, librarians, curators, and local guides, bringing together things that many a museum-goer takes to be part of Humboldt's 'authentic' collections. But, as in all parody, there is a lot more at stake than burlesque imitation. In his 59-minute film parodically named *Ensayo geopolítico*, the artist takes aim at the extractivist ambitions behind Humboldt's *Essai politique du royaume de la Nouvelle Espagne*. As he leaves through the archives of his nineteenth-century predecessor, holds between his hands the fragile pages of an herbarium, pores over the boxes filled with his minerals at the Muséum in Paris, or follows the curator into the storeroom of the museum in Dahlem where Humboldt's antiquities were held until recently, Kueva challenges the public to imagine the kinds of gestures, opportunities, and assumptions about collecting and museums that have made it possible for things to end up where they are today.

By reconstructing the trajectories of three objects Humboldt took back to Europe from his Mexican travels, we have posed a similar question: What kinds of social, material, and political configurations enabled Humboldt to collect things in Mexico and translate them into objects of knowledge and display, or into global commodities? On a most basic level, the three objects we study here could hardly be more different from one another: a mineral from a Mexican mine, thought by a Hispanic mineralogist to contain an unknown substance; the crushed body of an insect, laboriously cared for by Indian communities in Oaxaca; and vestiges of a city built long ago by an unknown people and vindicated as proof of the genius of ancient American civilisation by Creole scholars. Humboldt sought to insert these objects into the emerging languages of mineralogy, political economy, and archaeology albeit with limited success. The brown lead from Zimapán does not become a new metal without the report written by Del Río that Humboldt forgot to pass on to chemists in Paris. Cochineal production stands as a metonym for Amerindian capacity for time-consuming toil, which cannot be transferred into modern wage economies. He finds it easier to collect and translate Xochicalco, ironically given its massive size, for consumption by philosophers of history on the other side of the Atlantic. Like the Indian who had to climb the Jorullo twice so that Humboldt's flask of volcanic gases could eventually arrive safely in Paris, the knowledge produced over centuries and generously shared with the Prussian was and is an extractable and vanishing resource at the service, or not, of his larger, comparative enterprise and of those who would follow in his footsteps.

Notes

1 Alexander von Humboldt, 'Volcan de Jorullo,' *Tagebücher der Amerikanischen Reise*, IX, 105v. All translations are ours unless otherwise noted.
2 Humboldt provides an extensive description in *Essai politique sur le royaume de la Nouvelle-Espagne* (Paris: F. Schoell, 1811), v. II, 137–148.
3 Alexander von Humboldt, 'Travaux que j'ai fait à la Secrétarie [sic] de la Vice-Royauté de la Nouvelle Espagne,' *Tagebücher der Amerikanischen Reise*, VIII, 118v; 128r.
4 See the chapters by José Enrique Covarrubias, Peter Mason, and Mark Thurner, in this volume.

5   The image of the sage in his cabinet is the other side of the coin of the equally powerful and widely entrenched image of Humboldt as the heroic explorer of wild and untamed nature, explored in several contributions to this volume.

6   Alexander von Humboldt and Aimé Bonpland, letter to the Institut national de France, in *Annales du Muséum d'histoire naturelle* 3 (an 12 [1803]), 396–404, here 400.

7   Gregory Cushman, *Guano and the Opening of the Pacific World. A Global Ecological History* (Cambridge: Cambridge University Press, 2013), 26.

8   See José Antonio Amaya's contribution in this volume.

9   Humboldt suggested these samples be distributed for analysis among a number of scientists associated with the Institut National: Vauquelin, who had analysed the guano samples Humboldt had sent from South America; René Just Haüy (1743–1822), who was at the time developing crystallography as a method for classifying minerals; Jean-Antoine Chaptal (1856–1832), whose most important work focused on the agricultural and industrial applications of chemistry; as well as Claude Louis Berthollet (1748–1822), Louis-Bernard Guyton de Morveau (1737–1816), and Antoine-François de Fourcroy (1755–1809), who had collaborated in the standardisation of a chemical nomenclature. See Oliver Lubrich and Thomas Nehrlich, eds., *Alexander von Humboldt: Escritos* (Mexico City: Herder, 2019), 176–84.

10 In the letter to members of the Institut National, Humboldt and Bonpland refer specifically to erythronium. See, Humboldt, 'Géographie minéralogique du Mexique,' *Tagebücher der Amerikanishen Reise*, VIII, 144.

11 Hyppolite-Victor Collet-Descotils, 'Analyse de la mine brune de plomb de Zimapan, dans le royaume du Mexique, envoyée par M. Humboldt, et dans laquelle M. Del Rio dit avoir découvert un nouveau metal,' *Annales de chimie* 53:30 (nivoise an XIII, 1805), 268–271.

12 The broad outlines of the discovery have been studied by Lyman R. Caswell, 'Andrés del Río, Alexander von Humboldt, and the Twice-Discovered Element,' *Bulletin for the History of Chemistry* 28:1 (2003), 35–41; and José Alfredo Uribe Salas, 'Historia del vanadio, 1801–1831. Disputa por la autoría del descubrimiento,' *Asclepio* 72:2 (2020), 322.

13 https://en.wikipedia.org/wiki/Vanadium, consulted September 24, 2021.

14 For this brief biographical sketch of del Río we draw on Caswell and Octavio Puche Riart, *Andrés Manuel del Río Fernández, 1764–1849* (Madrid: Fundación Ignacio Larramendi, 2017).

15 Francisco Omar Escamilla González and Lucero Morelos Rodríguez, 'Bringing Werner's Teachings to the New World: Andrés Manuel del Río and the Chair of Mineralogy in the School of Mines of Mexico (1795-1805),' *Earth Sciences History* 39:2 (2020), 246–261.

16 Francisco Omar Escamilla González and Lucero Morelos Rodríguez, *Escuelas de minas mexicanas: 225 años de la fundación del Real Seminario de Minería* (Mexico City: Facultad de Ingeniería, UNAM, 2017), 92–107.

17 Humboldt, *Essai politique*, v. II, 146.

18 See the list of instruments in Humboldt, *Tagebücher der Amerikanischen Reise*, VIII, 138.

19 Humboldt, *Tagebücher der Amerikanishen Reise*, VIII, 97.

20 Alexander von Humboldt and Aimé Bonpland, 'Correspondance,' *Annales du Muséum d'histoire naturelle* 3 (1804), 396–405.

21 J. Joaquín Izquierdo, 'Las ciencias modernas en la primera etapa del Seminario de Minería en México (1792–1811),' *Memorias de la Academia mexicana de la historia* XXIII:3 (1964), 255.

22 Humboldt, *Tagebücher der Amerikanischen Reise*, VIII, 97. See also Escamilla González and Morelos Rodríguez, *Escuelas de minas mexicanas*, 88.

23 Humboldt, *Tagebücher der Amerikanischen Reise*, VIII, 8; 363.

24 Del Río, unfortunately, left no account of their interactions, with the exception of a few letters to Humboldt, which he published years later.

25 For a similar case involving Humboldt and José Celestino Mutis, see Amaya, this volume.

26 Andrés del Río, *Elementos de Orictognosia* (Mexico City: Mariano Joseph de Zúñiga y Ontiveros, 1795), v. I, 167–168.
27 Ramón Gil de la Quadra, 'Introducción a las tablas comparativas de las sustancias metálicas,' *Anales de las ciencias naturales*, v. 6 (1803), 25–38, here 38.
28 Dietrich Ludwig Gustav Karsten, *Tablas mineralógicas dispuestas según los descubrimientos más recientes*, trans. Andrés Manuel del Río (Mexico City: Mariano Joseph de Zúñiga y Ontiveros, 1804), 62.
29 Andrés del Río, 'Discurso sobre las vetas,' *Anales de las ciencias naturales* 7 (1804), 30–31.
30 Collet-Descotils, 'Analyse de la mine brune,' 271.
31 Louis Cordier, 'Extrait d'un article de M. André del Rio, sur la découverte du chrome dans le plomb brun de Zimapán,' *Annales de mines* 4 (1819), 499–500.
32 Andrés del Río, 'Carta dirigida al Sr. Barón de Humboldt,' *Mercurio de España* 1:2 (1819), 172.
33 Del Río, 'Carta dirigida al Sr. Barón de Humboldt.'
34 Andrés Manuel del Río, *Elementos de orictognosia* (Philadelphia, PA: Imprenta de Juan F. Hurtel, 1832), 485. Italics in the original.
35 Lorraine Daston, 'The Sciences of the Archive,' *Osiris* 27 (2012), 160.
36 Margaret Schabas, *The Natural Origins of Economics* (Chicago, IL: University of Chicago Press, 2005), 79–125.
37 Barbro Dahlgren, *La Grana cochinilla* (Mexico City: Universidad Nacional Autónoma de México, 1990).
38 Humboldt, *Essai politique*, v. III, 245.
39 Carlos Marichal, 'The Spanish-American Silver Peso: Export Commodity and Global Money of the Ancien Regime, 1550–1800,' in Steven Topik, Carlos Marichal, and Zephyr Frank, eds., *From Silver to Cocaine: Latin American Commodity Chains and the Building of the World Economy, 1500–2000* (Durham: Duke University Press, 2006), 25–52.
40 Exiled Jesuit historian Francisco Javier Clavijero reports that Europeans insisted cochineal was a grain even though one could see with the naked eye it is an insect and despite testimony to the contrary, by both learned Spanish naturalists and by Indians. See Francisco Javier Clavijero, *Storia antica del Messico* (Cesena: Per Gregorio Biasini, all'Insegna di Pallade, 1780), v. I, 114.
41 Guillaume-Thomas Raynal, *Histoire philosophique et politique des établissements & du commerce des européens dans les deux Indes* (Amsterdam: [s.n.], 1770), v. III, 59–60.
42 Amy Butler Greenfield, *A Perfect Red: Empire, Espionage, and the Quest for the Color of Desire* (New York: Harper Collins, 2005).
43 Nicolas Joseph Thiéry de Menonville, *Traité de la culture du nopal, et de l'éducation de la cochenille dans les colonies françaises de l'Amérique; précédé d'un Voyage à Guaxaca* (Au Cap-Français: Chez la veuve Herbault, libraire de Monseigneur le Général, & du Cercle des Philadelphes. À Paris, chez Delalain, le jeune, libraire, rue St. Jacques. & À Bordeaux, chez Bergeret, libraire, rue de la Chapelle St. Jean, 1787).
44 Humboldt, *Essai politique*, v. III, 245.
45 Dahlgren, *La grana cochinilla*.
46 A facsimile was published by the Archivo General de la Nación de México in 1981.
47 Alzate used these documents in his 'Memoria' so it is difficult to know whether Humboldt, who relied extensively on Alzate, actually read conde de Tepa's or Ruíz de Montoya's archives.
48 Humboldt was able to report that in a 'normal year,' cochineal exports from Veracruz yielded 12,000 livres. In 1802, New Spain had exported 46,964 arrobas, an equivalent of 3,368,557 piastres; it appears that he continued to receive information on cochineal exports after he returned to Europe because he reported 29,610 arrobas (223,867 piastres) for the year 1805. See Humboldt, *Essai politique*, v. III, 243.
49 R. A. Donkin, 'Spanish Red. An Ethnographical Study of Cochineal and the Opuntia Cactus,' *Transactions of the American Philosophical Society* 67:5 (1977), 1–84.
50 Humboldt, *Essai politique*, v. III, 256.
51 Humboldt, *Essai politique*, v. III, 256.

52 Humboldt, *Essai politique*, v. III, 260–61.

53 José Antonio de Alzate y Ramírez, 'Zoología,' *Gaceta de literatura de México* 2 (September 20, 1791), 255–256.

54 Alzate, 'Zoología,' 256.

55 José Antonio de Alzate y Ramírez, *Memoria en que se trata del insecto grana ó cochinilla: de su naturaleza y série de su vida, como tambien del metodo para propagarla y reducirla al estado en que forma uno de los ramos mas útiles de comercio, escrita en Mexico en 1777* (Madrid: Sancha, 1795), 1; Gabriela Goldin Marcovich, 'Voix Créoles: les savants de la Nouvelle-Espagne entre Mexico et l'exil italien (1767–1814),' PhD diss, École des Hautes Études en Sciences Sociales (2020), 419.

56 Humboldt, *Essai politique*, v. III, 244.

57 Humboldt, *Essai politique*, v. III, 256.

58 Humboldt, *Essai politique*, v. III, 108.

59 Johannes Fabian, *Time and the Other. How Anthropology Makes its Object* (New York: Columbia University Press, 1983).

60 Humboldt, *Essai politique*, v. III, 127.

61 Humboldt, 'Xochicalco,' *Tagebücher der Amerikanishen Reise*, VIII, 88–89.

62 'Describe it after Alzate's 'Descripción,' Humboldt, 'Xochicalco,' *Tagebücher der Amerikanishen Reise*, VIII. 88. Humbolt is referring to Alzate's *Descripcion de las antigüedades de Xochicalco. Dedicada a los Señores de la actual Espedicion maritima al rededor del orbe* (Mexico City: Zuñiga y Ontiveros, 1791), published as a supplement to the *Gaceta de Literatura de México*, which Alzate edited.

63 Humboldt deposited the celt, a jade Olmec axe, exhibiting some of the oldest Mesoamerican writing, in the Prussian royal cabinet, later the *Königliches Museum für Völkerkunde* in Berlin. The celt disappeared in World War II, and Humboldt's lithograph of the object, which he identified as an Aztec axe (*hache aztèque*) in *Vues des Cordillères* are, together with a few plaster molds, the only extant traces of the celt today.

64 Humboldt, 'Sobre los pueblos primitivos de América y los monumentos que han quedado de ellos,' in Lubrich and Nehrlich, eds., *Escritos*, v. 1, 244–264, here 244.

65 Antonello Gerbi, *La disputa del Nuovo Mondo: storia di una polemica, 1750–1900* (Milano: R. Ricciardi, 1955); Jorge Cañizares-Esguerra, *How to Write the History of the New World: Histories, Epistemologies and Identities in the Eighteenth-Century Atlantic World* (Stanford, CA: Stanford University Press, 2001).

66 Miruna Achim, 'La literatura anticuaria en la Nueva España,' in Nancy Vogeley and Manuel Ramos Medina, eds., *Historia de la literatura mexicana* (Mexico City: Editorial Siglo XXI, 2011), 549–569.

67 For an exhaustive reference on Guillermo Dupaix's antiquarian work, see Elena Isabel Estrada de Gerlero, *Guillermo Dupaix* (Mexico City: INAH/UNAM, 2017).

68 Charles Minguet, 'Extractos del diario de viaje de Alejandro de Humboldt sobre su estadía en México,' in Leopoldo Zea and Mario Magallón, eds., *Humboldt en México*, 2nd ed. (Mexico City: Fondo de Cultura Económica, 2019 [1999]), 20–21.

69 Alzate, *Descripción*, 1.

70 It was with this purpose in mind that he dedicated the text to the members of the Malaspina scientific expedition. But as would be the case later with Del Río's reports on erythronium, Alzate's *Descripción* was not published in Europe.

71 Alzate, *Descripción*, 'Advertencia.'

72 Alzate, *Descripción*, 1.

73 Alzate, *Descripción*, 2.

74 Alzate, *Descripción*, 2.

75 Alzate, *Descripción*, 7.

76 Though it is difficult to prove so without further research, Alzate might have had in mind the fortifications built during the reign of Louis XIV by Vauban on hilltops, such as those at Briançon and Mont-Dauphin.

77 Alzate, *Descripción*, 8.

78 Achim, 'La literatura anticuaria,' 549–569, and Goldin Marcovich, 'Voix creoles,' 468–473.
79 Alzate, *Descripción*, 8.
80 Alzate, *Descripción*, 12.
81 Alzate, *Descripción*, 11–13.
82 Alzate, *Descripción*, 16–17.
83 López Luján, 'Los primeros pasos de un largo trayecto: la ilustración de tema arque-
    ológico en la Nueva España del siglo XVIII,' *Memorias de la Academia Mexicana de la
    Historia correspondiente de la Real de Madrid* 51 (2010), 203–263.
84 Pedro José Márquez, *Due antichi monumenti di architettura messicana illustrati* (Roma: Presso
    il Salomoni, 1804), III.
85 Márquez, *Due antichi monumenti*, 17–21.
86 Humboldt, *Essai politique*, v. II, 162.
87 For an analysis of the debates surrounding the Pelasgians and the possible compari-
    sons with ancient Americans, see Gabriela Goldin Marcovich, "Expliquer ceux-ci
    par ceux-là'. Savoirs mexicains et architecture ancienne à Rome au tournant du XIXe
    siècle,' in Charlotte Guichard and Stéphane Van Damme, eds., *Les Antiquités dépaysées:
    histoire globale de la culture antiquaire au siècle des Lumières*, Oxford University Studies in the
    Enlightenment (Liverpool: Liverpool University Press, 2022), 99–122.
88 Humboldt, *Vues des Cordillères, et monumens des peuples indigènes de l'Amérique* (Paris:
    Schoell, 1810–1813), vol. I, 132.
89 Humboldt, *Essai politique*, 71; Humboldt, *Vues des Cordillères*, 133.
90 Humboldt, *Vues des Cordillères*, 8.
91 Humboldt, 'Sobre los pueblos primitivos,' 252.
92 For a detailed discussion of the ways Humboldt's stay in Italy shaped his thinking about
    the American antiquity see Marie-Noëlle Bourguet, *Le monde dans un carnet. Alexander
    Von Humboldt en Italie, 1805* (Paris: Éditions du Félin, 2017).
93 Humboldt, *Essai politique*, 70.

10

HUMBOLDT'S MISREADING OF THE MERCANTILIST FACE OF NEW SPAIN

José Enrique Covarrubias

TRANSLATED BY MARK THURNER

In *Del gobierno considerado en sus relaciones con el comercio* (1805), French economist and civil servant F.L.A. Ferrier articulated a series of criticisms of the liberal political economy of his time, which was then on the rise. Ferrier was among those authors who sought to vindicate the great mercantilist writers (Montesquieu, Locke, Galiani, etc.) recently discredited by the celebrated political economy of Adam Smith, which J.B. Say had been promoting and spreading in France at that time. Ferrier defended the idea that the old economic knowledge remained valid, despite the contempt in which it was held by the new experts on wealth and the prosperity of nations.[1] Ferrier argued that several economic theories mistaken as new had been intuited or formulated on previous occasions by the mercantilists, albeit in another idiom or discourse. A similar claim was made in the 1960s with respect to the scientists of colonial Mexico whose contributions Alexander von Humboldt had used to treat important issues in his *Political Essay on the Kingdom of New Spain* (1811). This critique may be found, for example, in the edited volume entitled *Ensayos sobre Humboldt* (1962), a work that brought together the contributions of several of the leading Mexican historians and academics of the time.[2] More recently, with the resurgence of economic history, similar critical questions have returned concerning the scope and bent of Humboldt's understanding of the economy of New Spain. This chapter engages with this critical current, concentrating on the demographic and economic issues overlooked or insufficiently registered in the *Political Essay*.

There is a general consensus among Humboldt scholars that the Baron's economic ideas were guided by the Physiocratic–Smithian tradition, although some scholars link him more closely with the Physiocrats, and others with Smith.[3] This current or tradition is not limited to Quesnay, Mirabeau, and other representatives of the physiocracy, together with the author of *The Wealth of Nations* (1776), but also includes other, previous, and subsequent authors whose reflections revolved around notions of the individual, reason, and nature, including

DOI: 10.4324/9781003231479-11

Mandeville, Bentham, Malthus, and Ricardo.[4] This current postulates 'the harmony of interests' principle as the cause of the formation and preservation of society. This principle was of great importance in the emergence of economic liberalism, understood as something different from political liberalism. The latter would have originated in a modality of natural law and theory of the social contract and would affirm as necessary a conscious and explicit decision of the individual for their integration into the community, whereas in the former case the always present and implicit guide of the individual interest is sufficient. Important in the theory of the harmony of interests is the concept of social, public, or general utility. This concept of general utility resumes the benefits of the convergence or identity of interests in favour of the common or public good.

It is important to recall, however, that from a more detailed perspective the representatives of this current of economic thought do not subsume everything related to economic life in the scheme of the harmony of interests, nor do they exclude all government intervention as unnecessary for it. This economic liberalism does not entirely deny the need for government intervention, if only indirectly. Such, for example, was the case of Smith, whose expectation that the administration of justice should preserve the community and give security to the economy is noteworthy in this regard. Already in the Dutchman Bernard Mandeville, this concept is noticed, and it reminds the reader of certain basic principles of this current of thinking which are also found in Alexander von Humboldt.

In Mandeville, the scheme of the harmony of interests starts from the beneficial action of self-interest and the satisfaction of the needs of the individual, which are derived from pride and the search for social prestige. These last passions move in people everywhere, but it is in commercial societies in which the situation is most clearly drawn. In these societies, more course is given to the pursuit of self-interest, which time and again enhances the egotism manifested in pride and social aspiration. This is also where the consumption of expensive and luxury goods is most widespread, with a great stimulus to trade, economic innovation, and general prosperity. Mandeville is very emphatic on this point, which gives meaning to his conviction that by following their own interests, individuals end up acting for the general good: private vices end up being public virtues.[5]

What manner of government intervention do the proponents of the current of 'the harmony of interests' find necessary? The late Mandeville points to the need for deliberate promotion of employment by skilled governors or rulers, those wise men who know best how to transform vices into virtues. On this point, we can already anticipate the first question regarding those economic aspects of New Spain otherwise ignored by Humboldt in his *Political Essay*.

Harmony of Interests and Government Intervention

In Mandeville, there is the conviction that booming employment is a factor in the economy that brings great benefits to society.[6] No art of making a nation happy and flourishing compares to giving everyone a job opportunity. To achieve this, the main concern of a government must be the promotion of factories, arts, and crafts to the highest degree that the human mind can conceive. However, this goal is not achievable without the offices of the government. In this regard, the government must demonstrate its ability to turn vices into public advantages. We have here the notion of economic potential,[7] for Mandeville takes it for granted that if the taste for luxury and honours are promoted, a government can easily generate one job after another and thus make the economy perform to its advantage. This will be the best productive policy, all the more affordable as private consumption increases relative to the demand for luxury goods and symbols of prestige.[8] Mandeville thinks that the East India Company does great good by inciting its sailors to spend their wages as quickly as possible, for example. Goods and needs are no longer judged by their suitability and quality but by their quantity. In this model, a parallel is drawn between the quantity of jobs and the goods produced, with the factors of demand and affluent consumption involved, so that the productivity formula could not be simpler and more clearly geared towards mass generation of goods.

This sense of economic potential in Mandeville, manifested in the direct correlation between the quantity of jobs and the goods produced, refers us directly to Humboldt.

There is also a sense of economic potential in Humboldt, but this is expressed in a different way, closely related to the broader issue of prosperity. In this case, the starting point is that agriculture is the main economic branch with an impact on the prosperity of New Spain, since it sustains demographic growth, which is not negligible by any means.[9] Humboldt's idea of prosperity is related to the notion that subsistence production allows for conservation and growth of the population. If the number of inhabitants has increased significantly, this fact reveals an equally important prosperity in the country.[10] However, two major recurrent diseases (smallpox and matlazáhuatl), as well as hunger, in a segment of the population, mainly the indigenous peoples, constitute periodic obstacles to population increase in New Spain.[11] But the important thing here is that prosperity brings not only demographic increase but also a good standard of living among the inhabitants.

Regarding this last objective, New Spain presents a kind of vicious circle, because the population is growing faster than the subsistence base (agricultural production). The line of reasoning here is not Malthusian, however, since Humboldt considers it possible to remedy the situation by correcting the disproportion between the population employed in agriculture, mainly in corn, wheat, and potatoes, and the population employed in other activities, notably in the transport of materials and merchandise.[12] At this point the question of employment and the proportion between the number of people employed in different branches arises, in this case agriculture and transport or muleteers. In other passages, Humboldt the traveller refers to the poor condition of communications, and his reader is led to understand that this is an important factor in determining the large number of individuals and animals engaged in transport.[13] Humboldt points out that the government should pay more attention to the roads of New Spain, especially for the introduction of carting, in particular along routes that connect the interior plateau with the coasts.

If Mandeville calls for government intervention to create jobs, Humboldt asks for it to reduce the number of people employed on the roads of New Spain, thus stimulating greater, more spontaneous occupation in agriculture.

In New Spain, however, productivity is unleashed with the concatenation of activities around mining, since it is enough that a mineral is discovered for a town to emerge soon after, generating demand for agriculture in the immediate fields, whose products are then sold at inflated prices. Thus, the mere existence of a mine makes it unnecessary for the government to found colonies, since mining exploitation itself generates a previously non-existent agriculture. In this Humboldt sees fulfilled the principle that "the motives of mutual interest ... are the most powerful links of the society," in this case channelled directly to the profits obtained from the production of precious metals.[14]

Humboldt's emphasis on the desirability of occupation in agriculture confirms his agreement with Physiocratic–Smithian ideas regarding the advantages of this activity, the one that in the long term has the greatest impact on prosperity.[15] This long-term perspective is important when developing the theme of prosperity or wealth generation and population growth. Here Humboldt's 'macro' orientation, inclined to generalisations when considering employment, consumption, and well-being, is self-evident. In a similar fashion, Humboldt generically discredits the consumption of luxury products, notably on the rise among the mestizos of New Spain in recent years,[16] as unproductive. In his view, it does not contribute to food production sufficient to eradicate the periodic famines and misery that was still so widespread. This tendency to adopt a broad framework for the consideration of economic phenomena is revealed even more clearly in the operations typical of political arithmetic or social arithmetic, which Humboldt makes liberal use of.

Society, Demography, and Economy in Broad Perspective

Since we have already dealt more extensively with Humboldt and political arithmetic elsewhere,[17] it seems pertinent only to recall here that curiosity about the proportion of the population dedicated to one or another economic activity is already present in the founder of this field of statistical study. In his *Political Arithmetic* (1690), William Petty determined the basic and average proportions with respect to the number of individuals engaged in the main productive activities of France and the Netherlands. He was interested in determining even the number of individuals per surface of land where sedentary societies arise, from which he deduces a kind of universal basic proportion, subsequently altered by the factors of location and economic and political activity.[18] It should be mentioned that this type of universal reflection is also of interest to Humboldt in the *Political Essay*, where he shows that one of his main interests is to resolve the question of whether the population density is adequate within the political demarcations of New Spain (intendencies and, in the north, interior provinces). Another interest is to provide figures relative to the longevity, age of marriage, and similar kinds of data (aspects of the average individual age within the various racial groups) on the kingdom's diverse populations.[19]

Although on the first question his findings are expressed more conclusively, on the second his offerings are fragmentary and provisional. Humboldt concludes that the second question is a subject for the natural history of man, thereby betraying his ascription to a field in which phenomena are situated in the very long term and within a perspective that encompasses all of humanity and its relationship with nature.[20]

We referred previously to the current of thought on 'the harmony of interests,' based on notions of the individual, nature, and reason, and developed in different ways by its several representatives. Of particular interest relative to this broad, natural framing of the problem is again the figure of Mandeville, the author perhaps closest to Petty chronologically and ideologically, given their common disinterest in the social, economic, and political opinions of the individuals they study or describe in their texts. As Gaukroger points out,[21] Mandeville denies clear boundaries between moral conduct and the pursuit of self-interest, even in its crudest form. On this subject, his point of departure is the supposed behaviour of savages and their 'natural' morality, in which a virtue such as compassion is nothing more than a form of self-interest. Even when being compassionate, Mandeville's savages act out of praise and fear of reproach, which reveals the existence of external pressures (constraints) that move them to do so. In essence, in the most civilised societies the same thing happens, Mandeville maintains, with the exception that in these societies those external pressures have been internalised. What is decisive, however, is that these pressures, supposedly at the origin of virtue and vice, actually have an origin external to individuals, specifically in their competition for scarce resources in ancient times. The impact of this natural state continues to be felt so that talking about vices and virtues in an absolute sense does not make sense. Rather, what individuals do is regulate their behaviours based on the mentioned competition. The result is a regulation that, again and again, is renewed and progressively fine-tuned so as to maintain the appearance of respectability and propriety when, in fact, the underlying phenomenon is never altered. Much of individual reason focuses on covering these appearances without neglecting the ever-present natural struggle.

Petty had already postulated, as has been noted, a natural state based on the proportion between a certain number of individuals and a certain amount of territory with the characteristics to support them. The development of societies is thus conceived under the Baconian idea of the usefulness of rational (scientific) interventions in the processes of nature for the best use of the potential of human sustenance enclosed within it, especially by way of political interventions.[22] Thus, although the original ratio between individual and territory is always in force, human genius or 'art,' applied to commerce, inventions, politics, and so on may well modify the conditions of the returns of nature. The comparisons undertaken by Petty between the demographic and economic profiles of different nations (e.g. the Netherlands and France) indicate his method for studying the potential in these countries. Such a procedure, together with his famous projects of population transfer between Ireland and England for a fusion of the two peoples, also reveals a frame of reference broader than that found in Mandeville. Perry's approach supposes inter- and supra-national comparisons with multiple intertwining factors between art and nature, economic practices and type of environment, the potential of utility in the individual, and potential of wealth in the surrounding geography.

The similarity between Humboldt's approach in *Political Essay on the Kingdom of New Spain* and Petty's political arithmetic is hardly surprising, given its common recourse to the method of comparison and the application of knowledge of nature, or, more precisely, of man in nature, to government interventions and economic projects.[23] The natural history of man contemplated by Humboldt requires a framework as broad as that of Petty, who at one point he praises for his method, applicable in such fields as botany (as 'botanical arithmetic').[24] For the purposes of social and economic study, however, Humboldt aligns himself, as we have seen, with the current of 'harmony of interests' and its appeal to notions of the individual, nature, and reason. There thus arises a tension between the generalising views of Humboldt's economic study, willing to insert its data within a hemispheric or even global framework,[25] and the more focused or applied attention that requires judging the success or failure of economic policies in New Spain.

Of particular interest here is the question of employment in New Spain. Humboldt addresses this question based on the assumption that an increase in agricultural production would definitively solve the problem and break the vicious circle of recurrent famines and demographic pressure. The point is of importance for the economic history of Mexico and quite possibly for the rest of the Spanish American countries visited by Humboldt, where his writings also constitute a relevant and useful source.

Recent research in social and economic history suggests that a very broad vision such as Humboldt's, disinterested in the political and social achievements of Spanish mercantilist forms of administration, is likewise uninterested in the historical logic of the various local governments. The experiences, practices, and values that, over nearly three centuries, gave rise to this logic, and, in some cases, still resonated and held weight during the time of his travels in America are not of primary interest to Humboldt.

In this regard, a recent review of the relationship between the economy, public works, and the situation of urban workers in Mexico City between 1687 and 1807 is of particular interest.[26] In this study, Enriqueta Quiroz reviews various public works projects in the capital of New Spain. These projects demonstrate a logic and chain of economic events different from those presented by Humboldt in his *Political Essay*. One issue that stands out from Quiroz's reading is, that on previous occasions there had been moments of great demographic pressure on the indigenous population. The dynamics of that population, even from the point of view of job occupation, had already led to a move away from agriculture by a certain mass of the population, with a consequent increase in the ranks of what Humboldt calls, after the physiocratic idiom, the 'sterile,' which is to say, the not truly productive population, when referring to the muleteers and others engaged in transportation. Thus, towards the middle of the seventeenth century, when the so-called lands composition or auction and purchase of uncultivated lands by Spaniards intensified (a phenomenon similar, to a certain extent, to the English enclosures), a good part of the native population previously restricted to their lands migrated to haciendas or became urban labourers.[27] This latter phenomenon is of particular interest to Quiroz, who in her study addresses the financing by the Royal Treasury of public works that produced employment for this type of population in the urban environment, thereby achieving subsistence in a modest but safe way and under strict political control of the authorities. In this manner, population growth received significant support under classically mercantilist policies (job creation by the government), while emigration to cities—especially Mexico City—frequently strengthened the household as an economic unit in the country. Individuals could thus find mutual support and state protection during economically difficult periods, something that, a consequence of high prices, was still occurring during the years of Humboldt's stay in New Spain.[28]

Regarding the resistance of the indigenous population during the sixteenth and seventeenth centuries to the labour and political regime imposed by the Spanish colonial administration, Humboldt mentions resettlement to the mountainous and distant areas, a process that culminates in a kind of colonisation or original conquest of territory that recalls the 'context of origin' of such great interest to political arithmeticians for their demographic calculations.[29] But Humboldt fails to grant sufficient importance to urban emigration, according to the findings of social historians such as Quiroz. The employment generated in the urban environment by the authorities and the importance acquired in that environment by the family economic unit, with favourable consequences in terms of prosperity and social peace, are ignored. Humboldt could have registered or given clues regarding these facts when dealing with the highly populated city of Mexico, but instead, he limits himself to pointing out that the population calculation in a city like this cannot be adjusted to the proportions found in the provinces or districts of the interior.[30]

Humboldt's lack of interest in the precise conditions of the urban economy, to which the inquiry into the causes of the demographic dimensions of Mexico City could well have led him, also had repercussions in his lack of attention to the large recent increase in Mexican luxury products consumed by the mestizo population. As noted, Humboldt assumes this consumption to be unfavourable and he reaffirms this stance in his Physiocratic–Smithian emphasis on the advantages of agricultural productivity and the necessary priority of investment in this branch, with a consequent reversal of the luxury consumption of the mestizo mass, settled largely in urban areas.

The increased consumption of luxury products even in markets, such as the so-called Baratillo Grande (Big Cheap Market) in the capital, is well established.[31] But it is also true that consumption among the urban working population and members of popular household economic units was differentiated. This differentiated consumption resists the blunt conclusions or interpretations of Humboldt, which are based on broad numerical or other trends.[32] Following Quiroz, and considering Humboldt's ideas in relation to those of the late mercantilists, collected in a recent study by the present author,[33] everything points to the fact that by not incorporating the concept of the utility of individuals within the framework of the family or household unit, so present in authors such as Campomanes, Humboldt limits his understanding of the principles of government regulation of an economy and society like that of New Spain. Of course, it is certainly possible that the increased consumption of luxury goods militated against the interests of the masses in that it did not favour greater agricultural production and did not contribute to making life cheaper for growers and workers, especially those of the mixed, lower castes. But any judgement in this regard would also have had to consider the damping of these effects through the creation of employment by public works financed by the Royal Treasury, the proven knowledge and perhaps the impulse of the authorities, the solidarity and mutual support mechanisms of the urban family framework during difficult times or economic crisis and—something very important and more clearly economic in nature—the diversity and continuous instability of popular consumption at the micro level in cities.

The Economic Valuation of Goods

Baron von Holbach tries to synthesise in his *System of Nature* what seems to him to be the essence of the doctrine of 'the harmony of interests.' 'Human societies,' he notes, 'cannot subsist except by virtue of a continuous exchange of those things in which men find their happiness.'[34] Holbach apodictically concludes that 'interest, or enlightened self-love, is the foundation of social virtues.' According to Simon, these memorable lines from Holbach do not correspond to the precise idea of Mandeville, the founder of the harmony of interests current of thought. The Dutchman was much more concrete in his approach, connecting it directly to economic dynamics. As Koslowski points out,[35] the Dutchman had left behind the so-called *Philautia* or doctrine of enlightened self-interest or self-love, also often described as long-term interest. Mandeville emphasised an individual interest free of these qualifications or requirements since he identifies it without further ado with the factual needs of individuals and any form of self-interest. Nor does Mandeville's positive assessment of social inequality correspond entirely to an enlightened mentality. He considers this inequality conducive to the public utility and happiness of all, given the interconnections and mutual services that it continually guarantees. Simon sees behind these arguments a replacement of the social contract by the market, the foundation of the institutionalised and regulated social. In short, in Mandeville, we see the replacement or the greatest possible restriction of politics by the economy.[36]

This distancing and disavowal of politics by Mandeville also reveals itself in his attacks on Aristotelian morality and the old aristocratic estate society, as well as the traditional categorisation of goods, which had recognised in goods not only the useful (the only thing relevant to the economy and society, according to Mandeville) but also the beautiful, and the honest.

This aspect is most clearly manifested in his attack on the moral theory of the Earl of Shaftesbury, a contemporary of Mandeville, who proclaims three types of goods, subsumable in the three categories mentioned earlier: *utile*, *pulchrum*, and *honestum*. As is well known, in *Characteristics of Men, Customs, Opinions and Times* (1711) Shaftesbury presents this triple categorisation of goods, whose validity would be general and permanent in all humanity and, as such, express concepts or notions of a natural or universal morality. In contrast to the ethics propagated by contemporaries who exalted modern commercial society, Shaftesbury based his ethics on an altruistic and non-economic spirit. For Mandeville, such an ethic was a mere product of convention and tradition, viable perhaps for parasites and monks but not for industrious, work-hardened men. The Shaftesbury altruist is not the type of individual who could achieve recognition, resonance, and transcendence in mainstream society. In truth, Mandeville assures us, the only really existing good is the *utile*. The ideal of the *pulchrum*, that is, of a dignity and excellence inherent in certain goods and appreciable in any place and time, is simply vain. Universal appreciation would be the requisite for granting its existence, Mandeville thinks, but if experience shows anything, it is the inexistence of this general and homogeneous esteem. Great artistic works, for example, have no intrinsic value but only that of the market. The same, Mandeville says, applies to the 'honest good': with this notion he refers to a thing that in another country or place is seen as abominable. The honest and the beautiful, considered as a mark of the noble and the un-useful, are merely products of convention and the assessment of the majority, and this already excludes the relationship of these categories with reality. On the other hand, Mandeville grants an absolute reality to useful things since, in his view, general, everyday experience certifies them again and again. For Mandeville, Shaftesbury's biggest error lies in assuming the possibility of acting against one's self-interest. This is tantamount to believing in honour or in a fictitious morality such as Don Quixote's.

In Humboldt, the characterisation of the goods of the economy is consistent with Mandeville, although without going to the same extremes and expressing it in a friendlier way. Between strictly subsistence or luxury goods, utility is always somewhere in the middle, and the more these goods gravitate towards that average condition, the more benefits they will report to society.[37] Such criteria can be seen above all in Humboldt's appreciation of basic materials and everything that contributes to subsistence, which, in turn, allows population growth and contributes to prosperity, thereby becoming decidedly useful.[38] The recourse to the concept of the useful serves Humboldt to expand the repertoire of things or elements submitted to his consideration, thus also incorporating a number of natural factors related to the subject. For example, he takes into account the yields of useful plants, especially for their nutritional value. The famous pages of the *Political Essay* on the productivity of the plantain and its feeding capacity in proportion to the cultivated surface area, which produces a much higher utility than that of cereals, are probably the most eloquent proof of this.[39]

Nevertheless, Humboldt's agrarian perspective was not shared by all his contemporaries, and it is here that once again we touch on aspects that are simplified in his treatment of the economic. Reference is made here in particular to the disagreement with Fausto de Elhuyar, the famous mining official of New Spain in Humboldt's time, who like Humboldt was trained in metallurgical science at the Freiberg School of Mines. In his *Memoria sobre el influjo de la minería* (1825), Elhuyar asserts that mining, and not agriculture, should be seen as the principal economic activity of Mexico.[40] He primarily invokes historical reasons to prop up his idea in favour of mining, reminiscent of the historical method of Adam Smith, an economist well known to and cherished by Elhuyar, despite Smith's censure of the value assigned by Spain and its colonies to precious metals. Only mining could support the necessary and rapid impulse needed for Spanish colonisation in its beginnings, he asserts, such that it could progress through an early, continuous, and significant trade between New Spain and its metropolis, this not to mention the easy acceptance of gold and silver everywhere, which enabled their expeditious exchange for materials and other very useful products for the overseas establishment. Elhuyar thus explicitly highlights the importance of the circumstantial in the valuation of assets and thereby relativises the absolute scope of what is taken as useful in a particular place or time. The case implicitly shows that the useful good can reveal that same relativity which, in Mandeville's eyes, applies to the beautiful or honest good, the result being that its supposedly objective reality is dubious.

However, the value assessment of mining as an activity that has moulded and still moulds individual abilities, tastes, and refinements remains important to Elhuyar's argument, assuming as he does that it will continue to do so among independent Mexicans. He touches laterally not only the point about the creation of added value by mining and its connected industries (minting, silver- and goldsmithing, etc.) but also the concept of the economic good that, more than useful, may be noble, thus corresponding to the category of *pulchrum*, with a real and universal economic value. It is precisely for this reason, Elhuyar argues, that the authorities have for centuries granted 'protection and government' to mining in Mexico. In short, historical experience has proven the importance of its esteem in the economy. Over time, Elhuyar's ideas would be corroborated, as mining would continue to be the most important productive branch in Mexico into the second half of the nineteenth century, at least to the extent that it was fundamental for international trade and attracting investments, becoming the object of continuous and priority attention from the point of view of development.

Conclusion

This brief chapter has not sought to demerit Humboldt's work on New Spain. Instead, it has sought to give an idea of the scope and limitations of the *Political Essay* as a source and interpretation of the Mexican colonial economy. It has focused on the categories or inclinations underlying Humboldt's political arithmetic and curiosity on subjects of wealth and population. If it has chosen to make certain comparisons with Mandeville, which may surprise some readers, it has been with the idea of taking to the extreme the implications of some of the concepts present in Humboldt and in the theoretical current of the 'harmony of interests' to which they may be ascribed. No connoisseur of the *Political Essay* can deny that the ideas of nature, the individual, and reason are combined in Humboldt's reflections on the demographic and economic situation of colonial Mexico or that they permeate his most fundamental thoughts on its territorial and administrative panorama.

It is suggested that the critical approach applied here be expanded to encompass other data, perspectives, and problematisations of the degree to which Humboldt's account may be taken to reflect the realities of New Spain, including not only the socio-economic aspects discussed here but also the political, ethnographic, geographical, and cultural elements. For now, it seems clear that in this case it is evident once again that any selection of perspectives and concepts will always explain certain phenomena better than others. The general framework applied in political arithmetic and economic analysis, along with an interest in contributing to the natural history of man, explain Humboldt's contributions regarding the study of production, exports and imports, manufacturing, the circulation of precious metals, insertion of the territory within a continental and global dimension, its fiscal projection within the Spanish empire, and so on. However, such a panoramic view makes it difficult for Humboldt to evaluate, and or indeed be sufficiently interested in, how the precise social and economic reality of New Spain required specific economic and administrative policies and strategies. Although perhaps not so relevant in terms of international comparison, the establishment of universal proportions or the elevation of ideas to a macro or general level, the polices, principles, and strategies of government that Humboldt encountered in New Spain reflected a long, previous economic history that had responded on a good number of occasions to the imperatives of the survival and prosperity of the population, a matter that Humboldt considered not in any way minor.

Notes

1 On Ferrier in context, see Jean-Louis Billoret, "La affirmation et les polémiques du modèle consulaire", in Gilbert Facarello and Philippe Steiner, eds., *Le pense économique pendant la Révolution française. Actes du Colloque Internationale de Vizille, 6-8 September 1989* (Paris: Presses universitaires de Grenoble, 1990), 316–317. The cited title in Spanish corresponds to that of the Mexican edition, based on the translation of an expanded edition of Ferrier's work, which appeared in 2 volumes (Mexico: Ignacio Cumplido, 1843–1844).

2 José Miranda, Juan A. Ortega y Medina, Rafael Moreno and Luis González y González, eds., *Ensayos sobre Humboldt* (Mexico: UNAM, 1962).

3 Miranda and Ortega y Medina, for example, related it more to physiocratic thought. Ette, Labastida and the present author, with that of Smith.

4 Pierre-Jean Simon, *Histoire de la sociologie* (Paris: Presses universitaires de France, 1991), 147–158, places Mandeville, the Physiocrats, Smith, Malthus, and Ricardo in this mode of thought, considered by Simon to be a forerunner of sociology. Likewise, these same authors (except Ricardo) can be seen grouped in a current of thought characterised from the perspective of the history of science (social arithmetic), in Stephen Gaukroger, *The Natural and the Human: Science and the Shaping of Modernity, 1739-1841* (Oxford: Oxford University Press, 2016), 270–295.

5 The central thesis of his famous *The Fable of the Bees, Or Private Vices, Publick Benefits* (1714). According to the moral and economic theories of previous eras, this should not be the case, as the fondness for these items was taken to be pernicious (it would corrupt the morality and industriousness of the people) and thus contrary to a genuine prosperity. For commentary on the moral and political implications of Mandeville's break with previous thought, especially that of the Aristotelian type, see Peter Koslowski, *Gesellschaft und Staat. Ein unvermeidlicher Dualismus* (Stuttgart: Klett-Cotta, 1982), 174–184.

6 In *Gesellschaft und Staat*, 181–182, Koslowski reflects on the implications of Mandeville's point and presents his high regard for government-run employment as an interruption of a teleological view of ethics (that concerning the harmony of interests) to fall back on a mechanistic scheme in the style of Thomas Hobbes.

7 Koslowski, *Gesellschaft und Staat*, 181.

8 Koslowski, *Gesellschaft und Staat*. Factors such as investments are not of great importance in this scheme. The decisive factor is the consumption of goods, hence the little consideration given to savings or investment in Mandeville's model.

9 Humboldt, *Ensayo político sobre el reino de la Nueva España* (México: Porrúa, 1966), 42–43.

10 Humboldt, *Ensayo político*, 50.

11 Humboldt, *Ensayo político*, 44–50, although it is true that in the twenty years prior to his visit to New Spain there have been no epidemics and famines comparable to those of previous times.

12 Humboldt, *Ensayo político*, 47.

13 Humboldt, *Ensayo político*, 463.

14 Humboldt, *Ensayo político*, 238.

15 Humboldt, *Ensayo político*, 316, 445. The formulation of this point in these passages is directly reminiscent of Adam Smith's theory of investment priority in agriculture.

16 Humboldt, *Ensayo político*, 47.

17 José Enrique Covarrubias, 'Humboldt y la aritmética política,' in en José Enrique Covarrubias and Matilde Souto Mantecón, eds., *Economía, ciencia y política. Estudios sobre Alexander von Humboldt a 200 años del Ensayo político sobre el reino de la Nueva España* (México: Instituto Mora/Instituto de Investigaciones Históricas de la UNAM, 2012), 55–77.

18 William Petty, *Political Arithmetic, Or a Discourse Concerning the Extent and Values of Lands, People, Buildings... As the Same Relates to Every Country in General, but More Particularly to the Territories of His Majesty of Great Britain, and His Neighbours of Holland* (London: Robert Clavel y Hen. Mortlock, 1690), 3.

19 Humboldt, *Ensayo político*, 93, 108.

20 Gaukroger points out that the natural history of man developed in the eighteenth century is both comparative history and comparative geography. The comparative is constitutive of this discipline, and it assumes the relevance of all possible comparisons that shed light on human beings, from those of anatomy to those concerning religious issues, from data on the antiquity of rocks to statistics on births, deaths, and marriages. This author concludes that 'it is a way of building understanding both from the discernment of similarities and dissimilarities.' Gaukroger, *The Natural*, 231.

21 Gaukroger, *The Natural*, 276.

22 Ted McCormick, *William Petty and the Ambitions of Political Arithmetic* (Oxford: Oxford University Press, 2009), is the principal study of Petty from this perspective.

23 Covarrubias, 'Humboldt y la aritmética política', 62, 76.

24 Covarrubias, 'Humboldt y la aritmética política', 63, n. 16.

25 As in the famous map that appears in the geographical introduction to *Political Essay on the Kingdom of New Spain*, which indicates the circulation routes of precious metals extracted in America. When treating the construction of an interoceanic canal in the Mexican Isthmus or Central America that would be useful for many countries, Humboldt justifies it by invoking the 'true interests of the human race'; *Political Essay*, 17.

26 Enriqueta Quiroz, *Economía, obra pública y trabajadores urbanos. Ciudad de México: 1687–1807* (Mexico City: Instituto Mora/CONACYT, 2016).

27 Quiroz, *Economía*, 111.

28 Quiroz, *Economía*, 245.

29 Humboldt, *Ensayo político*, 67, 238.

30 Humboldt, *Ensayo político*, 131–132.

31 Andrew Konove, *Black Market Capital: Urban Politics and the Shadow Economy in Mexico City* (Berkeley: University of California Press, 2018), 40–41.

32 Quiroz, *Economía*, 233, writes:

> Trend calculations are widely used today in macroeconomics and in general they tell us how much the economy of such and such a country 'improved' or 'grew.' However, consumers generally do not appreciate these changes. So, what is really happening? Perhaps the reason is found in the daily reality of consumers, which is undoubtedly far from macroeconomic calculations.

33 José Enrique Covarrubias, *En busca del hombre útil. Un estudio comparativo del utilitarismo neomercantilista en México y Europa (1748-1833)* (México: Instituto de Investigaciones Históricas de la UNAM, 2005).
34 Cited by Simon, *Histoire*, 149.
35 Koslowski, *Gesellschaft und Staat*, 176–177.
36 Simon, *Histoire*, 149–150.
37 It is more than likely that on this point, Smith and his reference to subsistence goods, comfort and luxury are an influence. This influence is also seen in the division of consumption into productive and unproductive types, also present in Humboldt. On these classifications, see Marcelo Carmagnani, *Le isole del lusso. Prodotti esotici, nuovi consumi e cultura económica europea, 1650-1800* (Milan: UTET Librería, 2010), 91–92.
38 Humboldt, *Ensayo político*, 445.
39 *Ensayo político*, 243. Humboldt affirms there that 'I doubt that there is another plant on the globe that, in a small space of land, can produce such a considerable quantity of nutritive substance.'
40 Published in Madrid, Imprenta de Amarita, 1825.

11

BONPLAND'S CACTUS, OR TRAFFICKING IN EXOTICS AND IGNORANCE

Irina Podgorny

In this chapter, I trace the itinerary of an exotic plant and its delusive but profitable name. Associated with Aimé Bonpland, (1773–1858), the famous French botanist and travel companion of Alexander von Humboldt and gardener of the empress Josephine de Beauharnais, wife of Napoleon Bonaparte, the global career of the delusive species *Cactus speciosus* reveals the entanglements of science, ignorance, politics, self-promotion, and the booming business of exotic plants, greenhouses, and fertilisers in the nineteenth century, entanglements whose echoes are still with us today. This work can be read as part of that long history that Robert Proctor calls 'the cultural production of ignorance,' which, in our case, relates to the widespread habit of denying or ignoring knowledge produced in other languages.[1]

Aimé Jacques Alexandre Goujaud Bonpland's biography is well known.[2] He was born in La Rochelle, France in 1773 to a bourgeois family. His father had been trained in medicine and had investments in the Caribbean colonies. By 1791, Aimé joined his elder brother in Paris, where they both studied medicine, botanised, and attended courses at Paris's *Jardin des Plantes*. Following his travels with Humboldt in the Americas and after a period of working in France, he moved to Buenos Aires in 1816 with two assistants, his wife Adeline and Emma, her daughter from a previous marriage. In 1821, he established a colony at Santa Ana near the Paraná River for the specific object of harvesting and selling yerba mate. In December 1821, Bonpland was arrested as a spy and detained at Santa Maria, Paraguay until 1829, while Adeline, to secure his release, started a real odyssey through the South American nations, from the court of Pedro I in Brazil to Buenos Aires, Montevideo, Santiago, Lima, and La Paz. During his detention, Bonpland was granted freedom of movement and acted as a physician for the local population and the military garrison. In 1831, he settled at San Borja in the Province of Corrientes.

DOI: 10.4324/9781003231479-12

There, aged 58, he married the *correntina* Victoriana Cristaldo, future mother of his three younger children, making a living farming, trading in yerba mate, and working for Juan Pujol, governor of Corrientes.

In the 1810s, Bonpland had been attracted to Buenos Aires by the promises of the government to appoint him as a leading scientific figure in several new projects of the newly independent state: a museum, a botanical garden, an agricultural institute, and a professorship in natural history.[3] However, Bonpland had not anticipated the fragile political state of his new homeland, perhaps as unhappy as the one he and his family had left on the other side of the Atlantic. Upon his arrival in early 1817, he learned that the government that had offered him employment had ceased to exist, their commitments gone with the revolutionary winds.[4]

In the meantime, Bonpland met with local savants, including priests Saturnino Segurola (1776–1854) and Bartolomé Doroteo Muñoz (17??–1831), who were engaged in natural history and the establishment of libraries in Buenos Aires.

These men told Bonpland about the work and collections of padre Damaso Larrañaga (1771–1848) in Montevideo. One year after his arrival in Buenos Aires, Bonpland wrote a long letter to Larrañaga, who was then director of the Public Library of Montevideo, manifesting his admiration for his work and his desire to visit Larrañaga's 'beautiful collections of plants, insects, minerals, etc.' Bonpland not only praised the padre's work, but he also urged Larrañaga to publish and spread his ideas and pursuits in Europe. He conceded that life in the Rio de la Plata Provinces had proved more difficult than he had anticipated. Before departing, Bonpland filled his luggage with books, sure to be able to sell them at a good profit in the ports of arrival. Indeed, he had expected to sell them to the new Buenos Aires library, as he had been promised.[5] When things fell through in Buenos Aires, Bonpland went looking for new customers.

Montevideo had been under Portuguese rule since 1817, and Larrañaga had spent most of that year in the court of Rio de Janeiro as a deputy before the court, returning on January 5, 1818. Bonpland's letter reached him shortly thereafter, when the public library had ceased to function as planned. He knew that Larrañaga was an avid reader and consumer of manuscripts kept in the colonial archives but also of books that, via different agents, arrived from Europe, Hispanic America, and Rio de Janeiro.[6] Before taking his leave of Larrañaga and apologizing for his rudeness, Bonpland went to the point, namely, the business of selling books.[7]

Following Larrañaga's subsequent request, on April 2 Bonpland sent a list of books offering a 10% discount, as follows:

1   *Voyage de Humboldt et Bonpland. Zoologie*, 2 vol. en 4to. avec fig. colories ......42 Fr.
2   Humboldt *Tableaux de la Nature* en 2 vol... 4,4
3   Schol (sic): *Dictionnaire des Sciences Naturelles* 5 vol. en 8° avec deux Atlas[8]... 26
4   Desfontaines: *Tableau de l'école botanique*[9].. 4
5   Plumier: *Plantae Americanae*[10].. 6

Most of the books Bonpland had brought to Buenos Aires originated in the liquidation of the Nassau-Saarbrucken Friedrich Maximilian Schöll publishing house in Paris. Schöll had been the main publisher of Humboldt and Bonpland's *Voyage* but following the French defeat he entered into the Prussian service, his business almost ruined as Napoleon's European dream came to an end.[11]

Bonpland, who, according to Bell, was not planning to settle permanently in South America, also brought with him his experience as a gardener and botanist. In 1816, he embarked a collection of European plants for the rich in Buenos Aires or for the *chacras* or small estates in the surrounding countryside.[12] In a city crowded with British merchants and with so many ex-smugglers embarked in what Tulio Halperin Donghi called the career of the revolution and war, there would surely be some refined gentlemen and ladies with imperial tastes which by 1816, and in part thanks to Humboldt and Bonpland, now included tropical gardening. Bonpland hoped to recover what he had lost in a France shaken by the defeat of Napoleon and the Restoration of the Bourbons, followed by the death of his patroness Empress Josephine in 1814. Intendant of the gardens of Château de Malmaison from 1808 and, since 1810, of the Castle of Navarre in Normandy, Bonpland, with an annual salary of 12,000 French pounds, had been in charge of acclimatisation of the plants in the gardens of both residences. Among these was a beautiful specimen he called *Cactus speciosus*, which, according to Bonpland, Humboldt and himself had observed in April 1801 at an altitude of 360 m in Turbaco, near Cartagena, in present-day Colombia. Since 1811, thanks to the seeds sent to France by the Baron, the plant had flowered at Josephine's greenhouses. At least, this is what Bonpland suggested in his catalogue[13] (see Figure 11.1).

The cactus, however, was not new. It had been an old acquaintance of the botanists of the Old and New Worlds who, via the drawings associated with the work of the Spanish naturalist Francisco Hernández (1514–1587), had learned of the Mexican existence of a plant the natives called *nopalxoch*. Nevertheless, Bonpland decided to create a new species without knowing for sure where the seeds that he had germinated in France actually came from. When the plants flowered in 1811, the flowers made him think of the plant that he and Humboldt had seen in 1801 near Cartagena. As a result, the cactus was retrospectively promoted as a by-product of the famous voyage.

Much has been written about Bonpland's apparent casualness regarding systematics and publishing. Beginning with Humboldt, this opinion was resumed by George Sarton.[14] This adverse opinion created a problem, first for Humboldt himself and his publication plans and later for those historians wishing to demonstrate that Bonpland's achievements were overshadowed by the great success of the Baron.[15] Here I suggest that Bonpland's practice can also be explained in part as an entrepreneurial strategy that responded to the exigencies of the market in exotics.

The Cactaceae in Europe

This passage with strong Humboldtian resonances of the distribution of plants by altitude and latitude, introduced the cactus plants to potential eastern and northern European customers:

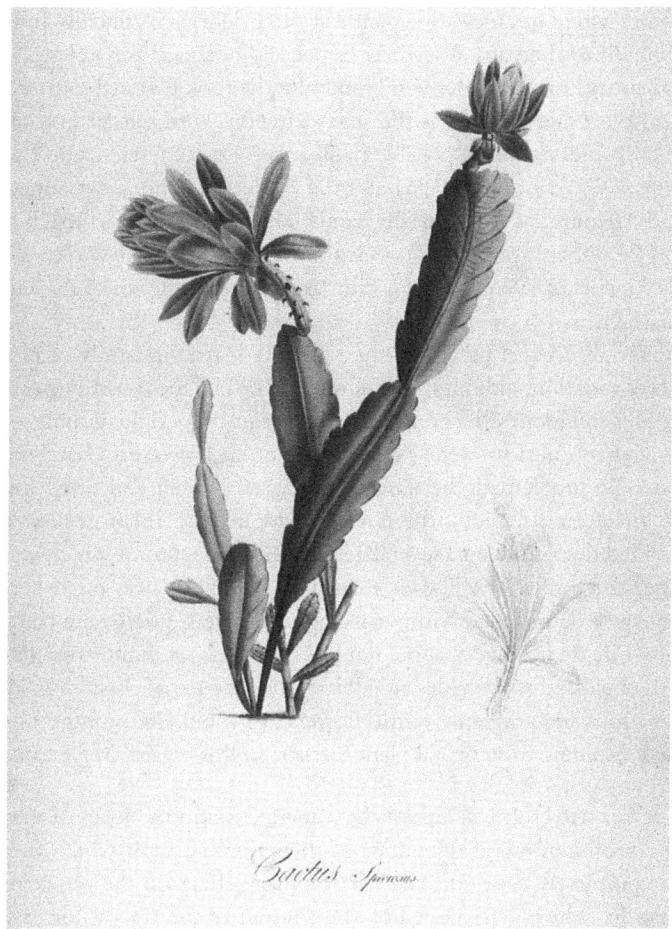

FIGURE 11.1   Cactus Speciosus. Aimé Bonpland, *Description des plantes rares cultivées a Malmaison et a Navarre* (Paris: Schoell, 1813), 8–10, plate 3.

With the exception of two species which have been found in East Asia, the Cactaceae belong to the New World, where they extend from 40° North latitude to 40° South latitude; and from sea level to the vicinity of the limit of perpetual snow. Most of these plants belong to the torrid zone, yet certain forms of them prevail in the temperate zones, and in tropical regions they grow on the mountains at higher elevations, where they find a cooler climate.[16]

It was published in German in 1836 as part of a chapter devoted to cacti in *Outlines of the Geography of Plants*. Authored by the botanist Franz Julius Ferdinand Meyen (1804–1840), this work would be translated to Swedish and English (1846). It was praised for its detailed reference to cultivated plants 'containing information that is worth reading for the layman.'[17] Born in Tilsit, East Prussia, Meyen had moved to

Berlin in 1821, where he studied medicine at the Friedrich-Wilhelms-Institut and dedicated himself to the natural sciences, especially botanical physiology. In 1830, under the patronage of Alexander von Humboldt, he took part in a journey around the world on the *Princess Louise* as the ship's surgeon, with special instructions to collect as many observations as possible in all areas of natural science in the coastal cities of Valparaiso, Arica, and Tacna and the Andean locations of Santiago, Lake Titicaca, and Arequipa, as well as in the Sandwich Islands, Manila, the China coast, and India. The publication of his travelogue, his physiological works, as well as Humboldt's support, earned him a position as associate professor at the University of Berlin in 1834.

In his textbook, Meyen presented the cactus as a striking family of plants very rich in species variation, although no more than 190 of these had been described in his day. He confidently reckoned that this number would double once the mountains of the Americas were carefully explored. However, classifying cactus had proven to be problematic because 'it was very difficult and often impossible to transport them, and, besides, the traveller seldom finds them in flower.'[18] The botanists divided the family into several genera, the characters of which were taken less from the structure of the flowers than from the differences of their forms so that the genus was determined without seeing the flowers. Cacti were hard to keep in herbaria, while their flowers, some tend to open only at night, were difficult to observe in the field and so were discarded from the practice of classifying. However, being conspicuous and gorgeous, cactus flowers became the most important selling point in the expanding ornamental plant market. In this regard, Meyen noted:

> Nature has tried to make up for the imperfection in the form of these plants by the profuseness and splendour of their flowers, with which they are so often completely covered. The extraordinary effect on the physiognomy of vegetation, which is produced by the contrast of the Cactus forms with the other groups of plants, is seen not only in nature, but also everywhere in our gardens. These would be without their gayest ornament if the Cactacea were wanting in them.[19]

Meyen wrote these lines when eastern and northern European gardens were starting to be graced by these flowers, thanks to the artificial heat of greenhouses that reproduced the climate conditions observed in the Torrid Zone. This, at least, was the pretension of the new industry of greenhouses, nurseries, and conservatories that proliferated for the northern modern gardener and horticulturalist since the beginning of the nineteenth century. The trend gave rise to what can be called 'the cultures of artificial nature' linked to the burning of coal both for producing energy and speedily transporting perishables, including the plants, soils, and fertilisers.[20] In this new cultural context, cacti became stove plants, crossing the 40th parallel and conquering northern Europe's horticultural aesthetics.

Meyen also remarked that of all the American plant families, the cacti were the most generally diffused in the Old World. In particular, the tunas (*Cactus ficus-indica* and *Opuntia*), which found their way first to Iberia and the rest of Mediterranean or southern Europe, where they were soon naturalised, were routinely used for fences and in military defences as chevaux-de-frise, producing abundant edible fruit known as *higueras de la Indias*, *fico indiano*, Fig of the Indies, or *Figuier de Barbarie*, Barbary Fig.[21] There were also attempts to cultivate the tuna and produce cochineal in an intensive way, although with little success until well into the eighteenth century.[22]

Inventing the Cactus

The Cactacea family, so common in the West Indies or Americas, was a total stranger, or nearly a total stranger, to early-modern Iberians and Europeans.[23] In the Americas or Indias Occidentales, cultivated and wild nopal cactus was widely dispersed, although it seems that only in Mexico was it cultivated prior to the sixteenth century. Native names for these 'Indian' plants were hybridised with the names of the 'unseen' plants of the Ancients, some of them thought to be native to ancient 'India.'[24] While *tuna* and *nopal* were derived from Taino and Nahua words, the Latin words of *cactus* and *opuntia* were subsequently lifted from Pliny's natural history and transferred to the plants 'of the Indias.'

Pardo Tomás and López Terrada have tracked the descriptions and names of American plants in the early 'Chronicles of the Indies,' and there is no mention either of 'opuntia' or 'cactus.' Peter Martyr d'Anghiera (1457–1526) was perhaps the first to describe the nopal without naming it in his account of the mythical foundation of Tenochtitlan. This was followed by Gonzalo Fernández de Oviedo (1478–1557), who described the tuna and another *Opuntia*. By the sixteenth century, prickly pear had become a common food among the Spaniards in the Americas. Álvar Núñez Cabeza de Vaca (1488?–1559), on the other hand, mentioned the different ways of preparing and consuming tuna among the gatherer tribes of northern Mexico, while Francisco López de Gómara (1511–1566) offered precise information on the denominations, varieties, and uses of tuna, employing the terms *nopal* and *nuchtli* to refer to the plant and the fruit, which he classified as red (encarnado), yellow, white, and 'picadillo', the tastiest being the white and cultivated one.[25] Oviedo, in 1526, registered the name of tuna for the fruit, the same used by Cabeza de Vaca, that Taino word imported from the Antilles that was later to be mentioned by Mattioli.[26]

As López Piñero and Pardo Tomás have pointed out, in his work on the plants of New Spain Francisco Hernández devoted fourteen chapters to cacti, all of which contained the word *tuna* and six of which used the Nahuatl word *nopal* or *nochtli*. According to Hernández, this plant 'spread across the Old World many years ago, causing great amazement because of its monstrous shape and its weird assemblage of thick, spiny leaves.'[27] Hernández described the most suitable climate for them and listed their medical properties, the places where they grew, when their seeds should be sown, and when they flower and fruit. He classified them by their flowers, leaves, and fruits, depicting the varieties, which via Recchi's version, were going to be used for the next two centuries as the main material basis for their descriptions in the most important seventeenth- and eighteenth-century botanical authorities.

According to Heinemann, the Flemish physician Mathias de l'Obel (1538–1616), Royal botanist to King James I of England, was the first to use the name 'cactos' for the New World thorny plants. In his *Stirpium adversaria nova* (1570/1571) he described the *Echinomelocactos* (*Melocarduus echinatus Indie occidus*) a spiny melon-thistle from the West Indies, a plant he observed at the London shop of Hugh Morgan, apothecary-in-ordinary to Queen Elizabeth.[28] 'Cacti' found its way into the published works of Pietro Andrea Mattioli (1501–77), Carolus Clusius (1526–1609), John Gerard (1564–1637), Matthias de L'Obel (1538–1616), John Parkinson (1567–1650), and Gaspard Bauhin (1560–1624), among others.[29] In these works, these New World plants, previously unknown in the Old World, were analogised to two Mediterranean plants and named after the ancient and controverted 'Opuntia' and 'cacti', the latter being the Latin name for Old World thistles, the artichoke included.[30] While the botanists of the British Isles kept the name 'Echinomelocactos', on the continent the name Melocactus or Melonendistel was adopted. As such, it would appear in the *Exoticorum* of Clusius (1605), accompanying the image of a specimen that had been taken to Holland.[31] L'Obel, on the other hand, described the South American thorny kind Cereus (Fackeldistel, i.e. torch-thistle) and *Indorum tune ficifera*, the tuna growing in Spain, Italy, and France, including Montpellier.[32]

Mattioli's influential commentaries on Dioscorides's *Materia Medica* (1544) had incorporated New World species including the so-called *Fico Indiano*, a kind of 'fig' from the Americas that was compared with the *Opuntia* of Greece and the fig of India as mentioned by Pliny after his understanding of Theophrastus:

> But the Indian figs are very different from ours. Theophrastus, in the fifth chapter of the 3rd book of the history of plants, said: India produces the Fig tree, which every year sends out the roots from the branches, not from the new ones, but from the one-year-old, & older ones… The leaves are not lesser than the plates, the fruits are never greater than the chickpeas, but similar to the figs, & for that reason the Greeks call this tree Fig. It bears very few fruits compared to its remarkable size. It grows around the Acesiva River. This is all written by Theophrastus, & the same almost written by Strabo in the fifteenth book of his Geographia, & by Pliny in the fifth chapter of the 12th book, who also writes in the second chapter of the 7th that of such greatness is this plant, that great armies of men on horseback stand under it in the shade. But this is different from another Indian fig tree, which has been brought in our times from the West Indies, but this one has neither in the trunk, nor in the branches, nor in the leaves, nor in the fruits, a veritable resemblance to the above mentioned. The Indians call TUNE the fruits of this one. The plant of which I believe is nothing other than the Opuntia

of Pliny, so called to be born near Opus, as Theophrastus writes with these words. Similar to the Indian Fig, even more marvellous is that plant, which is born around Oponte, & generates the roots from the leaves, to which it is given by nature, so that its fruits may be eaten, so that it may be sweet. We learn that, as can be observed, by taking off a leaf from the tree, & planting it in the ground as far as the middle, it not only makes the roots, but in a short time puts out the leaves, so that with this order, by rising the leaves from the leaves, that plant grows, like a tree, without trunk, without branches, & without germs, as can be clearly seen from the figure drawn here, so that this plant can be meritoriously listed among the miracles of nature. Its leaves are so big, that they exceed the size of an inch for the most part armed with long, & acute spines, if well in some of them in place of thorns we see some knotted petioles. It produces this plant the fruits at the top of the leaves almost similar to figs, but bigger, & crowned at the top, of a colour, which, in the green, is purple. Their flesh is as in ours, but redder, so that it paints the hands, as blackberries do, but eating many of them (as those who were already in that country did) makes red urine, like blood, which sometimes made the strangers very frightened with a lot of fun for the locals. A leaf with the fruits was given to me by the very kind M. Angelo Crotto, agent of the Count of Fiesco attached to the Imperator Ferdinand, with three not yet ripe fruits brought from Provence to Vienna. I have also seen a whole plant in the time that I was back from Goritia by the diligent pharmacist M. Giulio Moderato of Rimini.[33]

Giulio Moderato's garden, pharmacy, and advice were much appreciated by Ulisse Aldrovandi for example, with whom he travelled.[34] Thus, in the context of early modern materia medica, emerged the association between Indian figs (that is the figs from the Indies, east and west) and the plants that Pliny reported from Opunt. The plant mentioned by Theophrastus and Pliny was not, of course, the tuna that Moderato kept in his garden in Rimini.[35] Nevertheless, the tradition of attaching Pliny's description to the plant would survive via Mattioli's version of *De Materia Medica*, especially in northern Europe. In his *Species Plantarum* (1753), Linnaeus rejected the old generic names of *Cereus, Melocactus,* and *Opuntia,* as well as *Pereskia,* listing all under the generic name *Cactus.*

Given this long history of hybridisation and Latinisation, it is little wonder that Alexander von Humboldt's plan for the continuation of the publication of the botanical results of his travels to America was clearly noted: 'No synonyms.'[36] He seemingly wanted to regain the time lost by Bonpland's casual or perhaps strategic approach, a requisite that in this case would mean pulling a veil over the foundational work of earlier botanists, as if all the plants collected in the Americas had been discovered by the intrepid pair during their now-famous voyage. Bonpland, at least, did not acknowledge the long history of his *Cactus speciosus,* eluding well-known citations and derivations of Hernándezs' Nopalxoch, and those of the 1786 Royal botanical expedition lead by Martín Sessé y Lacasta

(1751–1808), a devoted Linnaean, who with Mariano Mociño had named it *Cactus phillanthus Linn*, a modern synonym for 'Nopalxoch cuez alticquizi 392 Hernández' (see Figure 11.2).

The plate in Figure 11.2, today kept at the Torner collection of the Hunt Institute in Pittsburgh, records an even more recent name: *Cactus phyllanthoides*. This name was invented in 1813 by the Swiss botanist and specialist in succulents, Augustin P. de Candolle (1778–1841) for a plant that he had seen flowering in Montpellier in 1811.[37] The timing of the flowering and publishing of de Candolle's and Bonpland's specimens suggest a coincidence. Most probably, Bonpland was unconcerned by the coincidence for he was selling, sowing, and enjoying flowers destined not so

FIGURE 11.2 Nopalxoch. Sessé, M., Mociño, M., Drawings from the Spanish Royal Expedition to New Spain (1787–1803) (1787–1803). Torner Collection of Sessé and Mociño Biological Illustrations. Courtesy of the Hunt Institute for Botanical Documentation, Carnegie Mellon University, Pittsburgh, Pa.

much for the glory of God and Science but for the greenhouses of the gentlemen and gentlewomen of northern Europe. His customers, wanting to be *à la mode*, no doubt worried more about the fate and adventures of the great explorers than the intricacies of ancient and modern plant nomenclature. Simply put, the old names of Mattioli, Oviedo, and Hernández did not sell while the new brands of Humboldt and Bonpland were flying off the shelf. These new brands resonated even more when combined with the seductive charms of 'the Antillean Empress,' Josephine. Little wonder then that Bonpland's *Cactus speciosus* should conquer Europe under the brand names of 'German Empress', 'Deutsche Kaiserin,' 'Giant Empress', and 'Drottningkaktus.'

Josephine's Heated Gardens and a Rose-Flowered Fig of the Indies

In 1811, Bonpland reported that some cactus seeds sent from South America by Humboldt had flowered at Josephine's heated greenhouses. He mentioned that, together with Humboldt, he had observed the same cactus in April 1801 in Turbaco, near Cartagena de Indias, in present-day Colombia, at an altitude of 360 m. *Cactus speciosus*, as Bonpland classified it, was soon depicted in full colour by the 'Raphael of the succulents,' botanical artist Pierre Joseph Redouté (1759–1840), whose fame guaranteed its commercial success.[38] A few years later, specimens of *Cactus speciosus* were indeed flowering in the conservatories and nurseries of England, receiving awards in several English gardening contests. It was among the most valued plants of the new industry of succulent plants and coal-heated greenhouses. In the meantime, Bonpland departed for South America. In the same year that his *Cactus speciosus* was blossoming in England, Bonpland was in Buenos Aires selling his books and trying to establish a greenhouse, in an effort to recover the wealth he had lost after Josephine's death.

Josephine de Beauharnais had bought the manor house of Malmaison, a rundown estate with 150 acres of woods and meadows about 12 km west of central Paris, in April 1799. Josephine endeavoured to transform it into 'the most beautiful and curious garden in Europe, a model of good cultivation,' where she located rare and exotic plants and animals. As Pierre-Yves Lacour recalled circa 1800, the collections of Josephine at Malmaison rivalled those of the Paris Muséum or Jardin des Plantes. Charles-François Brisseau de Mirbel was in charge of organising the botanical garden. English nurseryman John Kennedy was hired to bring plants from across the Channel. From 1804 until Josephine's death in 1814, nearly 184 species blossomed in France for the first time.[39] The French botanist Étienne Ventenat received 16,000 francs a year for naming and describing the new species she imported from Belgium, Holland, and England, Napoleon's blockade of Great Britain notwithstanding. Botanical artist Pierre Joseph Redouté received 18,000 francs for painting them, and his prints continue to sell quite well today. By 1808, she had paid 42,862 francs to Ventenat and 85,923 to Redouté.[40] Both of them, and later Bonpland, owed Josephine their own economic prosperity and social ease.

Bonpland was hired as the empress's gardener in 1810, after the death of Ventenat, who in the foreword to the first volume of his *Jardin de la Malmaison* (1803) underlined not only the utility of Josephine's endeavours but also the provenience of the flowers published in the catalogues, clearly linked to the political ambitions of her husband:

> You thought that the taste of flowers should not be a sterile study. You have gathered around you the rarest plants on French soil. Many, even those who have not yet left the deserts of Arabia and the burning sands of Egypt, have been naturalized by your care; and now classified in an orderly fashion, as we inspect them in the beautiful Garden of Malmaison, the sweetest memory of the conquests of your illustrious husband, and the most amiable proof of your studious leisure activities. You have been kind enough to choose me, Madam, to describe these different plants, and to make the public aware of the wealth of a Garden which is already equal to that which England, Germany and Spain offer us the most interesting of this kind.[41]

Not surprisingly, and against the trope of Malmaison as a place that embodied nostalgia for her native Antilles, most of the exotic plants listed and depicted came from Australia and the Cape of Good Hope (also sometimes glossed under 'the Indies'), originating in Nicolas Baudin's expedition (see Table 11.1).[42]

By 1800, the park was built upon a very extensive landscape plan, including a heated orangery with plants of the rarest kinds, maintained upon the same principles as those of Schönbrunn (Vienna) and Kew. 'But what is more elegant,' wrote a chronicler, is

> a saloon, adorned with charming paintings, that has been fitted behind the ampitheatre of plants, and in this recess, you may enjoy their fragrance with their aspect. It is to be regretted that this building is not contiguous to the mansion, of which it would be, particularly in winter, the most precious ornament.[43]

Five years later, she ordered the building of a greenhouse, heated by a dozen coal-burning stoves, a novelty that contrasted with the wood furnaces used in the Ancien Regime (Figure 11.3).

As Neil Safier has argued, a group of naturalists from the King's Garden in Paris attempted to reproduce in late eighteenth-century France what Spain had previously accomplished so successfully in the New World.[44] To do so, they reviewed their collections of seeds, samples and texts relating to the New World. These new greenhouses were to imitate the weather conditions of distant countries and allow exotic species to grow in a very different climatic zone.[45] This system replicated in the space of the garden two of the three climatic zones in which the globe was divided, that is the torrid and temperate zones. Hot houses or greenhouses (*serres-chaudes*) had been promoted in association with the sale of plants that could survive

TABLE 11.1 Plants cultivated at Malmaison and Navarre. Aimé Bonpland, *Description des plantes rares cultivées a Malmaison et a Navarre* (Paris: Schoell, 1813)

Species	Origin	Sent from/by
Peonia moutan	China (Ho-nan)	Joseph Banks (London, 1794), Bourseau (Paris)
Sida pulchella	Patria ignota (Aus.?)	Exp. Baudin (?)
Cactus speciosus	Colombia	H. & B., 1801
ambiguus	Patria ignota	
Metrosyderos saligna	Australia	Exp. Baudin
glauca	Australia	Exp. Baudin
pallida	Australia	Exp. Baudin
Silene chlorefolia	Armenia	
Goodenia grandiflora	Australia	
Lobelia fulgens	Mexico	H. & B., 1804
Melaleuca chloranta	Australia	Exp. Baudin
Peonia daurica	Siberia	John Bell (England), 1790, Malmaison 1810
moutan	China	
albiflora	Siberia	England, 1811
Erica grandiflora	South Africa	Malmaison and Navarre, 1775
edelinia	South Africa (?)	Navarre
concinna	South Africa	Navarre, 1810
purpurea	South Africa	Navarre
versicolor	South Africa	England, 1811 (Navarre)
vestita	South Africa	
patersonia	South Africa	
fulgida	South Africa	
Gompholobium furcellatum	Australia	Exp. Baudin
Correa viridiflora	Australia	
Eucalyptus diversifolia	Australia	Exp. Baudin
Eupatorium deltoideum	Patria ignota (South America?)	
Boehmeria caudata	Antilles	Île de France, Sonnerat, Malmaison
Linun trigynum	India	London, 1802, Malmaison, 1810
Linaria pauciflora	Patria ignota	Malmaison, 1810
Acacia linifolia	Australia	
curvifolia	Patria ignota (Mexico?)	H. & B. (?)
subulata	Australia (?)	Exp. Baudin (?)
armata	Australia	
Magnolia yulan	China	Malmaison, 1806
macrophylla	USA	
glauca	USA	
Pittosporum tomentosus	Australia	
Zieria smithii	Australia	

(*Continued*)

TABLE 11.1 (Continued)

Species	Origin	Sent from/by
Lopezia racemosa	Mexico	
hirsuta		
coronata		
pumila		
Rhexia pendulifolia	Guyana	M. Martin, Malmaison
glandulosa	Guyana	
Hibiscus sabdariffa	India	
Tristania neriifolia	Australia	England, 1808
Pimelea linifolia	Australia	
Heliotropum corymbosum	Peru	England
Chorizema ilicifolia	Australia	
Cotyledon tardiflorum	Patria ignota	
Lobelia surinamensis	Surinam and Guyana	
excelsa	Patria ignota	England, 1812
Banksia marcescens	Australia	Exp. Baudin
Ixora speciosa	India oriental	
Eleocarpus acuminatus	Australia	Exp. Baudin
Hovea celsi	Australia	Exp. Baudin (via François Cels)
Bossiea coccinae	Australia	
Duvalia oxalidifolia	Australia (?)	Exp. Baudin (?)
Liparia sphaerica	South Africa	
Solanum rostratum	Patria ignota	
Dalea mutabilis	Mexico	H. & B. (?)
bicolor	Mexico	H.&. B
Protea radiate	South Africa	1809
Mimulus luteus	Peru	
Begonia humilis	South America	H. & B. (?)
evansiana	China	
Digitalis purpurascens	Palatina	

only under conditions of artificial heating. These plants included the cacti, called 'raquettes' by French gardeners.[46]

Gardening of exotic plants was shaped by the 'language of precision' based on a good command of thermometers and hygrometers, climatological observations, heat control, and knowledge about the provenience of the plants. Thus, orangeries were devoted to plants from temperate climates, between the thirty-sixth and forty-third degrees of latitude; the hot greenhouses were used for the plants of the hottest countries,

> which find there not only a shelter from the cold, humidity, & the bad weather of ours, but the heat of their homeland in the air that surrounds them, & in the soil where they are planted; so that many take the same growth, & are the same productions as in their native soil, & hardly seem to feel their exile.[47]

FIGURE 11.3   The Hot House at Malmaison. Alexandre de Laborde, *Description des Nouveaux Jardins de La France et de des Anciens Châteaux*: (Paris: Delance, 1808), 65.

The object of the hot greenhouses was to replace by artificial heat 'the lack of natural heat of our atmosphere, & to preserve from its bad intemperance the Plants of the hotter countries.'[48] This approach contrasted with that of those who proposed to leave plants out in the open air even beyond the term when, during the nights, the temperature fell below that which they would enjoy in their countries during the coolest nights. For the propagators of greenhouses 'to accustom them & harden them to the cold, is by an absurd treatment to pretend to fortify them by altering their strength & make them healthy & vigorous by languor & insipidity.'[49] Artificial heat not only provided with a large number of exotic plants but also helped many species of European trees to yield long before their season of ripe and good quality fruit.

When Bonpland started working at Malmaison, he found seeds whose origin he could not track. Seeds had been dispatched to Paris from distant parts of the world, in part in accordance with gardeners' instructions, since most of 'those plants from the Torrid Zone, and a large number of those from less hotter countries cannot achieve the maturity of their seeds in the greenhouses, so it is necessary to bring them from their homeland.' Seeds collected in their perfect maturity, were embarked with precaution in boxes filled with earth, to preserve them from insects, from drying, and from contact with salt water during the crossing,[50] but they lacked labels and so it was impossible to tell where they were from. Thus, relying upon what was transmitted by Paris gardeners, Bonpland, for instance, in his *Plants from Malmaison* (p. 6), described *Sida pulchella* as obtained from seeds, but we do not know where the seeds were sent from. Mr. Noisette, one of our most distinguished nurserymen, whose establishment is worth a visit by foreigners, believes that it came from seeds brought from New Holland by Captain Boudin's expedition. Our nursery gardeners refer to it as *Sida pulchella*, and I thought I should keep this name, which is already known in the trade.

In fact, the 'Indies' trade shaped the naming of species, to such an extent that in commercial nurseries, plant names were 'too often false, or fanciful, or altogether elusive,' said a review of the Chiswick horticultural exhibition from the summer of 1855. Not only were names misleading, but 'countries' were also ascribed to plants, with the utmost confidence, although the latter never had the smallest intercourse with the former. Part of the confusion stemmed from the fact that during the early modern period Europeans often referred to 'the Torrid Zone' at large as 'the Indies,' and indeed trading companies that worked in these regions were called 'Indian companies.' All products from 'the Indies' east and west could be classified generically, in English, French, Dutch, Portuguese, and Spanish, as 'Indian.' Thus,

> [s]pecies from Mexico are fastened upon the Gold coast, Chinese plants are said to be Brazilian, well-known Nilgherry forms are asserted to have come from Western Africa, and so on. In these cases it would seem that the owners of such plants know less of geography than a parish schoolboy. In other instances, when a plant has half a dozen aliases, every one of these is used to sell it, although the owners must, or certainly ought to, know that they all stand for the same thing. Then again a plant already provided with a name in works accessible to anybody, is thrust forth into the world with a new one invented by some ingenious gentleman in the trade. Another mode of perplexing buyers is for every recipient of a new introduction, for which no scientific name has been found, to give it one of his own, too often with some magnificent meaning to entrap the unwary: hence the *splendidums*, and *glorosiums*, and *speciosissimums*, and *superbums*, and *maximums* in which so many modern productions are made to rejoice.[51]

The reviewer referred in particular to a very handsome scarlet cactus, which was produced and exhibited under the whimsical name of *Epiphyllum coeruleum grandiflorum*. Such practices, however, were not confined to England or France, quite the contrary they characterised the realm of exotic gardening, including the plants of lost provenience that flowered at Malmaison and were named after the habits of commerce.

Bonpland's *Description of Exotic Plants from Malmaison and Navarre* was seen as the continuation of the *Jardin de la Malmaison*, the second volume of a work initiated by the late Ventenat. The furnishings were the same: each plant was followed, after the generic name, by the family according to Jussieu, and the class and order of the sexual system, together with the quotations of the generic character, from Jussieu and Gärtner; then came the entire name of the plant with a Latin species identification; the indication of the country of origin; a more detailed description in French; furthermore, partly botanical remarks, partly notes on the introduction of the plant into the botanical gardens, its culture, and so on. Finally, the explanation of the copper plates, which were after Redoute's drawings, 'worked in the known manner, and of exquisite beauty of execution'. But in fact, botanists criticised 'Mr. Bonpland, famous for his travels with Baron von Humboldt' for not quoting Willdenow, for ignoring previous descriptions, and for misspelling scientific names. The reviewer insisted on the problems in the description of *Cactus speciosus*, transcribing Bonpland's Latin identification in extenso but remarking: 'To be added: inermis, crenaturis caullium fasciculato-pillosis.' The reviewer continued:

> The same plant had been described by Prof. de Candolle (Catalogus plantarum horti bota nic, Monspelliensis. 1813. pag. 84) under the name *C. phyllanthoides*, where Plukenet phytogr. tab. 247. fig. 5 et Almage stum. pag. 295, is given as a synonym. According to Mr. de Candolle, it had already bloomed in Montpellier in 1811, and is a very different plant from the extraordinarily splendid species, also known as *C. speciosus*, which was brought from Paris to the botanical gardens of Vienna and Berlin a few years ago, and which bloomed for the first time in the summer in the imperial and royal court garden of Schönbrunn.[52] This latter species has triangular and quadrangular trunks, which are decorated with short tufts of yellow spines on the notched edges. The flower is three times larger, high and shimmering crimson, and, what is particularly strange, it is almost flat, contrary to the habit of the other torch plants; it finally lasts several days in full beauty, which makes it an excellent ornamental plant. Willdenow (Enum. plant. horti berolinens. Supplem. pag. 31) distinguishes this last species, not determined enough, of the following masses: *C. speciosus erectus, leviter quadrangularis, angulis dentatis.*[53]

Nonetheless, Bonpland's *Cactus speciosus* made its way into the Botanical register of England. As Bell recalls, in the years 1814 to 1816, Bonpland made various trips to London, a centre of the Spanish American independence movements and propagandists. Bonpland had contact with South American revolutionaries from Bogotá, Caracas, and Buenos Aires, all competing for his services. In London, he met Bernardino Rivadavia, who was in Europe on a diplomatic mission on behalf of the United Provinces of the Río de la Plata. Rivadavia, ignoring or minimising the criticisms of Bonpland as a scientist, recommended Bonpland to the government of Buenos Aires, emphasising his merits and reputation as Humboldt's companion and as an accomplished botanist. With this recommendation, Bonpland left Europe with his family and assistants as well as with a multitude of seeds and two thousand living plants and books.[54]

But Bonpland had gone to London also for scientific and commercial reasons. His visits to London herbaria and his links with English botanists also involved the exchange of plants and, of course, the arrival to England of his *Cactus speciosus*. In 1818, *Cactus speciosus* Bonpland was published in the British Botanical register accompanied by a drawing taken in June that year in the conservatory of

Mrs. Gilbert, at Earl's Court, where the plant had been received from France the year before. The first time it bloomed in Europe, signalled the register, was in 1811, near Paris, in the garden of La Malmaison, then belonging to the Empress Josephine. It was supposed to have been raised from seed brought home by the celebrated travellers, by whom the species was first observed (Figure 11.1).

While Bonpland languished in South America, *Cactus speciosus* enjoyed a successful life as a transplanted migrant in England. By 1830 there were articles and booklets that explained how to grow and make it flower. Among others, 'On the Treatment of Cactus speciosus, speciosissimus', a letter by W.J. Shennan, Gardener to Mr. William Buchan, F.H.S., Gardener to Major Morrison, at Gunnersbury Park, told that since the mid-1820s, their *Cactus speciosus* and *speciosissimus* flowered freely in the park he nursed. Since a friend had some large plants of both sorts, but they never flowered, Mr. Shennan explained how he had solved that

> [w]e grow them in the stove until they get a pretty good size, or until we want them to flower, for they will flower at any age or size. In the month of June or July we turn them out of doors into a warm sheltered situation, and perfectly exposed to the mid-day sun; and there they remain till we take in our tender green-house plants, when we remove them to a shelf, or airy situation, in the green-house for the winter. In the spring we remove them into the stove or forcing-house in succession, as we wish them to come into flower. They will flower in the green-house; but the flowers are small, and the growth but slow, in comparison to those that are removed into a higher temperature. Their flowering depends, like most other things that flower upon their wood made the preceding year, upon its being well ripened and matured in the sun and air, and kept perfectly free from shade.

By 1828, *Cactus speciosus* fruited freely near London and ripened its fruit about three months after flowering. Others, such as Mr. John Nairn, gardener to Thomas Forbes Reynolds, Esq., of Carshalton, had grafted *Cactus speciosus* and *speciocissimus*, on cactus *triqueter*, making a singular plant, especially when the different species displayed their fine blossoms.[55] In the years to come, *Cactus speciosus*, sold by the most important horticultural houses of England, was presented with flower and fruits in many fairs and awarded several prizes, such as the Stirling Horticultural Society, Horticultural Society of Ireland, Devonshire (South Devon and East Cornwall Botanical and Horticultural Society), Sussex (The Chichester Horticultural Society, when in the Spring Meeting, April 15, a prize was awarded to Mr. Groundsell, gardener to J.J. Gruggen, Esq. for his Cactus speciosa and speciossima); Gloucestershir (Gloucester Horticultural Society); and Lancashire (Liverpool Horticultural Society).[56] By the 1830s, *Cactus speciosus* had become a very successful stove plant that had expanded in England linked to the nineteenth-century hot greenhouses explosion for a North European public eager to cultivate plants from a warmer climate. The Dry-Stove, which in design did not differ from the greenhouse, was chiefly devoted to the culture of succulents. The name and character of this structure were derived from the higher degree of heat generally kept in it and from the air being less moist than in the bark-stove, where more water was used, and consequently more vapor generated. Thanks to them, the horticultural fairs, the new horticultural journals devoted to a new public, and the concomitant expansion of the garden business, cacti became the pineapples of the nineteenth century.[57]

In June 1833, the specimens of *Cactus speciosus* exhibited at the Horticultural Society in London by John Green, gardener to Sir Edmund Antrobus, obtained the society's Banksian medal for their most unusual beauty. They had grown in a compost made of equal quantity of light turfy loam, pigeon's dung and one-third sheep's excrements, exposed the mixture one year to the influence of the summer's sun and winter's frost to mellow. The awarded cacti were about two years old, and to the amazement of the jury, Green's *Cactus speciosus* bore two hundred flowers, thanks to the energy of British Isles' solid waste.[58] By then, Bonpland's name had disappeared from the horticultural news and his cacti ran a life of their own, having wistfully forgotten that their birth as celebrities was due to a famous pair of travellers.

The Voyage and the Cacti

In May 1810, while Bonpland was sewing Cactus seeds, Alexander von Humboldt was desperate. The publication of the results of the American travels was nearing completion; the published parts sold well; Schöll, the editor, was quite generous, money abounded but … the botanical part, in charge of Bonpland, was left completely behind. Published in instalments, only one issue of the *Plantes équinoxiales* (first issue was published in 1808) and one of *Melastomacées* had come out in 1809–1810. Humboldt was resigned about the pace of those volumes but could not rely upon Bonpland for the further classification of the 5,000/6,000 species they both had collected. He was working for Josephine and for his cactus, which was going to flower in a few months. It was necessary to find a botanist with the suitable personality for processing those plant treasures brought from the Americas, among which, according to a rough calculation, about 1,200 species were going to be new. He needed a reliable botanist to achieve the work, and he requested the help of his old friend, Karl Ludwig Willdenow (1765–1812), director of the Berlin Botanical Garden. In that letter, Humboldt sketched the publication plan, the attribution and distribution of authorship, the language and format that would be adopted. Willdenow undertook the work, but he was forced to leave Paris after only a few months, and he died the following year in Berlin.

The work was then resumed by his disciple, Karl Sigismund Kunth (1788–1850), whom Humboldt willingly accepted. Thus, Kunth, 25 years old, came to Paris in 1813, where he remained until 1829 in untiring activity under a steadily growing reputation. In 1829, he left the French capital to follow a call as a full professor of botany in Berlin. As a result of his stay in Paris, Kunth published several large monographs, which were regarded as classical works in systematic botany, but he also left valuable material for later editors in the extensive herbaria, which were created with great skill. He had the best sources available for morphological work and the study of systematics. In Paris, Kunth not only enjoyed the acquaintance with the country's most famous botanists, such as Joseph Jussieu, Richard, and Desfontaines, but he also had access to the large collections of the Jardin des Plantes and those of Benjamin Delessert as if they were his own. A trip to England also opened up the London herbaria to him through the favour of Robert Brown. Following the task initially set for him in Paris, two important works were created. The first, 'Mimoses et autres plantes Légumineuses du Nouveau Continent, recueillies par M.M. de Humboldt et Bonpland,' illustrated by Pierre-Jean-François Turpin, was published between 1819 and 1824. The second was one of his most important works, also illustrated by Turpin, 'Synopsis plantarum quas in itinere ad plagam aequinoctialem orbis novi collegerunt A. de Humboldt et A. Bonpland,' seven volumes published in folio and in 4o between 1815 and 1825, which gives an eloquent testimony of Kunth's extraordinary activity considering that in it more than 4500 plants collected by Humboldt and Bonpland, among them 3,600 new ones, are described. To the 700 copper plates accompanying the work, Kunth himself drew all analyses of the flowering parts. After completion of these seven volumes, an excerpt of the synopsis was soon published in four octavos, the last of which, after 4500 determinations of the described species, presents the results of Humboldt's geography of plants. Kunth had taken over a 'distribution méthodique de la famille des Graminées' after Bonpland, who had worked on the topic himself, had returned to South America. But it was not published until 1837. Apart from these listed publications, Kunth also organised magnificent plant collections. He created an herbarium, which was not only one of the richest possessed by a private person but also represented a great treasure of unpublished knowledge, since it contained a great number of original plants and was arranged according to natural families and critically examined genera and species. After Kunth's death, it was purchased by the Prussian government. It consisted of a general collection containing 44,500 species in 60,000 specimens, of which a large part was made up of plants from the Paris Museum, most of which (about 3000) had been collected by Humboldt and Bonpland, and another part was made up of plants dried by Kunth from the Jardin des plantes; a collection of dried plants from the Berlin botanical garden with 10,030 species; and a collection of foreign species. These collections, containing a total of 55,000 species, also bear witness to Kunth's diligence and reflect in the many analyses accompanying the specimens, documenting Kunth's extraordinary skill in placing unnamed plants. These collections today form part of the so-called General Herbarium of the Berlin Botanical Museum.[59]

Kunth's skills as systematic botanist, as reviewed by his biographers, seemed to be quite the opposite of Bonpland's casualness and may be one of the reasons for the vanishing of the Turbaco Cactus from his work. Indeed, *Cactus speciosus* does not appear in any of the seven volumes or in the herbaria. As Leuenberger underlined, fifteen species of Cactaceae (as 'Opuntiaceae') were treated by Kunth in the volume 6 (1823), based on the collections and observations gathered by Humboldt and Bonpland in what is today Venezuela, Colombia, Ecuador, and Peru. All were published under the Linnean generic name *Cactus*.[60] Leuenberger listed the Cactacea treated by Kunth, with the reference to the herbarium materials and original notes by Humboldt and Bonpland recorded in their *Journal Botanique*, a manuscript kept in Paris.

The general herbarium in Paris, states Leuenberger (p. 139), houses an additional specimen 'Humboldt mss. 1445', labelled by Bonpland as *Cactus phyllantus, avril 1801. Carthagène* and Turbaco, added below: 'However, 'Cactus' and 'Turbaco' are clearly added in Kunth's handwriting, which suggests that this specimen had remained at Paris but, due to doubts as to its identification, was not used for the 'Nova genera et species by Kunth (1823), where *Cactus phyllantus* was not treated'. On the label published by Leuenberger (p. 140, fig. 7), one sees the epithet *phyllantus* replaced by *alatus*, showing the doubts of Bonpland/Kunth regarding the correct determination of the plant. Kunth, apparently, used Humboldt and Bonpland's field notes almost textually for the detailed description and the new names proposed by Humboldt and Bonpland in the manuscript were usually respected (p. 142). Leuenberger devoted a long paragraph to the contested *Cactus alatus*, where he reports the situation as follows:

a)   Not type material. No material in the Humboldt & Bonpland collection, but specimens under this name present at Paris and Berlin-Willdenow. The original epithet on the label at Paris is crossed out and replaced by 'alatus', apparently by Bonpland.... Later on, Kunth only labeled it as Cactus. Even without flowers, the specimen no. 1445 from Turbaco, Colombia, can be recognized as belonging to a different genus and species described much later by Schumann (1903) as *Wittia amazonica* and currently classified as *Disocactus amazonicus*. ... The material at Paris had been correctly recognized and annotated by M. Kimnach in 1960 as *Wittia panamensis* Britton & Rose, now a synonym of the former name. The four representative pieces of stem at Paris and the single stem segment at B-W are typical of *Disocactus amazonicus*.

b)   The detailed note written by Bonpland on no. '1445. Cactus Phyllanthus' surprisingly contains details on flower characters, whereas flowers are not extant in the herbarium specimens at Berlin-Willdenow and Paris. The note mentions red tubular flowers, and an addition by Bonpland in French near the end points out that the outer perianth segments were violet ('les folioles exterieures du calyce sont plus violets que les interieures'). This fully confirms the identification of the sterile herbarium material as *Disocactus amazonicus* discussed above. The locality and year are given as Turbaco, 1801, flowering in April.

c)   *Cactus phyllanthus* was not treated by Kunth (1823) but is a name mentioned in Humboldt's diaries from Colombia, between Hato del Quemado and Pandi, and from Venezuela. This taxon is now classified as *Epiphyllum phyllanthus* (L.) Haw. There can be little doubt that Humboldt and Bonpland saw this species as well, but the collected material all belongs to *Disocactus amazonicus*.[61]

It is curious that Leuenberger does not mention *Cactus speciosus* as published by Bonpland in *Plantes of Malmaison*, where Turbaco and Cartagena were quoted as the original homeland of the species as well as where the cactus supposedly flowered in 1811. Maybe the flower characters mentioned by Bonpland were not observed in Turbaco but in Malmaison. The evidence collected by Leuenberger and Kunth's doubts, who certainly had to be aware of Malmaison's cacti, seems to indicate that Bonpland associated the seeds he germinated with the memories of the flower and plant seen in Turbaco. Bonpland wrote:

> *Cactus speciosus* has many analogies, by the appearance, with *Cactus phyllanthus* and *Cactus alatus*, with which it was originally confused. It is only since the month of March of 1811, when this new species of *Cactus* gave flowers at Malmaison, that I was able to establish the differences between these three plants. In this species the joints are in general less wide and less long; the serrations are closer together and prolonged in the upper part in the shape of teeth; the flowers are large, red and very remarkable for their beauty: they are white, long and slender in *Cactus phlyllanthus*; small and of a whitish green in *Cactus alatus*. It is likely that Cactus speciosus came from seeds that Mr Humboldt sent to France.[62]

Thus, Bonpland qualified the plant with a third epithet: it was neither *alatus* nor *phyllanthus* but *speciosus*, the name with which it would be a success in England. The collections and the notes were, no doubt, extraordinarily complex for Bonpland, for Kunth, and, of course, for Humboldt, who more than once, lost sight of them. In a letter to Meyen, he acknowledged:

> I have searched everything for the fucus (in vain) and since I do not have any dried plants, I am sure that I do not have the specimens. But I just remembered that I could have given them to Kunth for comparison. I will send them there. Please forgive me, my dear, for the mess I have caused. I still have a sheet of meteorological observations from Lima.[63]

Kunth was also certainly aware of the critics of the contemporaries to Bonpland's determinations, who, in the case of *C. speciosus*, claimed that the same plant had been described by de Candolle in his *Catalogus plantarum horti botanici, Monspelliensis*, 1813 under the name *C. phyllanthoides*, where Plukenet and Dillenius were quoted. But the critics could also have mentioned an even longer history connected to Francisco Hernández and his famous Nopalxochil, which, on the other hand, lead to the deep history of the plant in today's Mexican territory.

In 1838, Bonpland was in the present-day Argentine province of Corrientes, collecting plants and geological samples, writing letters and doing business. When he was a prisoner in Paraguay between 1821 and 1829, news about his release was propagated in Europe; thus, in 1832, where *Cactus speciosus* was already a shining star in the gardens of England, *The Athenaeum* published the translation of Bonpland's first letter reaching Europe regarding his botanical discoveries and collections. It was a letter addressed to the president of the Academy by Baron de Humboldt, at Berlin.

At length I have had the happiness of receiving direct news from him through the care of Baron Delessert. A letter from Bonpland, dated Buenos Ayres, the 7th May 1832, advises, that he had received a few lines, which I had forwarded to him at the close of July last year, whilst resident at Corrientes, near the confluence of the Parana and Paraguay, in January 1832. 'I have been crossed,' says he, 'in every labour I have projected since I quitted the soil of France. My ill stars have persecuted me for the last fifteen years; but I am fain to believe that my fate will prove more auspicious, now that I am out of Paraguay. Being once more restored to my friends, and having renewed my connexion with civilized Europe, I have resumed my former labours in natural history with the greatest activity, in order that I may be enabled to return to my native country as quickly as possible. The collections I formed in Paraguay and the Portuguese Missions ought to have reached Buenos Ayres ever since the month of March. I look for them with the greatest uneasiness, and shall forward them immediately upon their arrival, (which cannot be long delayed,) to the care of the Minister of Foreign Affairs at Paris, praying him to deliver over the cases to the Museum of Natural History. The *Jardin des Plantes* will receive, not only what I have recently collected, but such herbaria as I have put together at Corrientes and Buenos Ayres, and particularly my general herbarium, and the geological series of the route we pursued. To this collection I shall add the specimens of rocks which I have just collected, as well as such as I may succeed in procuring during my excursions to Monte Video, Maldonado, and Cabo-Santa-Maria ...

Such are the fertility of the soil and the richness of the vegetation in the Portuguese Missions, that I think it my duty to return to that quarter, and I am willing to believe, that those who kindly take an interest in my early return to Europe, will not disapprove this trip. It would be cruel to leave this clime without adding such a host of remarkable productions to our botanical stores. My collections will comprise two new species of *Convolvuli*, the roots of which possess all the healing qualities of the jalap. I am in hopes that the School of Medicine will likewise set some essays on foot as to the uses to which three extremely bitter barks, derived from three new species of a class belonging to the family of the *Simaroubaeae*, may be put. These barks are of the flavour of the sulphate of quinine, and are used with the most salutary effect in cases of dysentery and other gastric derangements. If, whilst here, I could but receive proper information on the efficacy of these barks, as it might appear from trials in Paris, I would endeavour to secure a supply of them for our hospitals before my departure.[64]

Bonpland did not forget the cacti. In 1838 he sent 'a small collection of cactus seeds to Mandeville, the British Minister at Buenos Aires, as a present made for your rich

English gardens.'[65] Bonpland, in fact, in the course of his excursions found some specimens, which he sketched and described in a short dossier, 'Cacti collectii, Corrientes 1838,' where he listed those cacti, devoting to each one, a short description in Latin, a sketch in pencil and some other notes mixing Latin with French (see Figure 11.4).[66]

For instance, he annotated in French: 'D'après le rapport qu'on me fait le fruit de ce Cactus devient de couleur brune et se mange. Semina ovata, ombilico longo. … Le cordon ombilical qui unie le tempes semble former la pulpe qui enveloppe les graines' Cerei erecti and coccineus) floyelliformes, articulate, melovavti (..) Peut-être même le Cactus … le fruit de l'un et de l'autre.' And describing the fruit,

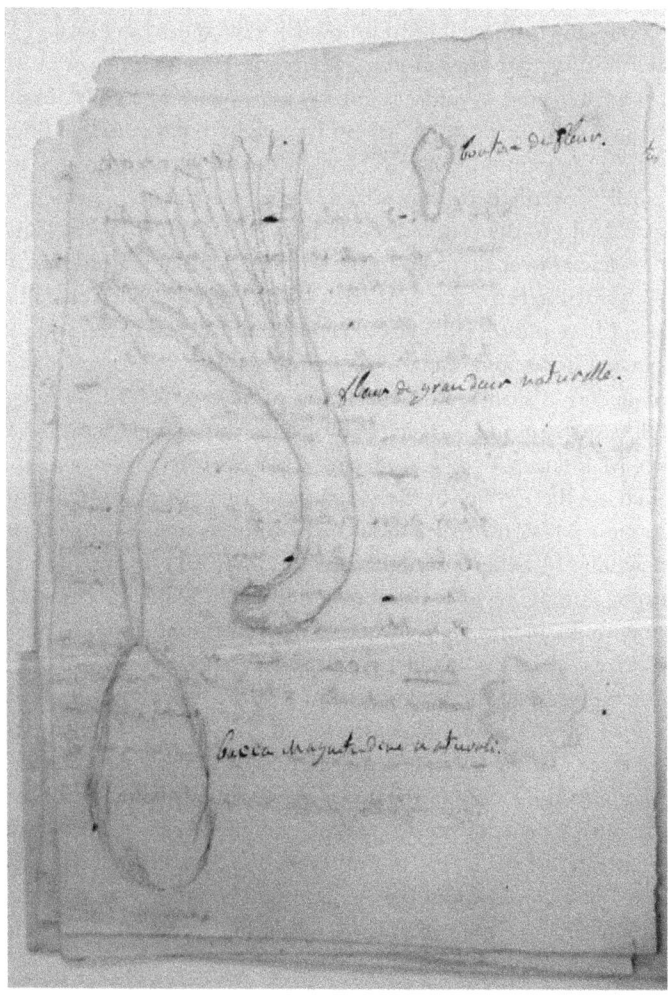

FIGURE 11.4 Cacti collectii, Corrientes 1838. Courtesy of Archivo del Museo de Farmacobotánica, Universidad de Buenos Aires, Argentina.

'on m'affirme qu'il ont observe Coccineus'. Was Bonpland referring to cochineal, the scaled insect, used for the production of red dyes, and thinking of a potential business? We do not know but by 1838, cacti, even for the French expatriated in distant Buenos Aires, were already associated to English gardens, and even Bonpland pretended to ignore that this association had French roots.

Bonpland wanted to present the consul the exotica he had collected in Corrientes. John Henry Mandeville, in 1838, was also the intermediary in charge of sending to the Museum in Paris the boxes with the samples of animals, seeds, rocks, and plants Bonpland had collected in the Rio de la Plata provinces. He was also sending back herbaria and minerals he had collected with Humboldt: sixteen boxes with the general herbaria and plants from all the Americas (toute l'Amérique) and the seventeenth and eighteenth with 'the collection of minerals from Peru, Mexico, etc. collected on the voyage made with Mr. Humboldt' (On trouve dans cette collection quelques morceaux qui n'appartiennent pas au Voyage).[67] By the late 1830s, when *Cactus speciosus* had made a career in the world of gardening, Bonpland seemed to have accepted his South American fate.

Cactus speciosus was one of the trading goods that Bonpland and Humboldt, willingly or not, helped consolidate as a by-product of their voyage, with profits from the publication and selling of books, letters, news, plates, and plants. Its history shows a commercial success in the early nineteenth-century realm of gardening, where the sale of ornamental plants, their depictions, and the implements to keep them would lead to the northern and eastern European cactus-mania, an obsession that continues even today. However, the supposed cactus from Turbaco had a much less linear development: kept in Humboldt and Bonpland's herbaria at the Paris Muséum national d'histoire naturelle, it was not treated in the botanical results of the expedition penned by the German botanist Karl S. Kunth, partly because it was impossible to determine, partly because of the confusion around this specimen.[68] *Cactus speciosus*, one can add, was also the result of the disorder that characterised the practice of accumulating data and facts in Paris, Berlin, London or elsewhere. In some contrast to the idea of order that is often attached to the so-called centres of calculation that the metropolitan scientific institutions would embody, the cactus described by Bonpland tells a different story. It is a story of disorder, fragmentation, ignorance, and confusion, where labels are lost and savants are unable to match things with notes, records with memories, dried leaves with living flowers, the past with the present, old books with present anxieties and profits.

In 1996 López Piñero and Pardo Tomás already pointed out that contemporary bibliography spread more errors than Bonpland's names. They referred to the 'unfortunate book' by the British writer Douglas Botting, at that time a best seller already translated into several European languages and widely distributed in Spain. It did not matter if Humboldt himself had affirmed that 'no European government has spent greater sums for the progress of botany than the Spanish government.' Botting wrote 'from the most complete ignorance' as he repeated the 'hackneyed cliché' of assigning to Humboldt the role of having been the first European scientist to study American nature.[69] In 2002, Botting's book was in its sixth German edition.

It formed part of the gifts that the Alexander von Humboldt Stiftung awarded to its fellows. Like *Cactus speciosus*, profitable nonsense survives in the business of exotica for those who know how to take advantage of the ignorance and tastes of their times.

Notes

1 Robert Proctor, 'Agnotology. A Missing Term to Describe the Cultural Production of Ignorance (and Its Study),' in Robert Proctor and Londa Schiebinger, eds., *Agnotology: The Making and Unmaking of Ignorance* (Stanford, CA: Stanford University Press, 2008), 1–36.

2 See Stephen Bell, *A Life in the Shadow: Aimé Bonpland in Southern South America, 1817–1858* (Stanford, CA: Stanford University Press, 2010) and the bibliography on Bonpland on pages 10–15. On Adeline Bonpland, see Alain Couturier, *Adeline Bonpland: voyage dans l'Amérique des Libertadores* (Paris: L'Harmattan, 2012), and *Adeline Bonpland: L'histoire cachée d'une aventurière du XIXe siècle* (Paris: L'Harmattan, 2018).

3 Bell, *A Life*, Chapter One.

4 On the Buenos Aires museum, see Irina Podgorny and Maria M. Lopes, *El Desierto en una vitrina. Museos e historia natural en la Argentina* (Rosario: Prohistoria, 2012).

5 The Catalogue of the National Library in Argentina (Biblioteca Nacional Mariano Moreno) contains most of Bonpland's publications, see https://catalogo.bn.gov.ar/.

6 Jorge Cañizares Esguerra, *How to Write the History of the New World: Histories, Epistemologies, and Identities in the Eighteenth-Century Atlantic World* (Stanford, CA: Stanford University Press, 2001); Irina Podgorny, 'Silent and Alone. How the Ruins of an Ancient City found Close to Palenque Were Taught to Talk the Language of Archaeology,' in L. Lozny, ed., *World Archaeologies: A Comparative Perspective* (New York: Springer), 527–553; and 'Mercaderes del Pasado: Teodoro Vilardebó, Pedro de Angelis y el comercio de huesos y documentos en el Río de la Plata, 1830-1850,' *Circumscribere* 9, 29–77. http://revistas.pucsbr/index.php/circumhc/article/view/5272.

7 In a letter to Larrañaga, Bonpland wrote:

> On leaving Europe, I thought it convenient … to bring a good number of books on Natural History. I have placed some of them in the library of this country, but I still have some left. Although I think that after your trip to Brazil your private library will have increased and that of your city too, I am sending you the list that will be sent to you by Mr. Cavallon. In the event that you decide to keep some works, I warn you that most of them may be reduced, that is, those that are for you or the library. I must also warn you that I am expecting a reply from Chile, where I sent a very considerable list of the works I had left to sell, and I therefore ask you to give Mr. Cavallon or myself an answer as soon as possible as to the choice you might make. I beg your pardon, sir, for speaking to you in my first letter of trade and natural history.

Escritos de don Dámaso Antonio Larrañaga, v. 3 (Montevideo: Instituto Histórico y Geográfico del Uruguay, 1924), 259.

8 Bonpland referred to the *Dictionnaire des Sciences Naturelles par plusieurs professeurs du Muséum national d'histoire naturelle et des autres principales écoles de Paris*. The first volume was published by F. G. Levrault and F. Schöll in Paris in 1804. Levrault in Strasburg resumed the Dictionary in 1816, when the subscription was offered at the following prices: 'Pour chaque volume du texte, papier ordinaire in-8.°, à 6 fr; Pour chaque cahier de planches en noir, format in-8.°, 5 francs, et format in-4.°, 7 fr. 5o cent. -Idem, enluminé, 12 francs, et format in-4°, 16 fr. Trente exemplaires sur papier velin: le prix du

volume est de 15 francs. The subscription would end at the 8th volume, then price of each volume would be 8 francs; and the figures in-8.° à 7 fr. 5o cent., et in-4°, 10 francs (*Dictionnaire des sciences naturelles*, prospectus, 1816, Strasbourg: Levrault). Bonpland was offering the books at a very good price: 26 (23.5 with the discount) instead of 40 francs.

9 René Louiche Desfontaines, 1750–1833, *Tableau de l'école de botanique du Jardin du roi* (Paris: Brosson, 1815).

10 French botanist Charles Plumier's *Description des plantes de l'Amérique* (Paris, 1693) contained 108 plates, while *Nova plantarum americanarum genera* (Paris, 1703–1704), forty plates and a compilation of one hundred genera, with about seven hundred species. Linnaeus adopted, almost without change, the genera published by Plumier. On Plumier and his relationship to the work by Francisco Hernández, see José María López Piñero and José Pardo Tomás, *La influencia de Francisco Hernández (1515-1587) en la constitución de la bótanica y la materia médica modernas* (Valencia: Instituto de Estudios Documentales e Históricos sobre la Ciencia, 1996).

11 On former Napoleon followers in the Río de la Plata, see Emilio Ocampo, *La última campaña del Emperador. Napoléon y la independencia argentina* (Buenos Aires: Claridad, 2007).

12 On the gardens in Buenos Aires, see Maxine Hanon, *Buenos Ayres desde las quintas de Retiro a Recoleta (1580–1890)* (Buenos Aires: El Olmo, 2000).

13 'Cactus. Icosandria monogynia Linn. Cacti. Juss. Species *CACTUS speciosus*,' in Aimée Bonpland, *Description des plantes rares cultivées à Malmaison et à Navarre* (Paris: Schoell, 1813), 8–10, plate 3.

14 Sarton, George, 'Fifth Preface to Volume XXXIV: Aimé Bonpland (1773–1858),' *Isis* 34:5 (1943), 385–399. Accessed January 14, 2020. www.jstor.org/stable/225737.

15 Alicia Lourteig, 'Aimé Bonpland,' *Bonplandia* 3 (1977), 16. Bell, *A Life in Shadow*.

16 Franz Julius Ferdinand Meyen, *Grundriss der Pflanzengeographie mit ausführlichen Untersuchungen über das Vaterland, den Anbau und Nutzen vorzüglichsten Culturpflanzen, welche den Wohlstand der Völker begründen* (Berlin, 1836), 168 (Cactuspflanzen).

17 Humboldt was particularly interested in promoting Meyen's work given the fact that it reflected well his ideas and proposal:

> On this busy day, I have only a few moments to tell you, my dear Professor, how deeply pleased I am that you, who as a botanist, circumnavigator and familiar with meteorological phenomena, have written a plant geography. The plan of your work has appealed to me all the more since the connection between botany and meteorology has recently been completely misunderstood and, in Schouw's hands, plant geography has degenerated into a purely meteorological treatise. Your treatise on *Zanichellia* will however appear in Brogniart Annales. When I left, it was in Mirbel's hands, who liked it very much. I understood from your letters at that time that I should leave them there for translation, since of German works— with the rather increasing than decreasing complete ignorance of the German language (nobody dares to read unprinted things)—only the drawings are looked at. I will soon write to Mirbel that he will return the original as soon as the translation is done.

Alexander von Humboldt to Franz Julius Ferdinand Meyen. Berlin, 24. Januar 1836, hg. v. Petra Werner unter Mitarbeit von Ingo Schwarz und Tobias Kraft, in Ottmar Ette, ed., *Edition Humboldt Digital* (Berlin: Berlin-Brandenburgische Akademie der Wissenschaften, 2019). https://edition-humboldt.de/v5/H0001224 On Meyen, see Petra Wener, 'Franz Julius Ferdinand Meyen: gefördert und frühvollendet. Zwischen Poesie und totem Zoo,' *HiN - Alexander von Humboldt im Netz. Internationale Zeitschrift für Humboldt-Studien*, [S.l.] 18:34 (2017), 148–165. http://www.hin-online.de/index.php/hin/article/view/247/469. Accessed January 09, 2020. http://dx.doi.org/10.18443/247. The Danish lawyer and botanist Joakim Frederik Schouw (1789–1852) was not translated into English. In 1822 he published his 'Plant Geography' (*Grundtræk til en almindelig*

Plantegeographie) in Copenhagen, translated into German in 1823 as *Grundzüge einer allgemeinen Pflanzengeographie*. In 1827 Schouw published in Copenhagen his *Beiträge zur vergleichenden Klimatologie*, where he used a term that normally is associated with Humboldt's publications on the Russian voyage. On this, see Michael Dettelbach, 'El último de los hombres universales: lo local y lo universal en la ciencia de Humboldt,' *Redes* 14:28 (2008), 112–113. In his work, Schouw defined 'Klimatologie' as 'geographical meteorology, or the study of the characteristics of the atmosphere in the different parts of the world. It is also a part of physical geography.' In 1832, Humboldt used 'Klimatologie' as the discipline that studied 'la distribution de la chaleur sur le globe' (the distribution of global heat), an idea from his 1817 memory 'Des lignes isothermes et de la distribution de la chaleur sur le globe' (*Mémoires d'Arcueil*, Paris: Perroneau), where the term 'climatologie' does not appear.

18 Meyen, *Outlines of the Geography of Plants: With Particular Enquiries Concerning the Native Country, the Culture, and the Uses of the Principal Cultivated Plants on Which the Prosperity of Nations Is Based*, translated by Margaret Johnson (London: The Ray Society, 1846), 142.

19 Meyen, *Outlines*, 142.

20 Jim Endersby, *Orchid: A Cultural History* (Chicago, IL: University of Chicago Press, 2016); Richard Drayton, *Nature's Government: Science, Imperial Britain, and the 'Improvement' of the World* (New Haven, CT: Yale University Press, 2000).

21 Meyen, *Outlines*, 142 and 147.

22 José Pardo Tomás, personal communication.

23 José Pardo Tomás and María Luz López Terrada, *Las primeras noticias sobre plantas americanas en las Relaciones de Viajes y Crónicas de Indias, 1493-1553* (Valencia: CSIC, 1993), 319; on the cultivated nopal, see Norma Angélica Castillo Palma, 'Cholula en sangre de grana. La destrucción de las nopaleras de cochinilla como resistencia indígena ante el agravio español,' *Historias* 49 (2001), 45–66, Roberto García-Morís, 'La Relación de lugares y pueblos donde se saca la grana de la Ciudad de Cholula' (México) de 1600. Alcance y contenido,' *Revista Española de Antropología Americana* 45:1 (2015), 75–89. Also see Gordon D. Rowley, 'The Sessé and Mociño cactus plates,' *Bradleya* 12 (1994), 8–31, and the bibliography cited there.

24 Similar processes have been analyzed by Irina Podgorny, 'The Elk, the Ass, the Tapir, Their Hooves, and the Falling Sickness: A Story of Substitution and Animal Medical Substances,' *Journal of Global History* 13 (2018), 46–68, and 'The Name Is the Message: Eagle-Stones and Materia Medica in South America,' in C. J. Duffin, C. Gardner-Thorpe, and R. T. J. Moody, eds., *Geology and Medicine: Historical Connections* (London: Geological Society Special Publications 452, 2017), 27, 195–210.

25 Tomás and Terrada, *Las primeras noticias*, 187–191, also Iris H. W. Engstrand, 'The Eighteenth Century: Enlightenment Comes to Spanish California,' *Southern California Quarterly* 80:1 (1998), 3–30.

26 Tomás and Terrada, *Las primeras noticias*, 319.

27 English translation by Simon Varey, *The Mexican Treasury: The Writings of Dr. Francisco Hernández* (Stanford, CA: Stanford University Press, 2000), 189.

28 Matthias L'Obel, *Stirpium adversaria nova* (London, 1571), 376.

29 Caspar Bauhin, *Pinax Theatri botanici sive Index in Theophrasti, Dioscoridis, Plinii et botanicorum qui à seculo scripserunt opera: plantarum circiter sex millium ab ipsis exhibitarum nomina cum earundem synonymiis & differentiis methodicè secundùm earum & genera & species proponens. Opus XL. annorum hactenus non editum summoperè expetitum & ad auctores intelligendos plurimùm faciens* (Basel, 1623), 457–458; John Parkinson, *Paradisi in sole paradisus terrestris, or, A garden of all sorts of pleasant flowers which our English ayre will permit to be noursed vp: with a kitchen garden of all manner of herbes, rootes, & fruites, for meate or sause vsed with vs, and an orchard of all sorte of fruitbearing trees and shrubbes fit for our land together with the right orderinge planting & preseruing of them and their vses & vertues* (London, 1629).

30 Horst Heinemann, '¿De dónde viene el nombre de cacto?,' *Cactáceas y Suculentas Mexicanas* 25 (1980), 27–33.

31 On L'Obel, apothecaries, and botany see Florike Egmond, 'Apothecaries as Experts and Brokers in the Sixteenth-Century Network of the Naturalist Carolus Clusius,' in Mordechai Feingold, ed., *History of Universities*, v. 23, n. 2 (Oxford: Oxford University Press, 2008), 59–91.

32 L'Obel, *Stirpium adversaria nova*, 453–454.

33 Pietro Andrea Mattioli, 'Fichi indiani & loro historia; Opuntia & sua historia,' in *I discorsi di M. Pietro Andrea Matthioli … nelli sei libri di Pedacio Dioscoride Anazarbeo della materia medicinale : hora di nuouo dal suo istesso autore ricorretti, & in più di mille luoghi aumentati : con le figure grandi tutte di nuouo risatte, & tirate dalle naturali & uiue piante, & animali, & in numero molto maggiore che le altre per auanti stampate : con due tauole copiosissime spettanti l'una à ciò, che in tutta l'opera si contiene, & l'altra alla cura di tutte le infirmità del corpo humano* (Venice: Malgrisi, 1568), 310–311.

34 The Italian original version of Mattioli is confusing. He wrote: 'Hebbine ancora una pianta intera nel tempo che mi ritrovavo in Goritia dal dilligentissimo semplicista M. Giulio Moderato da Rimini,' which can be understood as if Mattioli had seen the plant in Gorizia. But according to the secondary sources, Moderato's garden was in Rimini, cf. Timoteo Bertelli, *Sulla epistola di Pietro Peregrino di Maricourt e sopra alcuni trovati e teorie magnetiche del secolo XIII*, Memoria Seconda (Rome, 1868), 89–90; Mauro Ambrosoli, *The Wild and the Sown: Botany and Agriculture in Western Europe, 1350-1850* (Cambridge: Cambridge University Press, 1997), 114; Paula Findlen, *Possessing Nature: Museums, Collecting, and Scientific Culture in Early Modern Italy* (Berkeley: University of California Press, 1996), 173 and 177. I am indebted to Pepe Pardo and Guillermo Ranea for his help in these translations.

35 Heinemann, '¿De dónde viene el nombre de cacto?'

36 Letter from Alexander von Humboldt to Karl Ludwig Willdenow. Paris, May 17th 1810, hg. v. Ulrich Päßler unter Mitarbeit von Klaus Gerlach und Ingo Schwarz. In *Edition Humboldt Digital*, hg. v. Ottmar Ette (Berlin: Brandenburgische Akademie der Wissenschaften, 2019). URL: https://edition-humboldt.de/v5/H0006055.

37 A. de Candolle, *Catalogus plantarum horti botanici Monspeliensis* (Montpellier, 1813), 84.

38 Gordon Rowley, 'Pierre Joseph Redouté. Raphael of the Succulents,' *The Cactus and Succulent Society Journal*, The Cactus and Succulent Journal of Great Britain 18:4 (1956), 91–93.

39 Pierre-Yves Lacour, *La République naturaliste. Collections d'histoire naturelle et Révolution française (1789-1804)* (Paris: Muséum National d'Histoire Naturelle, 2014), 536–537.

40 William Thomas Stearn, 'Ventenat's 'Description des plantes de J. M. Cels', 'Jardin de la Malmaison' and 'Choix des plantes,' *Journal of the Society for the Bibliography of Natural History* 1 (1939), 7.

41 Etienne Pierre Ventenat, 'Dedicace,' *Jardin de la Malmaison* (Paris, 1803).

42 Of sixty-six species listed in Table 11.1, seventeen were from the Americas (25%) and six (9%) seem to have been sent by Humboldt and Bonpland. In Bonpland's times, the catalogue showed a preponderance of plants from Australia (31%), obtained from the Nicolas Baudin expedition (1800–1803) and from the Cape (15%). In the two volumes published by the former gardener Étienne P. Ventenat (*Jardin de la Malmaison*, Paris: Crapelet and Herhan, 1803–1804), the figures are as follows: from about 110 species listed, 18 (16%) are from the Americas (most from North America, 1 from the Antilles), about 30% from Australia, and 20% from the Cape, with several specimens originated in the Canary Islands and Egypt. Josephine's exotic plants came mostly from the Pacific, the Cape, and from the commerce with England.

43 Alexandre de Laborde, *Description des Nouveaux Jardins de La France et de des Anciens Châteaux: Mêlée d'observations sur la vie de la campagne et la composition des Jardins* (Paris: Delance, 1808), 65.

44 Neil Safier, 'Transformations de la zone torride Les répertoires de la nature tropicale à l'époque des Lumières,' *Annales. Histoire, Sciences Sociales* 66e année (2011), 143–172, here 152.

45 Safier, 'Transformations.'
46 See Emma Spary, *Utopia's Garden. French Natural History from Old Regime to Revolution* (Chicago, IL: University of Chicago Press, 2000); Lacour, *La République naturaliste*; and René Le Berryais, *Traité des jardins, ou Le nouveau de La Quintinye. 4e partie, Orangerie, serrre chaude* (Paris, 1788).
47 Le Berryais, *Traité*, 181.
48 Le Berryais, *Traité*, 275.
49 Le Berryais, *Traité*, 275.
50 See Marie-Noëlle Bourguet, 'La collecte du monde : voyage et histoire naturelle (fin XVIIe siècle-début XIXe siècle),' in C. Blanckaert, ed., *Le Muséum au premier siècle de son histoire* (Paris: Muséum National d'Histoire Naturelle, 1997), 163–196. Also see Ana Sevilla and Elisa Sevilla, 'Semillas andinas, invernaderos escoceses y herbarios londinenses en la red de Charles Darwin,' in Marisa Adriana Miranda, Miguel Ángel Puig-Samper, Rosaura Ruiz Gutiérrez and Gustavo Vallejo, eds., *Darwin y el darwinismo desde el sur del sur* (Madrid: Doce Calles, 2018), 257–276.
51 *The Gardeners' Chronicle* (Saturday June 23, 1855), 420.
52 On the controversial systematics of Berlin`s cacti, see Beat Ernst Leuenberger, 'The Cactaceae of the Willdenow Herbarium, and of Willdenow (1813),' *Willdenowia* 34:1 (2004), 309–322. Accessed January 20, 2020. www.jstor.org/stable/3997483.
53 Schöne Künste, 'Description des plantes rares que l'on cultive à Navarre et à Malmaison,' *Allgemeine Literaturzeitung* 105 (Friday 31 December, 1813), 1064–1069.
54 Bell, *A Life*, 22–26.
55 *The Gardener's Magazine and Register of Rural & Domestic Improvement*, 1826.
56 *The Gardener's Magazine*, 1830.
57 Also Friedrich Ludwig Finckh, *Die Cactūs, ihre Beschreibung, Cultur und Vermehrung ein Handbuch für Cactusfreunde* (Stuttgart: Löfflund, 1832); Berge, *Anweisung zur zweckmäßigen Behandlung der Cactus Pflanzen* (Stuttgart, 1832); Jacob Ernst von Reider, *Die Beschreibung und Kultur der Azaleen, Cactus, Camellien und Calla aethiopica* (Ulm, 1834). On the place of pineapples in the eighteenth century, see Lacour, *La République*; Spary, *Utopia's Garden*.
58 John Green, 'On the Management of the Cactus. In a Letter to the Secretary, Read June 4, 1833,' *Transactions of the Horticultural Society of London* 1 (1835), 401–402. This letter was reproduced in several periodicals and even translated into German, see *The Royal Lady's Magazine, and Archives of the Court of St. James* 2 (1834), 122–123; 'Über die Behandlung der Cactus, übersetz vom Herrn Theodor Nietner,' *Allgemeine Gartenzeitung: Eine Zeitschrift für Gärtnerei* 2 (1834), 309–310.
59 Ernst Wunschmann, 'Kunth, Karl Sigismund,' in Historischen Kommission bei der Bayerischen Akademie der Wissenschaften, *Allgemeine Deutsche Biographie* 17 (1883), 394–397.
60 Beat Ernst Leuenberger, 'Humboldt & Bonpland's Cactaceae in the herbaria at Paris and Berlin,' *Willdenowia* 32 (2002), 137.
61 Leuenberger, 'Humboldt & Bonpland's Cactaceae'.
62 Bonpland, *Description des plantes rares*, 9–10.
63 Alexander von Humboldt to Franz Julius Ferdinand Meyen. [Berlin], August 22nd 1834, hg. v. Petra Werner unter Mitarbeit von Ingo Schwarz und Tobias Kraft. In: edition humboldt digital, hg. v. Ottmar Ette. Berlin-Brandenburgische Akademie der Wissenschaften, Berlin. Version 5 vom 11.09.2019. URL: Appendix https://edition-humboldt.de/v5/H0001202.
64 *The Athenaeum* (1832) 635.
65 Bell, *A Life*, 122. Jean Baptiste Washington de Mendeville was the French consular agents at Buenos Aires.
66 Box 9, Dossier 13, Archivo de Farmacobotánica, Facultad de Farmacia, UBA.
67 Duplicats de l'État et contenu des caisses remises à Monsieur de Mandeville, Consul général de France à Buenos Aires pour être envoyées au Muséum royal d'histoire naturelle à Paris, Archivo Bonpland, Museo de Farmacobotánica, UBA.

68 See Leuenberger, 'Humboldt & Bonpland's Cactaceae,' 137–153, and Karl Sigismund Kunth, *Nova genera et species plantarum quas in peregrinatione ad plagam aequinoctialem orbis novi collegerunt Bonpland et Humboldt* (Paris, 1815–1825), 'Opuntiaceae,' 6 (1823), 52–56. https://www.biodiversitylibrary.org/item/271579#page/9/mode/1up.
69 Piñero and Tomás, *La influencia de Francisco Hernández*, 229.

12

HUMBOLDT'S COLUMBUS, OR THE IBERIAN WORLDS THAT HUMBOLDT IGNORED

Jorge Cañizares-Esguerra

Humboldt's *Examen Critique* (1836–1839) presents itself as a history of geography and cosmography in the sixteenth century, but it is really a study of the voyages and cosmographic ideas of Columbus and Amerigo Vespucci compiled by Spanish historians Juan Bautista Muñoz and Martín Fernández de Navarrete. Humboldt's text is prefaced by a review of late-medieval cosmographic speculation on the islands in the Atlantic along the route to Asia. Humboldt links this rising speculation to a tradition found in classical Greek and Roman texts. In short, Humboldt presents Columbus as a man deeply wrapped up in biblical, prophetic, medieval literature who was nevertheless capable of creating a new science, namely, earth physics.

In Humboldt's hands, Columbus becomes a solitary genius who despite his crusading and apocalyptic ideas was capable of cracking open the workings of ocean currents in the Atlantic. By using new instrumentation and by drawing correlations among isolated evidence in winds, magnetism, sea patterns, and oceanic flora, Columbus plotted new maritime routes to Asia. Humboldt thereby lionises Columbus as a global geographer, a master of instrumentation and the discoverer of unexpected patterns in the physics of the earth. Despite the meandering, deeply antiquarian structure of Humboldt's narrative, the unmistakable impression is that Columbus's science prefigured Humboldt's. Humboldt's Columbus is ultimately about Humboldt.[1] In this sense, Bolivar was not altogether mistaken when he cast Humboldt as a second Columbus; in *Examen Critique* Humboldt cast himself in this role.

In this chapter, I contrast Humboldt's self-fashioned *Examen critique* with the more deeply informed scholarship on Columbus of two enlightened Spanish historians, Juan Bautista Muñoz and Martín Fernández de Navarrete. Humboldt drew heavily on the work of these two scholars while at the same time ignoring their central arguments or interpretations. Together, Muñoz and Fernández de Navarrete devoted some eighty years gathering, collating, and publishing massive

DOI: 10.4324/9781003231479-13

documentary collections of the early-modern Hispanic expansion across the Indies. But Humboldt was not interested in the argument of Muñoz and Fernández de Navarrete, namely, that global modernity itself was an Iberian and 'Indian' project. For Humboldt, modernity was the result of the creativity and daring of individuals who broke with conventions. For the Prussian 'second Columbus,' Columbus had singlehandedly solved the puzzles of oceanic currents and earth physics. Heir to medieval physics and cosmographies and steeped in a religious vocabulary of crusades and prophecy, Columbus had nevertheless transformed the world by the sheer power of his intuitions and instruments, discovering the patterns of the Atlantic and charting new routes by observing subtle changes in the compass and the sea.[2]

Humboldt's selective use of Muñoz's and Fernández de Navarrete's scholarship sheds light on the wider invention of Humboldt that is the subject of this volume. The best-selling author Andrea Wulf has recently narrativised a dominant epistemological tradition that ascribes to Humboldt the heroic 'invention of Nature,' that is, the discovery of global ecological relationships. According to this reading, Humboldt was the first to see Nature as correlations of communities of plants, climate, and geology illustrated on profiles and maps. Wulf is surely not the first to present Humboldt as the founding father of ecology and many other things. She is merely the popular tip of an epistemological iceberg that has served to sink the Hispanic ship of knowledge.[3] Humboldt's reading performs a similar act. Columbus appears as a lonely pioneer who had to await his first enlightened scientific biographer. Columbus appears as a Humboldt *avant la lettre*.

Although Humboldt mined the scholarship of both Muñoz and Fernández de Navarrete he ultimately excises Columbus from the worlds that made his knowledge possible. This strategy was not new for him: Humboldt frequently acknowledged borrowing sources and ideas from intellectuals in the places he visited in the Americas and Spain while at the same time presenting the core of his scholarship as novel. Humboldt thus failed to acknowledge that key elements of his most precious insights on global physics were co-created or long anticipated. As the present volume makes clear, in his many publications, the New World appears in the form of objects and primary sources (documents, rocks, archives, waterfalls), not as the source of his ideas and interpretations. This chapter seeks to restore the traditions and genealogies of scholarship both Muñoz and Fernández de Navarrete implicitly or explicitly sought to rescue but that Humboldt ignored in his effort to self-fashion himself as the inventor of a new global science. Sadly, those traditions remain ignored and buried by the patronising condescension of much north Atlantic scholarship.

On Things Forgotten.

Muñoz was the founder of the Archive of the Indies in Seville as well as the collector of some 150 volumes of unpublished sixteenth- and seventeenth-century manuscripts on the Americas. Muñoz spent decades in dozens of royal and private documentary collections, reassembling the archive of the Hispanic monarchy in the Empire of the Indies that included America as well as islands in the Indian Ocean and Pacific Southeast Asia. Muñoz planned to produce a massive new history of

the Spanish conquest and colonisation but managed only to have one volume published, a history of the first three voyages of Columbus to the New World. In its
structure, Muñoz's volume resembles Antonio de Herrera's *Historia general de los
hechos de los castellanos en las Islas y Tierra Firme del Mar Oceano* (1601–1615). Like
Herrera, Muñoz delved deep into the immense official, unpublished paperwork of
the Indies as well as into the massive unpublished *Historia de las Indias* (ca 1529–
1557) by Bartolomé de las Casas, a year-by-year annal or chronicle of the conquest
from 1492 to the 1520s that drew on the vast archives of the monarchy and dozens
of members of the Dominican order. It was Las Casas who edited and secured for
posterity the only surviving copy of Columbus's diary. Herrera had access to Las
Casas's history and so too did Muñoz.[4]

Like Herrera and Las Casas, Muñoz went over the paperwork of Columbus's
voyages in great detail. All three authors offered painstakingly detailed histories
of Columbus's apparent geographical miscalculation, namely, the idea that he had
arrived in the Spice Islands of the eastern Indian Ocean. The three historians
chronicled Columbus's desperate efforts to transform his voyages into commercial
successes via the trade in Indian slaves in Spain and as porters in expeditions of
conquest in the Caribbean, or as slave laborers panning gold in the rivers of the
Caribbean islands. Las Casas, Herrera, and Muñoz also documented Columbus's
decrees to quell indigenous rebellions and to purchase gold and slaves through
trade. The three also documented Columbus's growing conflict with the Crown
and other colonists as an ever-larger number of restless Iberian settlers and Crown
and clerical bureaucracies embarked for the Indies, lured by Columbus's mounting
yet misleading observations on the botanical and mineral bounties of the islands that
presumably lay off the coast of China and India. Tellingly, this world of slavery and
emerging capitalism was a key element of Columbus's modernity that Humboldt
entirely overlooked or misread.[5] Humboldt saw Columbus the cosmographer as
entirely disconnected from the other Columbuses available in the historiography
since Las Casas: merchant, slaver, and failed colonial administrator. Unlike Muñoz,
who firmly grounded knowledge in social relations, Humboldt saw knowledge as
the disembodied work of genius.

Global Physics and the Two 'Planets' of the Hispanic Monarchy

Muñoz's insisted that the discovery of America and the South Sea was an epochal
mutation in global history. For Muñoz, Columbus's discovery of the western route
to Asia led to a radical new history of trade and commerce that transformed the
world forever.[6] If the changes had already begun with the Portuguese in the fifteenth century as they inched their way down the coast of Africa, then the voyages
of Columbus were of a different kind altogether. For Muñoz, the modernity of
Columbus was one of new knowledge. Columbus's new sea routes made possible a
whole new physics of the globe. An entirely new hemisphere or world and a new
sea (the South Sea) became known and with it a whole new planet. According to
Muñoz, Columbus's opening of new westward trading routes to Asia altered all that

was known since classical antiquity about the sphere, climates, and human politics, for America behaved like an entirely different planet than the one described in Aristotle's *Meteorology* and *Politics* or in Sacrobosco's *Sphera*. Unlike the Old World, in America, most of the population lived in the tropics, not the temperate zone. Here the tropics were temperate. There was no scorching heat in the American equator and there were no Blacks in its torrid zone.[7] Implicit in Muñoz's arguments on the impact of the discovery of the New World lay a history of Iberian contributions to the history of global physics, knowledge that Humboldt ignored.

From Aristotle's *Physics* to Ptolemy's *Tetrabiblos* to Sacrobosco's *Sphera* the earth had long been thought of as a web of interconnected organic relations. Consider classical meteorology: when the element water changes into the element air there is climate change; when solar fire warms the elements earth, water, and air, there is water circulation, rain, and oceanic currents.[8] Refraction of the light of stars affects terrestrial and bodily elements at different angles, depending on geographical location and relational interactions among fixed stars, zodiac constellations, and planets. In short, Ptolemaic astrology implied painstaking observation to uncover and graph hidden connections of global dimensions.[9] This Aristotelian or classical tradition of 'Humboldtian' global physics avant la lettre, in the sense that it sought to uncover hidden patterns and relations among supra and sublunar elements, was radically altered with the Hispanic 'discovery' of the New World. There were elements in the American or West Indian sphere that did not behave as predicted in Sacrobosco's *Sphere*. Humboldt presented Columbus as a medieval man who prefigured earth physics by dint of his genius, intuition, and instruments. But Columbus was no genius. He was a cosmographer of his age whose global physics were classical. It was the intuition and science of many others who, working in the expanding worlds of the Hispanic monarchy, reconfigured that tradition into a new earth physics that included a sphere that was missing in Ptolemaic cosmographies.

In his *Natural and Moral History of the Indies* (1590), José de Acosta presented the world as a circulatory machine of water, fire, and air to explain paradoxical New World or 'Indian' phenomena, including earthquakes, volcanic activity, mineralogical peculiarities, hurricanes, and rainy seasons that seemed to be inversions of Old World or Eurasian patterns.[10] According to Acosta, the Occidental Indies of America challenged the tendency of the ancients, particularly Aristotle, to draw false conclusions out of right premises. Classical physics–meteorology suggested that most individuals lived in the temperate climates while the tropics, particularly near the equator, were likely to be uninhabited due to the scorching sun. Acosta poked fun at these ideas and called attention to the difference of America, where advanced civilisations thrived within the tropics and the temperate zones were scarcely inhabited. The New World inverted what was known about the Old.[11] Acosta's ideas found an echo in Mexico in *Problemas y secretos maravillosos de las Indias* (1591), where Juan de Cárdenas deployed global meteorology and physics to interpret all sorts of natural phenomena. Like Acosta, Cárdenas deployed Renaissance meteorology to interpret 'secrets,' and mysteries of nature, including the peculiar taste of Mexican honey (which he attributed to the different migratory behaviour of bees) and the higher frequency of earthquakes in Mexico (which he attributed to the motion of subterranean air and water currents under its peculiar rock formations).[12] Although Cárdenas did not deal with biodistribution Acosta did. Acosta used the tools of Aristotle's *Physics* and Sacrobosco's *Sphera* to interpret the Andes as a microcosm. The inverted patterns of American meteorology, Acosta argued, were due to the Andes. This vast mountain range that ran the length of the New World was a global system of its own. Elevation, or verticality, could offset a climate zone within the sphere. One could find polar climates in the tropics or temperate ones, depending on the elevation. This tradition evolved in the Andean region over the subsequent centuries and was expressed in Francisco José de Caldas's maps of Andean biogeography, produced in Nueva Granada while Humboldt was visiting Bogotá and Popayán and in José Hipólito Unanue's scientific vision of Peru.[13]

Muñoz drew on a long tradition of Iberian scholarship that represented an empire constituted by two 'worlds' or 'planets.' Both the Hapsburg and the Bourbon dynasts represented themselves as kings of Spain and emperors of the Indies, namely, as monarchs with sovereignty over two different spheres. The tradition of the two spheres began with the recognition of the different laws of the zones, climates, meteorology, and politics of America. It coloured most imperial or colonial discourses, particularly in Peru where it generated strong religious narratives of election and exceptionalism. As early as the 1560s, European courtly intellectuals sought to present Charles V as the emperor of two orbs. Spanish notions of universal monarchy broke new ground as Spanish Hapsburg monarchs appeared in print lording over two orbs, not one medieval sphere. The iconographic trope of the two spheres remained unchanged under the Bourbons. The iconography of the wars of independence in Spanish America made a point of representing political emancipation as the breaking of the chains that had long yoked the American orb to the ancient sphere of the Old World (see Figures 12.1 and 12.2).

Archives and Knowledge or Ignorance of the New World

On the empirical side, Humboldt's exposition in *Examen critique* was almost entirely built on the scholarship of Fernández de Navarrete, whose decades-long archival work Humboldt plundered. Like Muñoz, Fernández de Navarrete also scoured royal and private archives in Simancas, Seville, and other cities in Spain. Unlike Muñoz, however, Fernández de Navarrete did not seek to create an archive to rewrite the history of the Indies. His intention was to gather the original documentation of fifteenth- and sixteenth-century maritime expeditions.[14] Between 1825 and 1837, Fernández de Navarrete published five volumes of documents primarily on Columbus's and Vespucci's voyages to America and Magallanes's and Loaysa's to Malacca and the Philippines.[15] This is the documentation Humboldt used for his *Examen critique*.

In the introduction to his collection, Fernández de Navarrete made clear his larger intention, which was to create an archive of an epochal scientific and technological revolution that demonstrated the prowess of the nautical expeditions, including

Biblioteca Nacional de España

FIGURE 12.1 Charles V, Emperor of the two spheres. Enea Vico, etching. In Lodovico
Dolce's *Vita Carlos V* (Venice, 1567).

innovation in navigational tools, ship pumps, and steam engines.[16] His published
volumes were just the tip of the iceberg. The totality of Navarrete archive reached
thirty-two volumes, now housed at the Naval Museum in Madrid.[17] Fernández
de Navarrete's efforts to document a revolution in knowledge triggered by technological, maritime, and cosmographical breakthroughs did not have much of an
impact. His massive archival-historiographical endeavour was quickly buried and
superseded by Washington Irving's and Alexander von Humboldt's famous studies
of Columbus.[18] In addition, as Fernández de Navarrete was making his archive
available the Hispanic Empire in the Indies collapsed. The revolutionary upheaval
yielded new republics and with them a narrative that reinforced the Protestant,

FIGURE 12.2 Ferdinand VI, King of Spain and Emperor of the Indies, in Fernando de
Araya, *Conclusiones mathemáticas* (Manila, 1758).

northern European account of Spain and her Indies as a sanguinary empire of
authoritarian cruelty and ignorance. For the most part, the Hispanic American
republics reinforced this account, inventing the trope of 'the colonial legacy' as a
burden to political and scientific modernity.[19]

Despite attempts to erase them from the history of science and knowledge, the hundreds of Iberian overseas expeditions Fernández de Navarrete documented built on
forms of embodied, technical knowledge of sailors and shipbuilders that had radically
transformed global geography, cosmography, and commerce. Fernández de Navarrete's
documentary collection took decades to assemble because it drew on a 'secret' or official
archive. Muñoz introduced his 'new' history of Columbus's three first voyages with

a lengthy evaluation of this archive. In it he outlined the epistemological structure of early modern Iberian colonial paperwork. This vast paperwork had been produced by societies and learned circles that for the most part did not seek to make their knowledge public. It was a world in which most documentation emerged piecemeal and out of vertical petitioning, grace, hierarchical supplication, secrecy, contracts, factional contest, litigation, and bureaucratic investigation. Most of these documents were never meant to be published.[20] As a result, the complex transformation of the mine and city of Potosi into the worlds' leading producer of silver, lacks a historiography that depicts it as an industrial and scientific revolution of global scope. Potosi's industrialisation relied on experimental knowledge. Innovation occurred in secret forms of communication between vassals and the Crown. Grace or 'mercedes' was the paper engine behind the innovation, as vassals sent petitions to crown officials to secure rewards. The need to communicate for rewards sparked a vast production of competing treatises and reports. Many of these reports were ciphered. The authors did not want their communication with the Crown stolen by others, including pirates.[21]

The New World was so far away from the monarch and his Council of the Indies that the Crown found itself promoting factions as a way of keeping any party from gaining too much knowledge and achieving too much control. As corporate factions emerged (representing conquistadors, encomenderos, friars, indigenous caciques, indigenous commoners, the secular church, bishops, cathedral chapters, and high court magistrates) conflict did too. Factionalism guaranteed the crown a permanent flow of information as parties sought to keep the king and the council constantly appraised. Perspectival biases plagued all documentation, rhetorical claims to objectivity notwithstanding. The epistemological conflict secured the crown as an honest broker in the mediation of disputes. It also secured a form of information in which knowledge and certainty was up for grabs.[22]

As Arndt Brendecke has shown, the goal of the Hispanic empire was information and triangulation, not knowledge.[23] It took the Crown a century to establish expertise and bureaucracies independent from local parties. To deal with the constant instability of faction-riddled information, the Crown ruled through forms of Rortian pragmatism, that is, majority consensus in metropolitan and viceregal councils as the only way out of structural doubt. The empire continued to be ruled via the legal, notarial practice of 'informaciones,' namely, forms of documentation that required parties to produce sets of witnesses, all implicitly unreliable. Every investigation, report, and evaluation created a trail of witnesses that could be easily countered by opposing testimonies. From inquisitorial inquiries to built-in investigations of ecclesiastical and lay local bureaucracies (*visitas*), the empire worked as a machine of paperwork producing structural doubt and therefore requiring mediation from above.

The many volumes of documentation on Columbus's travels assembled by Fernández de Navarrete were the product of forms of communication between vassals and monarchs that were not intended for the public sphere and its print culture, although such a sphere and culture did exist in the Hispanic Empire. Columbus's archive emerged out of his need to negotiate contracts, secure favourable testimonies of witnesses for Crown investigations, and for petitioning rewards. His diaries did not seek to record the secrets of nature but to compile data to defend and renegotiate contracts, and to persuade the Crown not to intervene in the Indies.[24] With the full support of the Crown, Muñoz and Fernández de Navarrete spent decades scouring private and royal archives rescuing for posterity paperwork that was not produced for publication. Secrecy was not an exclusive or deliberate policy, some sort of designed anti-public sphere, however. Secrecy was the very foundation of imperial governance and grace via petition. Vertical communication between the crown, its council, and the myriad factions is what made the Indies go round. Without the archival labours of Muñoz and Fernández de Navarrete in this world of secrecy and governance by petition, Humboldt's *Examen critique* would have never existed. The paperwork of Columbus, including his diaries, would have remained buried in private and Crown depositories.

Unlike Muñoz or Fernández de Navarrete, who sought to convey the difficulties of their labours and the endless peregrinations among the sea of paperwork, Humboldt failed to account for the radically different epistemological worlds of knowledge and trust in which Columbus and his Iberian heirs worked. This failure was a reflection of Humboldt's first-hand ignorance of the archival sources. In contrast, Fernández de Navarrete made public a world of knowledge in fifteenth- and sixteenth-century Iberian ships that had literarily transformed the world. As Humboldt himself noticed, it took some 2500 years since Aristotle and Ptolemy to fully map the Old World. It took only fifty years to map a new ocean and a new hemisphere. But in this Humboldt overlooked something that for Fernández de Navarrete was painfully clear: the engine of this revolution in knowledge production were the hundreds of motley expeditions of pilots, captains, artisans, commoners, and traders to purchase upward social mobility. These exploratory parties (*entradas*) and their pioneering leaders (*adelantados*) signed contracts with the Spanish Crown, seeking to carve out legal legitimacy against competing factions. Once in the Indies, these small enterprises multiplied as new arrivals launched new expeditions from the Antilles and the Canaries to Tierra Firme by signing new contracts with local crown representatives, including governors, audiencias, and viceroys. Unlike the Portuguese, the Spanish crown usually did not finance these expeditions but instead provided legal cover in exchange for a future promise of revenue on commodities (the 20% tax or quinto). The crown created a patrimonial bureaucracy of treasury agents and lay and church *letrados* to oversee local governance and distributive and redistributive justice, constantly demanding accountability from settlers and bureaucracies via special investigative functionaries (*visitas*, *residencias*, and inquisition). This was the engine that launched hundreds of maritime small parties to pursue riches and knowledge in unknown places.[25] These parties, in turn, assembled written information via diaries and questionnaires that Crown-appointed pilots and cosmographers, based at the *Casa de Contratación* in Seville and later in Cadiz, transformed into new geographical and commercial knowledge. The resulting archive of the *padrón real* included ever-growing geographical descriptions that together charted a new planet or sphere, namely, the New World.[26]

These were the maritime enterprises that Fernández de Navarrete sought to document by assembling paperwork that at the time lay scattered in the Crown, religious, and private archives. The purpose of his archival recovery was to reveal a pattern of knowledge production that had passed unnoticed. In effect, Fernández de Navarrete's project defied today's notion that scientific modernity was a product of 'the Enlightenment.' For Fernández de Navarrete, the cradle of scientific modernity lay in the fifteenth- and sixteenth-century Iberian ships that plied the waters of the Atlantic and the Pacific. The archive that Fernández de Navarrete assembled paid attention to patterns of knowledge formation and instrumentation that were very similar to Humboldt's but which Humboldt himself could not fully acknowledge. These patterns were not entirely secret, however. They had been made public in sixteenth-century Spanish manuals of navigation. A good example was Pedro de Medina's *Regimiento de Navegación* (1563), a pilot's guide to the investigation of new lands and seas.

In his pilot's manual, Medina offered a system of data collection to detect global patterns. Medina described five different types of instruments to gather five different types of data that yielded a new global representation of space, complete with the direction of wind currents, the elevation of the sun, the declension of stars, the movement of magnetic north, and the phases of the moon (see Figures 12.3–12.4).

(a) (b)

FIGURES 12.3 (a–e) Instruments to find one's bearings in the open sea. Pedro de Medina, *Regimiento de Navegación* (Madrid, 1563).

(*Continued*)

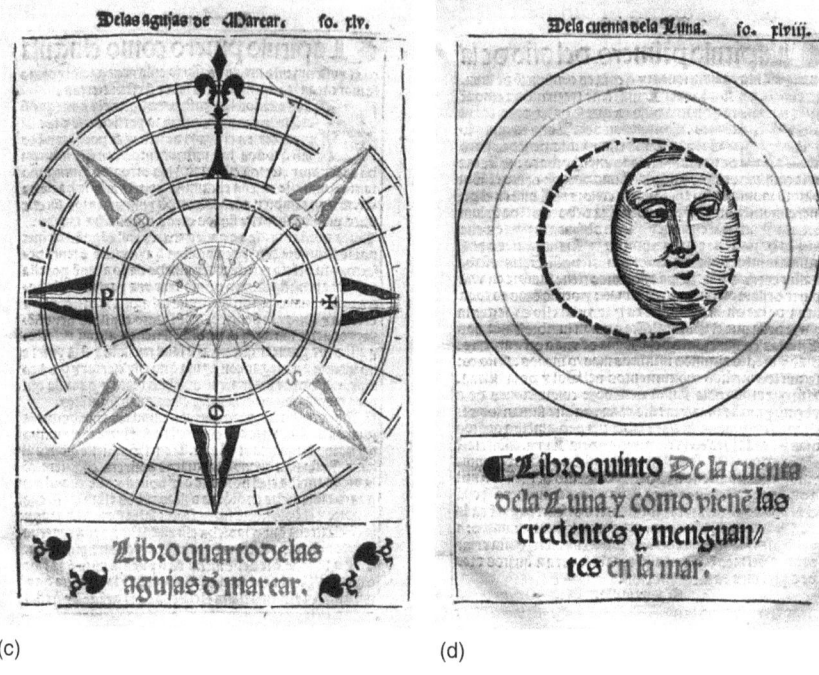

(c)   (d)

(e)

FIGURES 12.3 **(Continued)** (a–e) Instruments to find one's bearings in the open sea. Pedro de Medina, *Regimiento de Navegación* (Madrid, 1563).

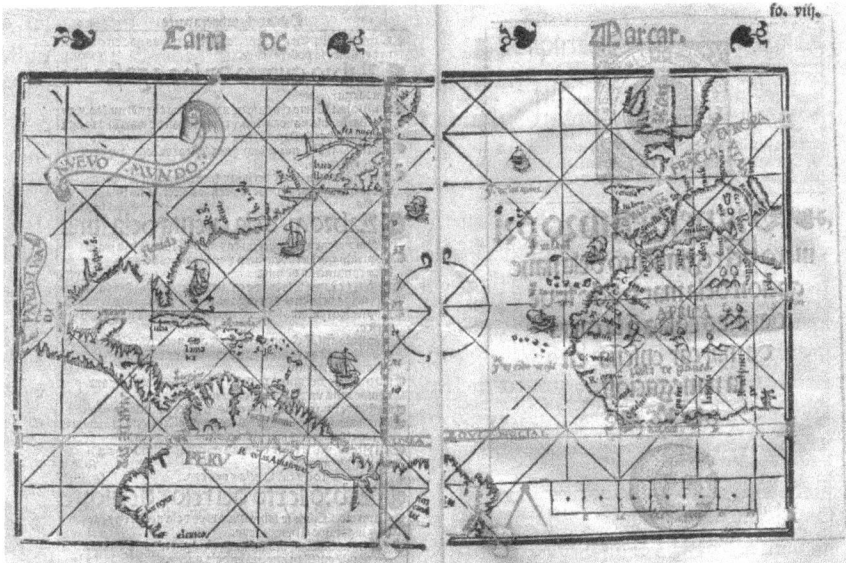

FIGURE 12.4 World map of coastlines. Pedro de Medina, *Regimiento de Navegación* (Madrid, 1563).

These five technologies allowed Medina to offer a new interconnected global representation of the entire earth, an accurate visual rendition of all continents and oceans.

The things that Humboldt attributed to Columbus were not really his but instead ways of interacting with nature through instrumentation to detect new lands and commercial routes in the Hispanic Indies. Humboldt made much of Columbus's use of the declination of the compass in relationship to the earth's magnetic north as means to detect changes between two worlds, east and west, of an imaginary meridian line hundreds of nautical miles to the west of the Azores. The use of the compass to calculate declension and thus longitude was one of the instrumental techniques offered by Medina.

But Fernández de Navarrete not only presented primary sources like Columbus's diary, illustrating the use of new instrumentation. He also included the paperwork of entire expeditions, thus reconstructing the pattern of collective expertise that had created the new commercial routes and thus a new earth physics. These were ignored in Humboldt's account. The collective study of earth physics is precisely what had allowed the Hispanic monarchy to launch the colonisation of South-East Asia from New Spain or Mexico. A case in point was the Spanish and Mexica exploration of the Pacific in the 1520s to the 1560s, when mixed crews sought to cross the Pacific Ocean from Acapulco to Malacca, then controlled by the Portuguese. In 1527, and as soon as he had conquered Tenochtitlan, Cortés sent his relative Alvaro de Saavedra Cerón with indigenous crews to cross the ocean to bring back cloves to Mexico for cultivation. The expedition arrived in Cebu but could not return.[27]

That same year the expedition of García Jofre de Loaysa departed from Spain for the spice islands. The survivors were captured by the Portuguese in Timor. Andres de Urdaneta was one of the survivors who returned to Spain in 1537. During his ten years in Asia, he had learned to navigate the islands. Urdaneta became an Augustinian and went to Mexico to teach cosmography at the newly founded University of Mexico. Another expedition was that of Ruy López de Villalobos in 1542. Like Saavedra's expedition, Ruy departed from Mexico but could not return. Dozens of survivors of this expedition were also captured by the Portuguese in Malacca and like Urdaneta shipped back to Lisbon. In 1558, the survivors of Loayza and Villalobos's expeditions living in Mexico, including Urdaneta (Loayza) and Juan Pablo de Carrión (Villalobos), were summoned by the Viceroy Velasco to organise a new expedition to the Spice Islands. The junta of 1558 summoned by Velasco offered on paper a route to return from Malacca and the Philippines; it was the route that Legaspi and Urdaneta followed in 1564. The key to getting back to Mexico from the Philippines was knowledge of seasonal changes in the Indian Ocean and the South China Sea. As the Indian Ocean warmed, currents slowly moved north into the South China Sea and beyond, propelling ships from the Philippines into northern latitudes near Japan. It was only near forty-two degrees latitude north that ships could catch the currents that would carry them into the American Pacific northwest. To sail back from the Philippines to Mexico, one had to understand the seasonal nature of currents from India to Taiwan to Japan. It could only be done once a year.

In short, the return route from the Philippines was the result of the accumulated expertise of several survivors of failed Spanish expeditions. Urdaneta and Carrión had unified cosmography with their first-hand knowledge of the changing seasonal nature of maritime currents in Asian waters. They accumulated this knowledge via contact with Southeast Asian sailors. This new insight into earth physics allowed Legazpi to settle the Philippines from Mexico.[28] The key to the new Iberian knowledge of earth physics was not only new instrumentation and the painstaking and systematic recording of pilots' observations. It was also knowledge of nautical practices accumulated through the expertise of local pilots. Urdaneta and Carrión did in the Spice Islands what Humboldt had done all over Spanish America, namely, draw on local expertise and knowledge to create a new earth physics. Humboldt, however, largely ignored the historical context that had produced that knowledge.

In the Footsteps of Humboldtian Ignorance

Humboldt's *Examen critique* was an exercise in ignorance as much as it was of knowledge. His study of Columbus was wholly reliant upon the documentation painstakingly assembled by two Spanish scholars, Muñoz and Fernández de Navarrete. Humboldt did not have their considerable archival experience, however, and thus ignored the context that had produced the knowledge and the archive. Humboldt's ignorance of the archive was reflected in his interpretation of Columbus, which in many ways served as a mirror for his own enterprise. Humboldt transformed

Columbus into an enlightened, romantic genius, his own alter ego, in the process decontextualising his hero. It is perhaps not a coincidence that Wulf's recent best-selling biography follows in his footsteps, similarly decontextualising Humboldt's knowledge.

But Muñoz and Fernández de Navarrete had scientific agendas of their own. Their projects were more original and interesting than anything Humboldt could possibly offer. They introduced archival evidence that enables a better understanding of the global history of geography. Whereas Humboldt used Columbus to self-fashion the persona of the Romantic genius whose intuition and personal expertise explain the origins of a new earth physics, Muñoz and Fernández de Navarrete presented Columbus to shed light on Iberia's precocious modernity. This global modernity was characterised by exploratory, bottom-up capitalism and slavery, factionalism and paperwork, and the massive accumulation of geographical knowledge through instrumentation and the translation of local knowledge. These Iberian endeavours created economies of scale, global trade, mercantile capitalism, and radically new forms of knowledge. Indeed, Humboldt's knowledge was largely a product of these endeavors.

Notes

1 Alexander Humboldt, *Examen critique de l'histoire de la géographie du nouveau continent : et des progrès de l'astronomie nautique aux 15 me et 16 me siècles*, 5 v (Paris: Gide, 1836–1839). The literature on Humboldt's Examen critique is surprisingly thin. See Karl Kohut, 'Alejandro de Humboldt, historiador. Un modesto homenaje a propósito del 250° aniversario de su Nacimiento,' *Inflexiones* 4 (2019), 9–35; Charles Minguet, 'Colón y Vespucio en la visión geohistórica de Alejandro de Humboldt,' in Leopoldo Zea and Mario Magallón, eds., *De Colón a Humboldt* (Mexico City: Instituto Panamericano de Geografía e Historia and Fondo de Cultura Económica, 1999), 9–20; Ottmar Ette, 'Entdecker über Entdecker: Alexander von Humboldt, Cristóbal Colón und die Wiederentdeckung Amerikas,' in Titus Heydenreich, ed., *Columbus zwischen zwei Welten. Historische und literarische Wertungen aus fünf Jahrhunderten*, v. 1 (Frankfurt am Main: Vervuert Verlag, 1992), 401–439.

2 Humboldt, *Examen critique*, v. 2. Juan Bautista Muñoz, *Historia del Nuevo Mundo* (Madrid: Ibarra, 1793). Martin Fernández de Navarrete, *Colección de los viajes y descubrimientos que hicieron por mar los españoles desde fines del siglo XV con varios documentos*, 5 v (Madrid, 1825–1837).

3 Andrea Wolf, *The Invention of Nature: Alexander von Humboldt's New World* (New York: Knopf, 2015). For Humboldt as a forerunner of ecology, see Aaron Sach, *The Humboldt Current: Nineteenth-Century Exploration and the Roots of American Environmentalism* (New York: Viking Books, 2006) and Laura Dassow Walls, *The Passage to Cosmos: Alexander von Humboldt and the Shaping of America* (Chicago, IL: University of Chicago Press, 2011).

4 Las Casas, *Historia de las Indias [1529-1557]*, 5 v (Madrid: Ginesta, 1875–1876); Antonio de Herrera y Tordesillas, *Historia General de los hechos de los castellanos en las Islas y tierra firme del mar Océano*, 4 v (Madrid: Imprenta Real, 1601–1615). On Muñoz, see Canizares-Esguerra, *How to Write the History of the New World* (Stanford, CA: Stanford University Press, 2001); Nicolás Bas Martín, *El cosmógrafo e historiador Juan Bautista Muñoz, 1745-1799* (Valencia: Universitat de Valencia, 2002). On Herrera see Richard Kagan, *Clio and the Crown: The Politics of History in Medieval and Early Modern Spain* (Baltimore, MD: Johns Hopkins University Press, 2009). On Las Casas, see Bernat Hernández, *Bartolomé de las Casas* (Madrid: Taurus, 2015).

5 On early Caribbean indigenous slavery, see Erin Woodruff Stone, *Captives of Conquest: Slavery in the Early Modern Spanish Caribbean* (Philadelphia: University of Pennsylvania Press, 2021) and Molly Warsh, 'Enslaved Pearl Divers in the Sixteenth Century Caribbean,' *Slavery and Abolition* 31 (2010) 3, 345–362.

6 Muñoz, *Historia*, Libro 1.

7 Jorge Cañizares-Esguerra, 'De la esfera a los dos planetas: las Indias como planeta alternativo desde la colonia a la independencia,' in Ramon Mujica, ed., *Forjando la Nación Peruana: El incaismo y los idearios políticos de la República siglos xviii–xx* (Lima: Banco de Crédito del Perú, 2021), 19–40.

8 Craig Martin, *Renaissance Meteorology, from Pomponazzi to Descartes* (Baltimore, MD: John Hopkins University Press, 2011).

9 Robert Westman, *The Copernican Question: Prognostication, Skepticism, and Celestial Order* (Berkeley: University of California Press, 2011).

10 José de Acosta, *Historia natural y moral de las Indias* (Seville, 1590). In Book 1, Acosta develops his epistemological critique of evidence and reason. In Books 2 and 3, Acosta seeks to explain the many circulatory mechanisms that transform the Torrid Zone into the most temperate and most inhabited climate in the Indies, the opposite of Euro Asia. In Book 4, he offers a comprehensive overview of the unique mineral, botanical, and animal resources of the land.

11 Acosta, *Historia natural*.

12 Juan de Cárdenas, *Problemas y secretos maravillosos de las Indias* (Mexico, 1591).

13 Jorge Cañizares-Esguerra, 'How Derivative Was Humboldt? Microcosmic Narratives in Early Modern Spanish America and the (Other) Origins of Humboldt's Ecological Sensibilities,' in Cañizares-Esguerra, *Nature, Empire, and Nation: Explorations of the History of Science in the Iberian World* (Stanford: Stanford University Press, 2006), 112–128. See also Mark Thurner and Jorge Cañizares-Esguerra, 'Andes,' in Mark Thurner and Juan Pimentel, eds., *New World Objects of Knowledge: A Cabinet of Curiosities* (London: University of London Press, 2021), 217–224; and in this volume, see the chapters of Gómez and Thurner.

14 Carlos Seco Serrano, 'Estudio preliminar,' in Martín Fernández de Navarrete, ed., *Obras de d. Martín Fernández de Navarrete: Edición y estudio preliminar de D. Carlos Seco Serrano*, v. 75 (Madrid: Atlas, 1954); and Jesús Fernando Cáseda Teresa, *Martín Fernández de Navarrete y la literatura de su tiempo* (Logroño: Instituto de Estudios Riojanos, 2000).

15 *Colección de los viajes y descubrimientos*.

16 *Colección de los viajes y descubrimientos*, v. 1 (Madrid, 1825), LII–LV and CXXIII–CXXV. See also, María M. Portuondo, 'Finding "Science" in the Archives of the Spanish Monarchy,' *Isis* 107:1 (2016), 95–105.

17 Martín Fernández de Navarrete, *Colección de documentos y manuscritos compilados por Fernández de Navarrete*, 32 v (Nendeln, Liechtenstein: Kraus-Thomson, 1971). *Índice de la colección de documentos de Fernández de Navarrete que posee el Museo naval* (Madrid: Instituto Histórico de Marina, 1946); Dolores Higueras Rodríguez, 'La colección Fernández de Navarrete del Museo Naval,' *Cuadernos monográficos del Instituto de Historia y Cultura Naval* 24 (1995), 35–60.

18 Washington Irving, *A History of the Life and Voyages of Christopher Columbus*, 4 v (London: John Murray, 1828) and Humboldt, *Examen critique*.

19 Lina del Castillo, *Crafting a Republic for the World. Scientific, Geographic, and Historiographic Inventions of Colombia* (Omaha: University of Nebraska Press, 2018); Jorge Cañizares-Esguerra and Neil Safier, 'Natural Histories of Remembrance and Forgetting: Science and Independence in the Spanish and Portuguese Americas,' in Marcela Echeverri and Cristina Soriano, eds., *Cambridge Companion to Latin American Independence* (forthcoming).

20 For an example of the conceptual struggles to fit in early-modern Spanish cartography into the world of print and public sphere, see José María García Redondo, *Cartografía e Imperio: El Padrón Real y la representación del Nuevo Mundo* (Madrid: Doce Calles, 2018).

21 Jorge Cañizares-Esguerra, 'Bartolomé Inga's Mining Technologies: Indians, Science, Cyphered Secrecy, and Modernity in the New World,' *History and Technology* 34:1 (2018), 61–70.

22 Arndt Brendecke, *The Empirical Empire: Spanish Colonial Rule and the Politics of Knowledge* (Berlin/Boston: De Gruyter, 2016).

23 Brendecke, *The Empirical Empire.*

24 Jorge Cañizares-Esguerra, 'La conquista como batalla de contratos,' in Carmen Alveal and Thiado Dias, eds., *Espaços coloniais. dominios, poderes, e representaçoes* (Rio de Janeiro: Editorial Alameda, 2019), 265–284.

25 James Lockhart, *The Men of Cajamarca A Social and Biographical Study of the First Conquerors of Peru* (Austin: University of Texas Press, 1972); Rafael Varon Gabai, *Francisco Pizarro and His Brothers: The Illusion of Power in sixteenth-century Peru* (Norman: University of Oklahoma Press, 1997); Eugene Lyon, *The Enterprise of Florida: Pedro Menendez de Aviles and the Spanish Conquest of Florida, 1565-1568* (Gainesville: University Press of Florida, 1983).

26 José María García Redondo, *Cartografía e Imperio: El Padrón Real y la representación del Nuevo Mundo* (Madrid: Ediciones Doce Calles, 2018).

27 Cañizares-Esguerra, 'On Ignored Global Scientific Revolutions,' *Journal of Early Modern History* 27 (2017), 1–13.

28 Patricio Hidalgo Nuchera, 'La embajada de Juan Pablo de Carrión a la Corte en 1558 y el conocimiento colectivo del tornaviaje,' *Anais de história de além-mar* 15:1 (2014), 52–78; Cañizares-Esguerra, 'On Ignored Global Scientific Revolutions.'

INDEX

Page numbers in **Bold** indicate tables.